T0340382

Risk Management
and Simulation

Risk Management *and* Simulation

Aparna Gupta

CRC Press
Taylor & Francis Group
Boca Raton London New York

CRC Press is an imprint of the
Taylor & Francis Group, an **informa** business

A CHAPMAN & HALL BOOK

CRC Press
Taylor & Francis Group
6000 Broken Sound Parkway NW, Suite 300
Boca Raton, FL 33487-2742

First issued in paperback 2019

© 2014 by Taylor & Francis Group, LLC
CRC Press is an imprint of Taylor & Francis Group, an Informa business

No claim to original U.S. Government works

ISBN-13: 978-1-4398-3594-4 (hbk)
ISBN-13: 978-0-367-37988-9 (pbk)

Library of Congress Cataloging-in-Publication Data

Gupta, Aparna.
  Risk management and simulation / Aparna Gupta.
    pages cm
  Includes bibliographical references and index.
  ISBN 978-1-4398-3594-4 (alk. paper)
    1. Risk management. 2. Risk management--Simulation methods. I. Title.

HD61.G86 2013
338.5--dc23   2013007014

Visit the Taylor & Francis Web site at
http://www.taylorandfrancis.com

and the CRC Press Web site at
http://www.crcpress.com

To my parents, Amar-Sneh

Contents

List of Figures

Foreword

The concept of risk has existed since the beginning of human experience. Almost any human choice or endeavor has potentially unfavorable outcomes that may include losses with varying degrees of severity. From very early on, humankind learned to recognize risks, to protect or hedge against them, and to limit losses, financial or otherwise, resulting from decisions with adverse consequences. Our understanding of risk has been a slow process that picked up speed in the last few centuries, and rapidly accelerated in the last few decades. Today, we understand the factors behind any adverse event with a lot more clarity than ever before. We have developed analytical tools and risk management frameworks to effectively address, and limit the losses of, any unfavorable outcome, from the minor and routine to the most devastating catastrophe in our realm of possibilities.

Of all the areas of human endeavor, risk is of paramount consequence to modern finance. In today's financial markets and financial institutions, risks and their prudent management have never been more important. Risks and their management are at the core of the role that financial markets and institutions are called to perform in the increasingly complex and dynamic present and future global economic setting. The challenges the current environment for risk management poses have required a new breed of professionals that are cross-trained in traditional finance disciplines and sophisticated quantitative and analytical disciplines.

Were humankind to regress to an economic 'stone age,' whereby it would have little need for the present day financial markets, institutions, instruments, and regulations, the practice of finance would most likely need nothing more than elementary arithmetic and a healthy dose of genuine gut feeling. However, in the absence of such unlikely regression, today's reality remains a ubiquity of risks, wide variety of risk types, complex channels for risks to flow, and an ever-growing need for sophisticated tools, methods, and instruments to manage the risks. Impact and importance of risks and their management are significant to all - individuals, households, financial firms, non-financial corporations, business and commercial enterprises, not-for-profit enterprises, and governments.

Although the regulatory thrust and a large segment of risks, specifically pure risks, either require or maintain a defensive stance, an all-inclusive and a robust response to risks must not be in defensive terms alone. Society, and in fact humanity, will be much better served if risks faced by it are viewed in proactive terms and managed for opportunity and reward they may bring. Therefore, a more prudent development of risk management is one that adopts and supports a comprehensive, proactive view using appropriate models, methods, tools, and frameworks to support assessing, decision making, monitoring, and control of risks.

In light of the above observations, it is a pleasure to note the contributions *Risk Management and Simulation* makes. This book strives to construct a comprehensive view of risks, yet maintains an exposition level that is accessible to a beginner in the training or practice of risk management. The author begins the book with building the basic concepts and fundamental framework for risk, decision making for risk, and risk management. Her introduction to the subject is accompanied by rigorous development of models and computational techniques for effective management of a variety of risk types, including market risk, equity risk, interest rate risk, commodity risk, currency risk, credit risk, liquidity risk, strategic, business, and operational risk.

Using initial development as the foundation, the author advances the topic by introducing the reader to a series of topics in modeling and computational techniques useful for risk management. The emphasis in computational techniques adopted in this book is simulation modeling and analysis, which is a particularly powerful suite of computational techniques broadly applicable and particularly useful in large-scale and complex contexts of risks. In the second half of the book, the author delves deeper into many of the important risk types, identifying specific issues related to each risk-type and advancing the models and risk management framework for the risk-type, beginning with various market risks - equity, interest rate, currency, and commodity risks.

I am particularly impressed by the fact that the book not only develops credit risk and credit risk management approaches, but it also ties in strategic, business, and operational risk considerations from the perspective of financial as well as non-financial firms. For financial institutions, strategic and business risk management translates to an involved process of asset-liability management, which the chapter on this topic aptly includes. Unlike other books in this space, the author takes the operational risk theme further by also studying the use of insurance as a mechanism for risk management and risk transfer. This seamless connection of topics is much needed given today's increasing blur between segments of the financial services industry. In summary, this book is a unique and wonderful combination of risk management and financial simulation concepts that will set the reader on a strong footing for a rewarding career in risk management.

Drawing on her experience in teaching risk management and computational technique courses to undergraduate and graduate students for a decade and a half, the author offers an accessible book on an increasingly important topic. Her explanation of the complex models and computational techniques is well thought-out and well-motivated, which should make reading the material both inviting and appealing. Whether you are reading this book as a textbook for a course in risk management or computational finance, or using it

as a self-study guide as a beginner practitioner, the exercises and MATLAB®
guidance the book provides should additionally aid your learning.

Our ability to manage risks well is the key to our success as individuals,
firms, governments, and society. Therefore, mastering risk management skills,
with a comprehensive understanding of various types of risk, and the modeling
and quantitative techniques for each of the risk types, as well as an awareness
of present day challenges, are a must for any finance professional. I wish you
well in your efforts of learning and training using this book, and hope it opens
doors for further exploration and enquiry for you.

Emmanuel (Manos) D. Hatzakis, *Ph.D.*, *CFA*, *FRM*
Risk, Structuring and Analytics Expert, UBS

Preface

I am delighted to introduce this book on mathematical modeling and computational development of risk management. Instead of narrowly defining risk management as a defensive activity, my take on it is positioning as a proactive opportunistic view of risk and its management. Moreover, it has been my effort in this book to keep the material at an approachable level for undergraduate seniors and master's level students.

The book is structured to offer a comprehensive view of issues in risk management, scoping a broad range of risk types both in speculative and pure categories. Management of risks in different contexts and in different kinds of enterprises can have varied goals and objectives. These seemingly disparate goals and objectives can be served well by creating a framework for risk management. Implementing the framework utilizes a rigorous, quantitative analysis approach, but also necessitates an individual and organizational awareness and responsibility toward risks and their management.

The rigorous, quantitative analysis to support risk management needs ever more sophisticated modeling and computational techniques. In this book, we begin with elementary models for risk, and then develop dynamic models for risks evolving over time. Sophisticated, complex models for risk rely on computational techniques for obtaining insights and for solving problems in risk management. Our emphasis throughout the book is based on computational techniques that help solve the problems of risk management.

Part I of the book focuses on developing the concepts and framework for risk management. In the first chapter, after defining risk and distinguishing its properties from what we label as uncertainty, a detailed typology of risk is developed. We begin developing formal models for quantitative definition of risk, fundamental to the models for risk management developed in later chapters. In this part of the book, we also develop the necessary constructs to be able to assess the exposure and management of risks under a formal risk management framework. A detailed discussion of each stage of the risk management framework and appropriate tools for each stage are discussed. Lastly, in order to give a practical perspective for the risk management context, an overview of regulatory systems in place and their historical evolution for some key segments of the US financial sector are included in this part of the book. The intention here is not to give a comprehensive description of the regulatory environment, but to present some salient features of it to motivate the importance of the contexts of risk management problems, issues, and models developed in the rest of the book.

In Part II of the book, the emphasis is on modeling and computational techniques for solving risk management problems. We begin with developing the simulation framework crucial for computationally solving risk management problems. We set down the principles to follow for constructing a simulation framework, where each step must be given its due importance, including model development, verification, validation, designing simulation experiments, and conducting appropriate output analysis. Many types of risks in the risk typology need to be considered and managed in a dynamic setting. To facilitate this requirement, in the rest of this part of the book, we develop time-dependent models for evolution of risk and methods to solve these models.

The largest part of the book, Part III, is devoted to specific issues of risk management in the range of risk types. The interest rate risk and equity risk component of market risk are discussed at length, utilized to advance the issues and approaches adopted for risk management. Following market risk, the attention is shifted to another important risk type, credit risk, where risk management issues and modeling approaches specific to credit risk are developed. Strategic, business, and operational risks are addressed, leading from the speculative to the pure risk types in risk management considerations. Finally, insurance as a mechanism for risk management and risk transfer is studied.

Part IV develops the computational thrust of the book by pushing it one large step forward. It is devoted to looking at some advanced concepts and techniques for risk management, along with developing methods that improve computational efficiency when solving risk management problems.

Suggestions for how this book may be used for courses in Financial Engineering/Quantitative Finance curricula follows. For a course on risk management, the instructor may focus on Parts I and III of the book, with limited modeling oriented topics picked from Part II. For a computational finance or a financial simulation course, the selection will focus on Parts II and IV of the book. Since the application context must be used as motivation for the modeling and computational techniques, portions of chapters in Part I and III would need to be identified to serve this purpose. All chapters are accompanied by extensive review questions and detailed exercises. Almost all exercises are quantitative or extensively computational in nature, where use of MAT-LAB mathematical software is encouraged. In support of this, algorithms are described and MATLAB functions are suggested throughout the book. The author also makes MATLAB code available on her website for supporting the development of examples and exercises in each chapter.

I have taught risk management and simulation for more than a decade in both undergraduate and graduate level courses. The material presented and discussions developed in this book are an outcome of interactions with and participation of many students in these courses. I would like to acknowledge the participation of Ali Acilar, James Burnes, Mecit Cetin, Hyunwoo

Cho, Chaipat Lawsirirat, Lepeng Li, Zhisheng Li, Harry Ma, Adam Petrie, Vicky Sharma, Sasidhar Sigampalli, Natasha Yakovchuk, Lingyi Zhang, and Xin Zhong. I am grateful to many students who challenged me to motivate theoretical and mathematical concepts in more practical terms relevant to risk management. I would like to acknowledge the role James O.Aram, Vincent Anderson, Omar Azhar, Matthew Cobb, John Cucinelli, Michael Fifield, Michael Gruet, Daniel Hutchison, Liu Hong, Manjia Huang, Dan Jerke, Graylin Kim, Rachel McCracken, Glenn Van Moffaert, Jill Muldoon, Ravin Ramsamy, John Rekstad, Alex Richman, Peter Ryan, Michael Shimazu, Saad Ullah Usman, Jingning Wang, Di Wu, Shijie Yang, Qingshan Zeng, Zhi Zeng, and others have played in hopefully making this book more readable by the intended audience than it would have otherwise been.

I wish to extend a warm thank you to my editor, Mr. Bob Stern, for both asking me to write this book and for his patient support through the process of writing it. My sincere thanks to an anonymous reviewer for the valuable comments that have significantly improved the presentation of material in this book. I also wish to thank my doctoral advisor, Professor William F. Sharpe, for his advice and comments on this book project. I wish to gratefully acknowledge my Project Editor, Ms. Judith Simon, and Ms. Jeanne Washington for their painstaking proofreading of the manuscript. I, of course, bear the responsibility for any remaining errors.

Last, but not in the least, I would like to thank Sreekanth for the infallible tea service through many wakeful nights of writing, and to Anirudh for being a constant source of encouragement.

Symbol List

Symbol Description

Ω	The sample space of a random variable.
μ	Mean of return of an asset or portfolio.
σ	Standard deviation of return of an asset or portfolio.
ρ	Correlation in the return of two assets or portfolios.
Π	Value of a portfolio.
Δ	Change in value of what follows, or a small increment in value.
s.t.	Short for 'such that.'
w.p.	Short for 'with probability.'
max	Short for 'maximum of' or 'maximize.'
min	Short for 'minimum of' or 'minimize.'
sup	Short for 'supremum.'
inf	Short for 'infimum.'
$\forall$	For all values of a quantity that follow the symbol.
$\in$	The quantity preceding is contained in the object following the symbol.
$\cup$	Union of sets.
$\cap$	Intersection of sets.
$\subseteq$	The set preceding is contained in the set following the symbol.
$\mathcal{I}_{\{A\}}$	An indicator of function of whether event A has occurred or not.
$\sum$	Summation of elements that follow the symbol. This is often indicated with the range of summation.
$\rightarrow$	Indicates either 'maps to' or 'tends to.'
$\simeq$	The two quantities are approximately equal to.
$\sim$	The random variable follows the distribution given by what follows.
∂	Partial derivative with respect to one of the independent variables of a function.
mod	The modulo function that produces the remainder after dividing the first number by the second number.

Part I

Risk and Regulation

Chapter 1

Defining Risk

The awareness that happenings of the future cannot be determined with certainty must have arrived quite early to prehistoric man. However, completely grasping the notion of risk remains an ongoing process. The reason for this is over the millennia, and especially in the past few centuries, we have created an increasingly complex and interdependent habitat. Man no longer fends for himself and his small family unit by growing food and building shelter. Instead we depend on people and their efforts across the globe for every single and simplest of our needs. Therefore the happenings of the future at far and wide locations of the globe that cannot be determined with certainty hold the potential of affecting our happiness, well-being, and even subsistence.

The gravity of the situation has not gone unnoticed. Human enquiry has made significant efforts to grasp the implications of non-certainty of happenings of the future. That is, while certainty is unattainable, what can we comprehend and say about the future happenings that can help improve the condition. In fact, humanity has thrived through the ages due to its capacity and capability to create mechanisms to manage risks. The efforts to understand and tame risk have been nontrivial, defining fields and disciplines of study that have engaged our energies for centuries. From mathematician Daniel Bernoulli's pioneering work and contributions in probability and statistics [9] to the contributions of some of the eminent Nobel laureate economists of the 20th century, such as, Harry Markowitz, Bill Sharpe, Franco Modigliani, Merton Miller, Fischer Black, Myron Scholes, and Robert Merton – all have attempted to grasp the implication of risk and significance of its management in human enterprise.

The non-certainty of future happenings and risk is not all bad. In fact, quite the contrary. Every new age, every new development, every new technology has posed, and will continue to pose, a new challenge to humanity of tackling a new spectrum of risks. Tackling risks and management of risk creates opportunities for growth and reaching new heights of prosperity. Since risk is ubiquitous and permanent, humanity's continued success lies in its ability to develop skills to out-smart the risks and make decisions in light of the risks that are sustaining and thriving.

From the above discussion, it may appear that we are suggesting the non-certainty of happenings of the future to be the definition of risk. This may be a workable definition, but to make the definition more effective, we will choose to make an additional qualification through the definition of risk. By non-certainty we imply that a certain specific thing may or may not happen in the future. But the key question we would like addressed is, if a certain specific thing may or may not happen, what are the things that can happen?

In many cases, we may have a pretty good idea of what all may happen in the future, even if nothing is certain. But in some cases our mental cognition or our collective experience doesn't allow for us to imagine some specific outcomes that can happen. For instance, a devastating meteor decimating life as we know it on our planet earth tomorrow. This brings us to the distinction between possibility and probability. A devastating meteor is a possibility, since at a stretch of imagination one may be convinced that it can happen. However assigning it a believable probability will not be that easy.

Twentieth century economist Frank H. Knight was the first to make the distinction between risk and uncertainty [49]. Those future outcomes that are measurable and their corresponding probabilities of occurrence can be assessed ex-ante are termed as risk. While those future non-certain happenings whose outcomes are only known as possibilities, and ex-ante probabilities are hard to assess, are defined as uncertainty. Then there is also the domain of '*unknown unknowns.*' These are happenings of future that we are not even capable of identifying as possibilities. We will not be able to devote much attention to the '*unknown unknowns*' in this book.

Risk. We simply define risk as variability that can be quantified in terms of probabilities. With this definition of risk, the tools for management of future non-certainties can be most effectively applied. But all bets are not off when we must wander out of the domain of risk into the territories of uncertainty. In today's increasingly connected and complex world, also evidenced by recent financial crises, there is a lot of grey region between risk and uncertainty. Surprises spring up every so often, and we must acknowledge the presence of and be ready to deal with uncertainty. Tools based on simulation help us also assess the impact of uncertainty, and thus, provide insights for better management of uncertainty. We will bring the reader's attention to this wherever appropriate. Predominantly, however, we will be concerned with risk in our discussions of measuring and management of non-certainties of future.

If risk is the variability in future happenings that can be quantified in terms of probabilities, one distinction is needed in terms of subjective and objective probabilities. Each one of us, based on our beliefs, historical data and evidence can have a subjective view of probabilities of future outcomes. These are subjective probabilities. Objective probabilities are those that can be backed with observational evidence, not just subjective estimates. The objective probability assigns a likelihood for an event to occur based on an analysis of recorded observations and data. Unavailability of relevant or insufficient data can often result in the boundary between objective and subjective probabilities to become rather thin. We will need to bear this in mind and visit this distinction where relevant.

Management of risk, and for that matter uncertainty, in any context requires understanding the sources for this risk. A full grasp of the sources of risk is greatly helped if the risks are classified by a meaningful taxonomy. In the next section, we will develop a structure of additional definitions to support developing this taxonomy of risks.

1.1 Types of Risk

Studying risks in a specific context requires understanding the context and identifying quantities to observe to develop a measurable quantification of the risk. In our context, the impact of risk will uniformly be measured in financial and/or monetary terms, although sometimes assigning a financial measure to a risk becomes extremely tricky. For instance, what is the right financial/monetary value of a life of a 10-year-old killed by an intoxicated, reckless automobile driver. For the management of risks even such tragic events must be evaluated, and the financial/monetary value assigned will end up depending on from whose perspective the assessment is being done.

Understanding the sources for risk exposures is an essential prerequisite for a viable and effective management of the risks in any context. A full grasp of the sources of risk is greatly helped if the risks are classified by a meaningful taxonomy. Taxonomy of risk needs to be developed depending on the nature of the entity developing a risk management plan and strategy. The sources of risk from an individual's and a household's perspective would be quite different from those of a multinational manufacturing enterprise or a large financial institution. The taxonomy, however, should aim to be versatile to benefit a variety of entities in their risk management efforts.

The first most important classification we will develop is on the basis of pure versus speculative risk. Once these are explained, we will elaborate on the next level of classification for each of the two risk types. As stated earlier, risk is not all bad. Risk has a fundamental linkage with reward, which makes us expose ourselves to risk in the first place. Rarely do people expose themselves to risk for the risk's sake. In these cases, the risk itself may be the seeker's reward. People take up jobs and professions, travel long distances, firms take up investment projects, engage customers, collaborate with other enterprise, each of which exposes them to new and additional sources of risk. They nevertheless engage in these activities in pursuit of benefits or rewards from these activities.

The nature of reward defines the first classification of risk we present next.

1.1.0.1 Pure Risk

Each of us as individuals or as a part of a larger group, organization or enterprise identify a certain mode of operation to be the normal condition. We expect this to be the status quo for our activities of normal living, in order to perform our base level of functions of life. When we return home from work, we expect to find our house more or less how we left it in the morning. When we take a flight or a road-trip, we expect to arrive at the destination intact and roughly in some expected duration of time. This normal status quo outcome is the best case scenario that lets us perform the activities of our life peacefully and conveniently.

However, mishaps happen, events can happen that can throw us off the normal status quo. When we get thrown off the status quo in such cases, it is due to the occurrence of an unhappy event resulting in losses. For instance, the house may be found ransacked by burglars, the automobile may breakdown resulting in extended delay in arriving at the destination, the flight may have to be rerouted, or worse yet, crash due to equipment malfunction or very rough weather. These are examples of variability in outcome that categorically result in either normal status quo to be maintained or suffer loss.

Pure risk. We define pure risk as variability quantified in terms of probabilities that can either result in realization of losses or no losses. Occurrence of damage due to earthquake, floods, hurricane, fire, theft, sickness, accident, death, unemployment, fraudulent action of an employee, machinery malfunction, and computer hacking, are all examples of loss events that would qualify for pure risk.

1.1.0.2 Speculative Risk

Risk has an integral connection with the hope for reward. This linkage is at the core of the definition of speculative risk. Any risk we choose to expose ourselves to that can yield both attractive, profitable outcomes as well as loss outcomes is speculative in nature. Individuals, firms, and governments voluntarily expose themselves to speculative risk in an attempt to create possibilities that help improve their condition, while at the same time signing up for the possibility they could end up being worse off. But then again, there is no gain for no pain.

Speculative risk. We define speculative risk as variability quantified in terms of probabilities that can result in realization of both gains and losses. Firms launch new products and services in the hope of attracting new customers and increasing sales, but constantly face the risk of weak demand or competitors beating the firm to market. This is an example of speculative risk, since both gain and loss outcomes can occur. The firm must choose and manage the risk wisely so the outcomes and their likelihood are most favorable.

Individuals must decide to allocate their savings in investment vehicles, such as, stocks, mutual funds, saving deposits, real estate, etc. All these investments can work out to provide high returns to the investor, but can also suffer losses due to a variety of reasons. This is quintessential speculative risk. Investors should choose their risk exposures that yield the desired level of reward for the right level of risk.

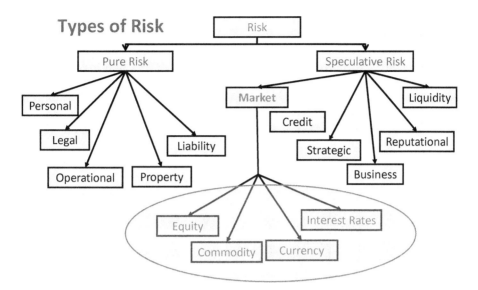

FIGURE 1.1: Classification structure for types of risk.

1.1.1 Classification of Pure Risk

After the definition of pure and speculative risk, we need to extend the classification further to build the taxonomy of risks. A taxonomy is a critical tool for effective risk management. This remains true for pure risk, therefore we consider classification of pure risk into personal, property, liability, operational and legal categories. A summary of the taxonomy is provided in Figure 1.1.

Personal risk. This is the risk from events that can cause loss to one's person. These constitute a variety of individual risks. It can be a loss of health due to illness or accident. Premature death of the head of a household can cause distress to the household due to unfulfilled obligations. Loss of regular income due to unemployment can cause disruptions to the life of an individual and his/her dependents. When an individual's savings for financial support through retirement starts to fall short, it can result in considerable hardship.

Property risk. Persons or firms own property that is exposed to risk of damage or loss due to various causes, such as, fire, earthquake, lightning, tornados, vandalism, or theft of property. This can result in direct loss due to property being damaged, and indirect loss due to the property not being available for use by the owner.

Liability risk. One is legally liable if something we do results in bodily harm or property damage to someone else. This is a very real risk for individuals and firms. Moreover, there is no upper limit on the amount of loss for liability risk, since value of someone else's property and harm can be arbitrarily high. To settle the determination and indemnification of losses very high legal costs may also be incurred as an indirect effect. This can result in a lien being placed on the individual's income and financial assets to satisfy the legal judgment.

Operational risk. Losses can be incurred by a firm due to inadequate systems, management failures, and faulty control. Losses can also be incurred by an organization due to fraudulent behavior of its employees, or simply due to lapse in judgment or human error. In today's high-tech workplaces, computer system breakdown or malfunction can result in significant losses, including human errors by misuse of technology, such as pushing the wrong button, accidentally destroying files or entering wrong values for parameters. In banking there are numerous examples where a trader or an employee falsifies or misrepresents risks incurred in a transaction, resulting in very heavy damage to the firm, both monetary and reputational.

Legal risk. A firm or an individual can incur significant losses due to knowingly or unknowingly not following the rule of law or regulatory requirements for a variety of reasons. A customer or a counterparty can reasonably or unreasonably conduct legal action against a firm, resulting in significant losses from legal fees. A counterparty may not have legal or regulatory authority to engage in a transaction, and the counterparty may sue to avoid meeting its obligations on losing money in the transaction.

Finally, pure risk can be of particular kind or of fundamental kind. Particular pure risk affects a single individual or a small group of individuals, while fundamental pure risk has much more macro-level effects. A stray attack in a dark alleyway on the way back from work is a particular risk, while genocide due to 'ethnic cleansing' is a fundamental risk. A tornado passing through a community and causing damage to property is an example of particular risk, while global warming is a fundamental risk. This important distinction is an indication of the size and enormity of loss, meriting a fundamentally different approach to manage the risk.

1.1.2   Classification of Speculative Risk

Speculative risk arguably includes a larger class of risks. The risks here are also of a vast variety, covering the range of contexts where risk management is a worthwhile activity. In banking and financial institutions, the fundamental role of the organization is to help manage the risks of a society's economic activities. Banks offer credit to individuals and firms in the economy to support investment, innovation, and growth. In doing so, they expose themselves to a variety of speculative risks, which they should manage very deliberately.

Insurance companies take over pure risks underlying a society's economic activity, in exchange for a fee, of course. The insurance firms must appropriately allocate these accumulated funds from fees to remain viable, thus exposing themselves to a range of speculative risks. Markets are 'places' where agents exposed to various speculative risks come to trade the risks, and in the process discover the economic value of the risk. Finally, all non-financial firms and individuals must deal with speculative risks in their day-to-day decisions, as well as long-term decisions, as they navigate the risk-reward terrain.

The speculative risks can be categorized as market risk, credit risk, liquidity risk, strategic, business, and reputational. Each of these risks can be further broken down into further subcategories, for instance, subclassification of market risk can be seen in Figure 1.1. We describe each of these risks in more detail next.

Market risk. As markets attempt to discover the economic value of risks that agents trade in any market, there is a high degree of variability in the value of instruments being traded. This is termed market risk, that is the risk that changes in market prices and/or rates will reduce or increase the dollar value of a security or an investment portfolio. In order to get a sense of the aggregate market risk, instead of per security, we track general market risk in terms of the market as a whole (in terms of a market index) falling or gaining value. The aggregate market risk can affect all securities in the market, not just because all the securities in the market make up the aggregate, but also because sentiments towards the aggregate can filter to specific securities.

Financial markets can be broken down into four major segments: equity market, fixed-income (interest rate) market, commodities market and foreign exchange market. These markets are for financial securities or assets of four primary types, designed for specific objective of activity in the global economy. Other specific types of additional risks that a financial product or an entity may be affected by are, for example, fixed income products have risk of default of the issuer. This risk links market risk to credit risk. An investment fund may be tracking the performance of a certain aggregate market benchmark, and in doing so may be exposed to tracking error risk. In some cases, basis risk may be important that measures the breakdown in relationship between price of a product and the price of an instrument used to manage the risk of the product.

We discuss the subclassification of market risk in a little more detail, since these are the most important risks we will study in this book.

Equity risk. Simply put, this is the volatility in stock prices, and sensitivity of an instrument or portfolio value to change due to changes in the value of a single stock or the level of aggregate stock market indices. We label specific or idiosyncratic risk of equity to be the characteristics specific to the firm, its line of business, quality of its management, or breakdown in its production process. The general equity market risk is termed systematic risk, since according to the result of portfolio theory, this risk cannot be eliminated through diversification, while specific risk can be diversified away.

Interest rate risk. Simplest form of interest rate risk is seen when the value of a fixed-income security, like a bond, falls due to increase in market interest rates. In a more complex portfolio of fixed-income instruments, differences in maturities, nominal values, reset dates of instruments, and cash flows (longs and shorts) can add more variety of interest rate risks. For instance, curve risk arises due to the shifts in the term-structure of interest rates that affect instruments with different maturities differently, and as a result, can create imbalance in long-short positions in a portfolio. Basis risk, which was mentioned earlier, can arise when instruments with the same maturity respond to interest rates differently causing a mismatch due to rates being imperfectly correlated. Gap risk arises as a result of difference in sensitivities of assets and liabilities to changes in interest rates.

Foreign exchange risk. Increasingly more firms must transact in different currencies due to global operations or markets for their goods. Currency or foreign exchange risk arises from open or imperfectly hedged positions in a particular currency. This exposure can arise as a natural consequence of business operations, not necessarily due to explicitly taking a position in a currency. It can have a pretty drastic impact on a firm by potentially sweeping away profits. It can place a firm at a competitive disadvantage, can generate huge operating losses, and in the end, inhibit growth and investment. Foreign exchange risk arises from an imperfect correlation between currency prices, international interest rates, and domestic interest rates.

Commodity risk. The variability in price of raw materials that are crucial for the production economy constitute this risk. Price risk of commodities behave differently from interest-rate and foreign exchange risk, because commodity prices are often controlled by few suppliers, which can magnify volatility. Fundamentals affecting commodity prices include market liquidity (or illiquidity), degree of ease, and level of cost of storage of the commodity. Due to these reasons, commodity prices have higher volatility and display sudden large changes or jumps. Commodities can be classified into the following groups, based on some common shared properties.

Hard (nonperishable): Precious metals (gold, silver, platinum), base metals (copper, zinc, tin)

Soft (perishable short shelf life and hard to store): Agricultural products (grains, coffee, sugar, orange juice)

Energy: Oil, gas, electricity

Credit risk. When banks issue a loan, they expose themselves to the risk that either the interest or the principal itself may not be paid, or may not be paid on time. This is credit risk. While default is the extreme case (i.e., a counterparty is unable or unwilling to fulfill its obligation), credit risk also includes changes in the credit quality of a counterparty, since this affects the value of the loan or portfolio of loans. Institutions are also exposed to the risk that a counterparty, which may be another institution or a government, is downgraded by a rating agency.

Credit risk is an issue when the position has a positive replacement value (i.e., is an asset), where either all the market value of the position is lost, or more commonly, part of the value of the asset is recovered after a credit event. This defines the terms *recovery value* or *recovery rate*, and the amount that is expected to be lost is the *loss given default* amount. Credit risk is often called counterparty risk when the security in question is other than bonds and loans.

Liquidity risk. Liquidity is having access to cash when needed. Liquidity risk is identified as either funding liquidity risk or asset liquidity risk. Funding liquidity risk refers to a firm's ability to raise necessary cash to meet its immediate cash needs, such as, for rolling over its debt, or to meet the cash, margin or collateral requirements of counterparties. Asset liquidity risk is when an institution cannot convert the value of an asset to cash. This would arise if the institution is not able to execute a transaction at the prevailing market price because at the time there is no appetite for the deal on the other side of the market. Liquidity risk can result in substantial losses, but this risk is hard to quantify. It affects an institution's ability to manage and hedge market risk, and capacity to fund shortfalls in funding by liquidating assets.

Business risk. This is the classic risk of conducting business arising due to demand uncertainty, fluctuation in prices, cost of production, supplier costs, and availability. These are managed through core tasks of management, by choice of channels, products, suppliers, marketing, etc. It is important to connect business risk management within a formal enterprise risk management framework, by integrally combining market risk, credit risk with business risks.

Strategic risk. Strategic risks are those arising when generating a long-term plan for business growth and new product introduction. Plans for new business development and growth often require very significant investment, for which there may be a high degree of uncertainty in store. On the other hand, it carries the potential of great profitability. Examples are incorporating new unproven technologies in products, offering products to new market segments, and taking up a massive advertising campaign to generate a new brand image.

Reputational risk. As firms conduct their business, they create a reputation by the quality of their products, how they treat their customers, and how ethically they manage their affairs, both financial and managerial. Accounting scandals that defrauded bondholders, shareholders, and employees of many major corporations during the late 1990s have made this risk prominent. Corporate scandals like those at Enron, WorldCom, and Global Crossing are devastating for the image and the value of a firm. Reputation is also very important for banks and financial institutions, where trust and confidence of customers, creditors, regulators, and the marketplace are important.

With the taxonomy of risk built, we are now ready to advance to the next level of sophistication necessary for management of risk. We need to begin developing models to characterize risk, where the models would belong in a framework built to perform risk management.

1.2   Getting Started with Modeling Risk

In the previous section, we made a distinction between risk and uncertainty observed in the world to establish risk as being measurable and quantifiable. Our goal in this section is to take this definition further by giving a well-grounded basis for a mathematical assessment of risk. The analysis and management of the various types of risks identified under the general categories of pure and speculative risks in Section 1.1 are greatly facilitated by developing some abstractions. We will call these abstractions *models* for these risks.

For developing a model abstraction for a risk, we first need to determine the possible realizations the risk can have. This collection of possibilities is often called the *sample space* for the risk. A typical and simple example of the sample space is if we were interested in comprehending the possibilities for the top-face upon rolling a fair die on a flat surface. These possibilities will obviously be the six faces of the die – $\{1, 2, 3, 4, 5, 6\}$. We would call this the sample space for the risk of what might happen when we roll a die. For the risk in the price of your favorite stock a year later, an appropriate sample space may be $[0, \infty)$, even if the stock price exceeding a certain high-enough value may be practically zero.

Based on the model describing a risk, one often wants to identify specific values or sets of values of interest as *Events*. For a variety of events and set of events, it is natural to want to assess the likelihood of their occurrence. Events and the probability of their occurrence is fundamental to assessing risks. For instance, the stock of your interest exceeding a set target in a year is an event, one for which you would like to assess the likelihood. Similarly, the risk of your stock falling below the price at which you bought it is an event. Defining models for risk, their sample space, events set, and likelihood with mathematical rigor create a language to develop risk management frameworks in the rest of the book. We will spend the rest of this chapter laying down the essential constructs of this language.

1.2.1   Random Variable and Probability

A random variable is a variable representing a risk that takes on a certain set of values, but every time one makes an evaluation (observation) of the variable, it takes on the values randomly. The *sample space*, Ω, as described earlier, is the collection of possibilities for the risk. While elements in a sample space may not be numeric, such as, the two sides of a coin on the toss of a coin, we want random variables to take on numeric values. Hence, we denote a sample space to be a domain for the random variable, where the random variable maps the (possibly non-numeric) elements in Ω to a set of real values ($X : \Omega \rightarrow \mathbf{R}$). Consider a dartboard in a room. When I throw a dart, all possible points of impact make up the sample space, Ω. This includes every point of the dartboard, much of the wall, and even parts of the floor or ceiling, if I am not good with darts. My score $X(\omega)$ ($\omega \in \Omega$), however, is a finite set of values between 0 and 60. The symbol '$\in$' reads as 'belongs to', or simply, 'in.'

Suppose instead of a single coin, n coins are tossed, where n could be 10, 20, or even 100, if you have the patience. Since each coin has only two faces, 'Head (H)' or 'Tail (T),' the sample space for a single coin is, $\Omega_1 = \{H, T\}$. A random variable that assigns a numerical value to the outcomes of a toss is set to be, $X(H) = 1$, $X(T) = 0$. Now, if Y is the total number of heads shown by n coins tossed, the sample space, Ω_n, will be built of elements with a sequence of n H's and T's. In this case, for each $\omega \in \Omega_n$, $Y(\omega) \in \{0, 1, 2, \cdots, n\}$. If none of the n tosses land on heads, $Y(\omega) = 0$, while if all the tosses result in heads, $Y(\omega) = n$. A random variable with such a discrete set of outcomes is often called a *discrete random variable*, to distinguish it from random variables that can take a continuum of values, such as the temperature in Timbuktu on a lazy summer noon. In the sub-Saharan summer heat of Timbuktu, there is not much else one would like to be up for than to be lazy. A random variable with a continuum of values making its sample space is called a *continuous random variable*. Besides being a discrete random variable, the n coin tosses random variable is also finite, i.e., its sample space is finite.

Once a random variable is defined to describe a risk, the natural next question to ask would be, what is the likelihood of it realizing any of its various different possibilities. How likely is it for the n-coin toss random variable to get 7 'H's when the $n = 10$? For that matter, how likely is it for there to be 7 or more 'H's on the 10-coin toss random variable? Or how likely is it for me to score 40 or above when I throw the dart? To accomplish such assessment of likelihoods, we construct a special set of sets that denotes the category of values a random variable can take, called the σ-*algebra*, denoted by $\boldsymbol{A}$. We also define a *probability measure*, P, that measures the likelihood of the random variable taking on certain value or set of values. The three constructs presented so far are often written together as $(\Omega, \boldsymbol{A}, P)$, which is called a *probability space*.

The σ-algebra for the dartboard game would include all values on the dartboard and sub-intervals of the range of scores $[0, 60]$, while the probability measure for my scoring those values will depend on my skill level of throwing darts. In case of the n-coin toss random variable, the σ-algebra will contain all values and subsets of the set: $\{0, 1, 2, \ldots, n\}$. If the coin tossed and the tossing itself is done without any bias, the heads and the tails will be equally likely. The probability measure in that case will be a well-known distribution, the binomial distribution. This makes the n-coin toss random variable a binomial random variable. We will study this and other well-known random variable models later in this chapter.

As a well-defined mathematical construct, the σ-algebra defined above should satisfy the following properties:

1. $\Omega \in \mathbf{A}$.

2. If $A_1, A_2, \ldots, \in \mathbf{A}$, then $\bigcup_{i=1}^{\infty} A_i \in \mathbf{A}$.

3. If $A, B \in \mathbf{A}$, then $A - B \in \mathbf{A}$.

These properties essentially deliver a meaningfulness to the definition of a σ-algebra, in saying that if we insert some events in the σ-algebra, then all unions and differences of these events will also make for events that should belong to the σ-algebra. Hence, a σ-algebra should contain all possible events of interest for a risk. Correspondingly, the measure of likelihood of these events should satisfy certain properties for the probability measure to be well-defined. The definitional properties for the probability measure are:

1. $\forall A \in \mathbf{A}, P(A) \geq 0$.

2. If $A_1, A_2, \cdots, \in \mathbf{A}$ are pairwise disjoint, then $P(\bigcup_{i=1}^{\infty} A_i) = \sum_{i=1}^{\infty} P(A_i)$.

3. $P(\Omega) = 1$.

Properties 1 and 3 are intuitive in saying that likelihoods of all events should be non-negative and the event of the entire sample space is a certainty, since the sample space exhaustively includes everything that can happen for a risk. Property 2 yields an important characteristic to the probability measure in terms of likelihoods of events that have nothing in common, in that their likelihoods can be added to determine the collective likelihood of all the events.

It should be noted that as an implication of the above properties of σ-algebra and probability measure, $P(\emptyset) = 0$. By properties 1 and 3 of σ-algebra, $\Omega - \Omega = \emptyset$ and $\emptyset \in \mathbf{A}$. The rest can be deciphered from properties 2 and 3 of probability measure, since $\emptyset \bigcap \Omega = \emptyset$ and $P(\emptyset \bigcup \Omega) = P(\Omega) = P(\Omega) + P(\emptyset)$. This allows defining for any event $A \in \mathbf{A}$, if $P(A) = 1$, we say that the event A occurs *almost surely* (a.s.) or *with probability 1* (w.p. 1). By a similar derivation, for any event, $A \in \mathbf{A}$, $P(A) = 1 - P(A^c)$, where A^c is called *A-complement*, and is defined as $A^c = \Omega - A$.

1.2.1.1   Summarizing Random Variables

For a random variable, X, we seek a comprehensive summary of its probabilities of assuming different values in a single function called the probability density or mass function. Density is relevant when the random variable is a continuous random variable, and mass is relevant for discrete random variables. Therefore, for $X : \Omega \to \mathbf{R}$, we define $f(x)$ as $P(X(\omega) = x)$. This definition is good for discrete random variables, but strictly speaking, it is inaccurate for continuous random variables. For a continuous random variable, at points of continuity of the density function, $P(X(\omega) = x) = 0$. Therefore, for continuous random variables, the summary probability description is accomplished using cumulative probability function.

Cumulative density (or distribution) function or cumulative mass function does exactly that, it accumulates probability from lowest to higher values of a random variable. Therefore, $F(x)$ is defined as $\sum_{x_i \leq x} f(x_i)$, where $\{x_i\}$ are all the values of the discrete random variable, X. For a continuous random variable, $F(x) = \int_{-\infty}^{x} f(y)dy = P(X(\omega) \leq x)$ is the accumulation of probabilities up to x, summarized in the cumulative density (distribution) (CDF) function. By the properties of a probability measure, we have $F(-\infty) = 0$ and $F(\infty) = 1$. In the continuous case, the derivative of the CDF, $\frac{dF}{dx}$, is the probability density function. Moreover, since $P(X(\omega) = x) = 0$ at points of continuity of the density function, $F(x) = P(X(\omega) \leq x) = P(X(\omega) < x)$ for such points.

The probability density function (PDF) and CDF provide a comprehensive summary of the probabilities of a random variable, and can be used to compute probabilities for all events concerned with the random variable. However, sometimes we are interested in a single number summary of the properties and characteristics of a random variable. This is achieved by taking weighted 'averages' of the random variable, where the weights are the probability of the outcomes of the random variable. This is defined by the notation and the formula as follows,

$$E[X] = \int_{-\infty}^{\infty} x f(x)dx, \tag{1.1}$$

$$= \sum_{\{x_i\}} x_i f(x_i), \tag{1.2}$$

where Eqn. (1.1) is for continuous random variables and Eqn. (1.2) for discrete random variables. The expected value or expectation of a random variable, $E[X]$, is called the mean of the random variable, and is often depicted by the symbol, μ. More generally, we would like to find expectation (weighted average) of any function, $g(x)$, of the random variables (risks). This can be achieved by the following modification of Eqns. (1.1) and (1.2),

$$E[g(X)] = \int_{-\infty}^{\infty} g(x) f(x)dx, \tag{1.3}$$

$$= \sum_{\{x_i\}} g(x_i) f(x_i), \tag{1.4}$$

where $g(X)$ is a function of the random variable, X. One specific function useful to measure the deviations of the random variable from its expected value is, $g(x) = (x - E[X])^2$. This defines the variance of the random variable. The square root of the variance is called standard deviation of the random variable, often denoted by the symbol, σ. In formulas the variance (and standard deviation, σ) of a random variable are defined by,

$$\sigma^2 = \int_{-\infty}^{\infty} (x - E[X])^2 f(x)dx, \tag{1.5}$$

$$= \sum_{\{x_i\}} (x_i - E[X])^2 f(x_i). \tag{1.6}$$

We are often interested in the expectation of higher order polynomial functions of random variables, specifically $g(x) = x^k$ or $g(x) = (x - E[X])^k$ for positive integer, k. Expectation of these functions of the random variable are called $k-$th order moments and central moments, respectively. These higher order moments help characterize random variables beyond their mean and variance. For instance, the third-order central moment indicates the degree of asymmetry in the distribution, or skewness, while the fourth-order central moment measures 'heaviness' of tails of the distribution. Tail of a distribution is the shape of the density function at the far extreme values, and if these are somewhat high, the tail is said to be heavy or fat. The distribution is called a heavy-tailed or fat-tailed distribution.

1.2.1.2   Several Random Variables and Correlation

In risk management, more often than not, we would need to handle many risk factors simultaneously, therefore we need to develop the vocabulary of joint properties of several random variables, $X_i; i = 1 \dots n$. To begin with, we will take $n = 2$, and define joint density function, $f(x_1, x_2)$, for the pair of discrete random variables, (X_1, X_2), as $f(x_1, x_2) = P(\{X_1(\omega_1) = x_1\} \cap \{X_2(\omega_2) = x_2\})$. As in the case of single random variable, definition on the exact same lines for continuous random variable will not work. Hence, we will need to define the joint cumulative distribution function, $F(x_1, x_2)$, for a pair of continuous random variables as,

$$F(x_1, x_2) = \int_{-\infty}^{x_1} \int_{-\infty}^{x_2} f(y_1, y_2) dy_1 dy_2, \tag{1.7}$$

$$= \sum_{\{x_1(i) < x_1\}} \sum_{\{x_2(j) < x_2\}} f(x_1(i), x_2(j)), \tag{1.8}$$

where Eqns. (1.7) and (1.8) are for continuous and discrete random variables, respectively.

For higher than one-dimensional random variables, some additional properties must be defined. One such measure is a joint centrality measure that captures how a pair of random variables move relative to each other, i.e., when one takes higher values than its mean, whether the second tends to take values higher or lower than its mean. We will call this joint centrality moment the covariance of the random variables, defined by,

$$Cov(X_1, X_2) = \int_{-\infty}^{\infty} \int_{-\infty}^{\infty} (x_1 - E[X_1])(x_2 - E[X_2]) f(x_1, x_2) dx_1 dx_2, \tag{1.9}$$

$$= \sum_{\{x_1(i)\}} \sum_{\{x_2(j)\}} (x_1(i) - E[X_1])(x_2(j) - E[X_2]) f(x_1(i), x_2(j)). \tag{1.10}$$

We can also write this in more compact notation as, $Cov(X_1, X_2) = E[(X_1 -$

$E[X_1])(X_2 - E[X_2])]$, where expectation is taken as a double integral or summation using the joint density function. Expectation of other functions of the two random variables can be similarly determined. The covariance is used to create a normalized measure of correlation between a pair of random variables, defined by,

$$\rho_{X_1, X_2} = Corr(X_1, X_2) = \frac{Cov(X_1, X_2)}{\sigma_{X_1} \sigma_{X_2}}, \qquad (1.11)$$

where σ_{X_1} and σ_{X_2} are standard deviations of random variables X_1, X_2, respectively.

From the joint density of multiple random variables, the distribution function for one of them can be extracted, which is termed as the marginal distribution function, defined by,

$$F_{X_1}(x_1) = \int_{-\infty}^{x_1} \int_{-\infty}^{\infty} f(y_1, y_2) dy_1 dy_2, \qquad (1.12)$$

$$= \sum_{\{x_1(i) < x_1\}} \sum_{\{x_2(j)\}} f(x_1(i), x_2(j)), \qquad (1.13)$$

where Eqns. (1.12) and (1.13) are for continuous and discrete random variables, respectively.

For several random variables, one important property is that of independence. Knowing that two or more risks are independent gives the benefit of studying them and making decisions for them independently. Independence of two (or more) random variables implies that each of them takes values independent of each other. In terms of the probabilities and distribution function, this means $P(\{X_1(\omega_1) \leq x_1\} \bigcap \{X_2(\omega_2) \leq x_2\}) = P(\{X_1(\omega_1) \leq x_1\}) P(\{X_2(\omega_2) \leq x_2\})$, and $F(x_1, x_2) = F_{X_1}(x_1) F_{X_2}(x_2)$ for all x_1, x_2. An implication of independence of random variables on expectation of functions of the random variables is that, $E[g(X_1)h(X_2)] = E[g(X_1)]E[h(X_2)]$, for any function $g(X_1)$ and $h(X_2)$. It is worth noting that this property arising from independence of random variables implies that the random variables have zero correlation, although zero correlation between two random variables may not imply independence of the random variables. Therefore, independence is a stronger inter-relational property between random variables.

1.2.1.3 Conditional Probability

Suppose we know that some event has occurred, perhaps defined in terms of a random variable, that will affect our judgment on the likelihood of other events, possibly defined in terms of a set of other random variables. Following the dart example, if we know that the dart I threw did land on the dartboard, then the probability assessment of my scores would become different, hopefully better! The modified probabilities on having extra information are the **conditional probabilities**. Let's say B is an event in the σ-algebra, $\boldsymbol{A}$, which could be defined in terms of a random variable X_1. For example, B could be that an even number of 'H's show up when 10 coins are tossed. This will modify our judgment on the probability of the event of getting 2 'H's. In this example the second event is defined in terms of the same random variable, however in general this may not be the case. The conditional probability can be computed as follows:

$$P(X_1(\omega_1) = 2|B) = P((X_1(\omega_1) = 2) \bigcap B)/P(B), \quad \text{where} \quad P(B) > 0. \ (1.14)$$

It is, however, possible that occurrence of an event A has no bearing on that of another event, B. Then $P(A|B) = P(A)$. In this case we say that the events A and B are **independent**. If A is an event defined in terms of random variable, X_1, and event B is defined in terms of random variable, X_2, and X_1 and X_2 are independent random variables, then this implies events A and B would be independent. This is the direct implication of independence of random variables on conditional probability of events defined by the random variables.

This allows us to define the conditional probability density (mass) function, $f(x|B)$, defined by the Eqn. (1.14). We can see that $f(x_1|B) \geq 0$, and show that $\int_{-\infty}^{\infty} f(x_1|B)dx_1 = 1$. The conditional density function allows us to define **conditional expectation**, as follows,

$$E[X_1|B] = \int_{x_1} x_1 f(x_1|B)dx_1, \tag{1.15}$$

where the event B could be defined in terms of a second random variable X_2 or in terms of random variable X_1 itself. Additionally, the following properties can be shown for relationship between conditional expectation and (unconditional) expectation.

$$E[X_1] = E[X_1|B]P(B) + E[X_1|B^c]P(B^c), \tag{1.16}$$

$$E[X_1] = \sum_i E[X_1|B_i]P(B_i) \text{ where } \bigcup_i B_i = \Omega,$$

$$P(B_i) > 0, B_i \bigcap B_j = \emptyset, i \neq j, \tag{1.17}$$

$$E[X_1|A \bigcap B] = E[X_1|A] \text{ where } A \subseteq B. \tag{1.18}$$

If we have a set covering of the sample space, Ω, as in Eqns. (1.16) and (1.17), the unconditional expectation of a random variable, X_1, can be computed in terms of conditional expectations based on the elements of the set covering. The last property in Eqn. (1.18) simply brings to light that when we condition on the event $A \bigcap B$, with $A \subseteq B$, only the occurrence of A matters.

After this general development of constructs to develop models of risk, in the next section we will look at some specific models of risk. These will be useful throughout the rest of the book.

1.2.2 Specific Models of Risk

Risks abound. In their management it is always beneficial to be able to characterize each risk by a model. A model is never perfect, but if one is able to capture the most important characteristics of a risk, relevant for the context of risk management, then this greatly facilitates the process of risk management. We devote this section to the discussion of properties of well-known standard random variable models that may be chosen to model various risks. This overview should help the risk manager get familiar with the first set of risk models available, although the space of random variables is vast, only restricted by the number of distribution functions a modeler can imagine.

Besides discussing the properties of the model in formulas and equations, we provide ample display of figures and graphs for a visual appreciation of the properties of these models. As a support for implementation of these models, we also provide suggestions for commands in the MATLAB mathematical software tool [20]. We begin with the widely used and celebrated bell-shaped distribution.

1.2.2.1 Normal Distribution

It won't be too much of an exaggeration if we said that normal distribution is the most popular distribution in finance and risk management. It is a distribution with the familiar symmetric bell-shape, with the tip of the bell defining the mean of the distribution, and the thinness of the bell defining the degree of variability summarized in variance. The probability density function of a normal distribution is given by,

$$f(x; \mu; \sigma) = \frac{1}{\sqrt{2\pi\sigma^2}} e^{-\frac{(x-\mu)^2}{2\sigma^2}}, \tag{1.19}$$

where $-\infty < x < \infty$, μ is the mean and σ is the standard deviation of the distribution. The PDF of the normal distribution looks straightforward, however it should be noted that finding its integral to determine the CDF for normal distribution cannot be done analytically. For this purpose, integration must be done computationally in tools like MATLAB.

This distribution is also called the Gaussian distribution, which is good to know since later in the book we will define a Gaussian process. When the mean is set to zero, $\mu = 0$, and the standard deviation is set to one, $\sigma = 1$, the distribution becomes standard normal, denoted by $N(0,1)$. In days when computing was done in paper-and-pencil, printed tables with CDF values for standard normal distribution were used to compute probabilities for a general normal distribution. This was facilitated by the fact that $\sigma X + \mu$ has a general normal distribution, $N(\mu, \sigma)$, when $X \sim N(0,1)$. Therefore, μ can be considered the location parameter of the distribution and σ is the scale parameter. A plot of PDF and CDF of normal distribution is provided in Figure 1.2, for different choices of μ in the PDF.

Due to the symmetry of the distribution about its mean, $P(X \leq x) =$

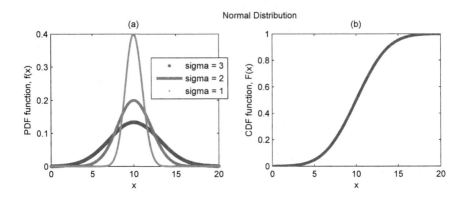

FIGURE 1.2: (a) Probability density function for normal distribution. (b) Cumulative distribution function for normal distribution.

$P(X \geq 2\mu - x)$. The bell-shape of the distribution centered at μ results in majority of the probability to be concentrated around the mean of the distribution, therefore the event $(\mu - \sigma \leq x \leq \mu + \sigma)$ has 66.67% probability for any choice of μ and σ. Therefore, if σ is small, the density function is a peaked curve, getting flatter as the σ increases, as shown in Figure 1.2. Beyond variance, the third-order central moment for normal distribution, which is a measure of asymmetry in the distribution, is zero for all values of μ and σ. Therefore, $E[(X - E[X])^3] = 0$. The tails of the normal distribution, measured by the fourth-order central moment (also called Kurtosis of the distribution), defines the benchmark for fatness. A distribution with a fatter tail than normal distribution is termed a fat-tailed distribution.

1.2.2.2   Uniform Distribution

The uniform distribution is the simplest possible distribution, and as the name suggests, its density function is a flat-level curve. We can construct a discrete version of the uniform distribution also, with finite, say n, number of outcomes equally spaced in an interval. The continuous uniform distribution has a density function given by,

$$f(x; a; b) \quad = \quad \frac{1}{b-a}, \text{ for } a \leq x \leq b, \tag{1.20}$$

$$= \quad 0, \text{ otherwise}, \tag{1.21}$$

where a and b are any real values. The distribution is denoted by $U(a, b)$, and standard uniform distribution is obtained for $a = 0$ and $b = 1$, which is $U(0, 1)$. The mean of the uniform distribution is the middle point of the support of the density function, i.e., where the density function is positive. Therefore, $\mu = \frac{a+b}{2}$. Standard deviation of uniform distribution is $\sigma^2 = \frac{(b-a)^2}{12}$. In Figure 1.3, the probability density and cumulative distribution function for

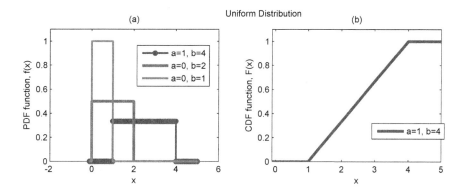

FIGURE 1.3: (a) Probability density function for uniform distribution. (b) Cumulative distribution function for uniform distribution.

uniform distribution is displayed for different choices of a and b. The uniform distribution may appear to be too simple to be useful. However, in Chapter 4 we will show that this distribution plays a fundamental role in the simulation of risks.

1.2.2.3 Central Limit Theorem

As a small detour, we introduce an important result that highlights the significance of the normal distribution. If we have N random variables, $\{X_i; i = 1, \ldots, N\}$, all of them have identical distribution, as well as mutually independent, we will call such set of random variables i.i.d., short for independent, identically distributed. Lets say they all have $U(a, b)$ distribution. We define a new random variable, $Y_N = \frac{\sum_{i=1}^{N} X_i}{N}$, which is essentially the sample mean of the N i.i.d. random variables.

The Central Limit Theorem, in short CLT, states that the random variable Y_N has an approximately normal distribution, with the approximation getting better as N gets larger. Therefore, in the limit (hence 'Limit' theorem) we have,

$$\frac{\sum_{i=1}^{N} X_i}{N} \quad \to \quad N\left(\frac{a+b}{2}, \sqrt{\frac{(b-a)^2}{12N}}\right), \text{ as } N \to \infty. \qquad (1.22)$$

This result is true for any set of i.i.d. random variables, no matter what their distribution. In fact, the result is stronger than we state here; the reader should refer to Billingsley [11] or Durrett [22] for more details.

In Figure 1.4, we display histograms for Y_N for increasing values of N. With increasing value of N, the histograms remain located at value $\frac{a+b}{2} = 0.5$, whereas the variance reduces significantly. The overlaid red curve is a fitted normal density function, which shows an improving fit with increasing value of N.

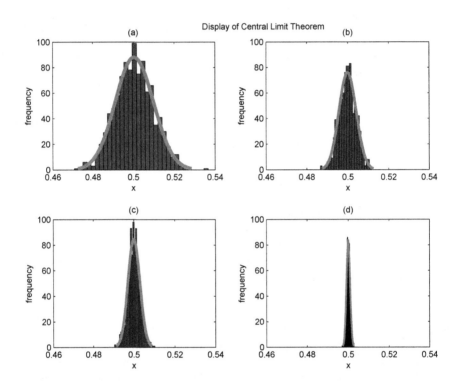

FIGURE 1.4: Display of Central Limit Theorem. (a) N = 1,000 (b) N = 5,000 (c) N = 10,000 (d) N = 100,000.

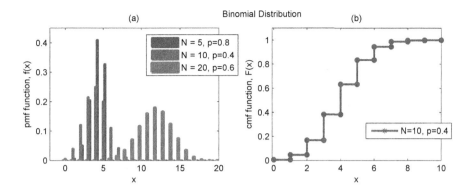

FIGURE 1.5: (a) Probability mass function for binomial distribution. (b) Cumulative distribution function for binomial distribution.

1.2.2.4 Binomial Distribution

This is the first discrete random variable we consider. Binomial distribution is built based on Bernoulli distribution, which describes a Bernoulli trial. Bernoulli trial is any risk that has only two possible outcomes - marked 'success' and 'failure'. For example, a bet on a toss of a coin will either make you win when a heads is realized, or lose when a tails shows up. In a duration of a year, a bond will either not default (labeled as 'success') or end up defaulting (labeled as 'failure').

Binomial distribution looks at a collection, say of size N, of Bernoulli trials, and measures the number of times success is obtained among the N trials. Therefore, the range of values for a binomial random variable is $\{0, 1, \ldots, N\}$. The probability mass function for a binomial distribution is given by,

$$f(x_i; N, p) = \frac{N!}{x_i!(N - x_i)!} p^{x_i}(1 - p)^{N - x_i}, \text{ for } x_i \in \{0, 1, \ldots, N\}, \quad (1.23)$$

where p is the probability of success in each of the N Bernoulli trials and $x!$ denotes x-factorial. The mean and variance of binomial distribution are Np and $Np(1 - p)$, respectively. The probability mass function and cumulative distribution function for a set of values for N and p are given in Figure 1.5.

1.2.2.5 Poisson Distribution

The most popular distribution for the study of queues is the Poisson distribution. Like binomial distribution, Poisson is a counting distribution, i.e., it is discrete and takes integral values from 0 to ∞. Therefore, while in theory it can assume an arbitrarily large value, we will observe that the probability of large values drops sharply. Study of queues is a vast and important topic, since queues are encountered in numerous contexts where people or things must wait to get jobs processed or service received. Efficiency of these pro-

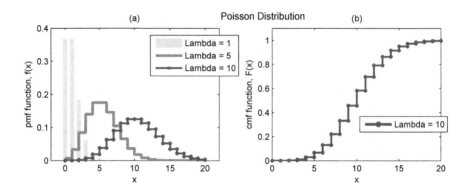

FIGURE 1.6: (a) Probability mass function for Poisson distribution. (b) Cumulative distribution function for Poisson distribution.

cesses and services is highly dependent on how well the queue structures are organized, which is studied in the discipline of queuing theory.

Simply put, Poisson distribution measures the number of arrivals or occurrence of events in a duration of time or space. Customers arriving at a bank teller or ATM, applications for auto insurance arriving for processing, number of errors showing up in an application for loan, etc. are all examples of risk where Poisson distribution may be useful. For our purposes, this will turn out to be an important distribution since in a later chapter we will develop the distribution into the Poisson process.

The probability mass function of a Poisson distribution is given by,

$$f(x_i; \lambda) = \frac{\lambda^{x_i} e^{-\lambda}}{x_i!}, \text{ for } x_i \in \{0, 1, \ldots, \infty\}, \tag{1.24}$$

where λ defines the rate of arrival or occurrence of the event and, as before, $x!$ denotes x-factorial. λ is a single parameter that defines this distribution, which also happens to be the mean and the standard deviation of Poisson distribution, and is stated as a rate. Figure 1.6 displays the probability mass function and cumulative distribution function for a set of choice of λ parameters. We notice that the probability mass function peaks around the mean value, λ, and decays rapidly on both sides of the peak. Hence, even though theoretically there can be infinite outcomes, the probability of the arbitrarily large values being realized is minuscule.

1.2.2.6 Exponential Distribution

We switch our attention back to continuous random variables, and look at a distribution that is closely related to the Poisson distribution. Exponential distribution is a positive valued distribution, and as it turns out, has a simple

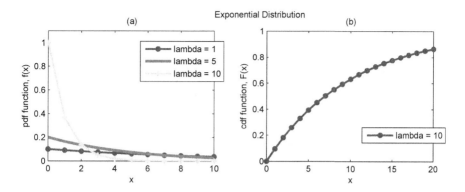

FIGURE 1.7: (a) Probability density function for exponential distribution. (b) Cumulative distribution function for exponential distribution.

probability density function given by,

$$
\begin{aligned}
f(x;\lambda) &= \lambda e^{-\lambda x}, \text{ for } 0 \le x < \infty, && (1.25)\\
&= 0, \text{ for } x < 0,
\end{aligned}
$$

where λ is the only parameter needed to describe the distribution. The mean and variance of the distribution are $\frac{1}{\lambda}$ and $\frac{1}{\lambda^2}$, respectively. In Figure 1.7, we plot the probability density and distribution function for a set of λ values. As can be seen from the probability density plot, exponential distribution is an asymmetric function with a faster decay for increasing value of λ, as is also evident from the functional form of the PDF in Eqn.(1.25).

The cumulative distribution function for an exponential random variable is equally straightforward, as shown in Figure 1.7 (b), given by,

$$
\begin{aligned}
F(x;\lambda) &= (1 - e^{-\lambda x}), \text{ for } 0 \le x < \infty, && (1.26)\\
&= 0, \text{ for } x < 0.
\end{aligned}
$$

For exponential distribution, the CDF is not only straightforward, it is also easy to invert it to obtain the inverse-CDF, given by $F^{-1}(y) = -\frac{\ln(1-y)}{\lambda}$. We will revisit the inverse-CDF of exponential in Chapter 4 when we describe methods for simulation of risks.

The relationship between the Poisson distribution and exponential distribution should be noted before moving on. The λ parameter used in the two distributions is in fact the same quantity. In Poisson distribution, the parameter describes the rate of occurrence of events, and in the exponential distribution it is the time or space between occurrence of events. That explains why the mean of the exponential distribution is the reciprocal of rate of occurrence in the Poisson distribution. Additionally, exponential distribution possesses the 'memoryless' property, namely $P(X > x + y | X > y) = P(X > x)$ for $x, y > 0$. Being memoryless here refers to the fact that knowing an event has

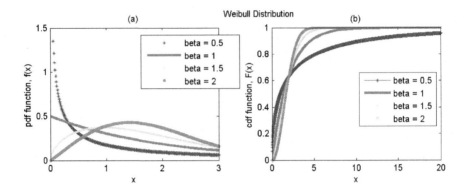

FIGURE 1.8: (a) Probability density function for Weibull distribution. (b) Cumulative distribution function for Weibull distribution.

not occurred for a certain amount of time provides no information on how much longer it might take for the event to occur.

1.2.2.7 Weibull Distribution

Weibull distribution is again a two parameter distribution, defined by its shape β and scale α parameter. It is most closely related to the exponential distribution, as will be evident from the probability density function. Like exponential, Weibull is also a positive valued random variable. Moreover, Weibull distribution with a shape parameter less than 1 is also an example of a heavy-tail distribution. The distribution is most commonly used for modeling risk of failure of machinery, and is also utilized in insurance models. The probability density of the Weibull distribution is given by,

$$f(x; \beta, \alpha) = \frac{\beta}{\alpha}(\frac{x}{\alpha})^{\beta-1}e^{-(\frac{x}{\alpha})^{\beta}}, \text{ for } 0 \leq x < \infty, \quad (1.27)$$
$$= 0, \text{ for } x < 0.$$

Therefore, when shape parameter $\beta = 1$, the PDF simplifies to the exponential distribution with $\lambda = \frac{1}{\alpha}$. In Figure 1.8, some probability density and distribution functions are displayed for different values of the shape parameter, β, clarifying why the parameter has earned the name. For $\beta < 1$, the density function has the shape of the letter J, racing to ∞ near 0. When shape parameter lies in the interval $(1, 2)$, the density approaches zero at $x = 0$, but has infinite slope at this point. The slope at $x = 0$ becomes finite for $\beta = 2$. And finally, for $\beta > 2$, the density is zero and has a zero slope at $x = 0$. The density of the Weibull distribution is unimodal, which means it has one point where it peaks. The mean and variance for the Weibull distribution are described in terms of the gamma function, $\Gamma(.)$.

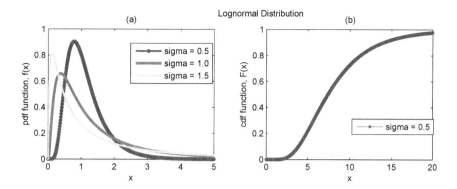

FIGURE 1.9: (a) Probability density function for lognormal distribution. (b) Cumulative distribution function for lognormal distribution.

1.2.2.8 Lognormal Distribution

Next to the normal distribution in terms of popularity for modeling risk in finance is the lognormal distribution. As the name of the distribution suggests, the lognormal distribution is related to the normal distribution in that natural logarithm of the random variable is normally distributed. Keeping with this relationship, a lognormal random variable, Y is denoted to have $lnN(\mu, \sigma)$ distribution, implying that the logarithm of the random variable has normal distribution with mean, μ, and standard deviation, σ. For lognormal distribution, μ continues to be a location parameter and σ, a scale parameter. Put another way, if $X \sim N(\mu, \sigma)$ then e^X is lognormally distributed. The probability density function, obtained by change of variable in the normal probability density, is given by,

$$f(x; \mu; \sigma) = \frac{1}{\sqrt{2\pi\sigma^2 x^2}} e^{\frac{-(\ln x - \mu)^2}{2\sigma^2}}, \text{ for } -\infty < x < \infty. \qquad (1.28)$$

As opposed to the normal distribution, precisely due to the nature of change of variable, the lognormal distribution is a positive valued random variable. This is a particular advantage in modeling risks that must be positive in value, such as equity prices. Moreover, for different choice of parameter σ, the distribution takes different shapes. This is demonstrated in the probability density plots of Figure 1.9. The mean and standard deviation, expressed in terms of mean (μ) and standard deviation (σ) of the underlying normal, are $e^{\mu + \frac{\sigma^2}{2}}$ and $(e^{\sigma^2} - 1)e^{2\mu + \sigma^2}$, respectively.

1.2.2.9 Chi-Square Distribution

The sum of random variables may or may not have the same distribution as the individual random variables. This property is an important one, since in many risk management contexts the quantities of interest are sums of random

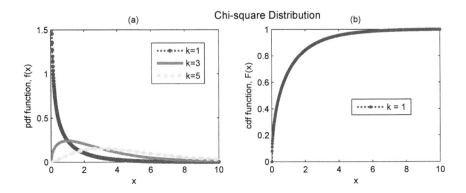

FIGURE 1.10: (a) Probability density function for Chi-square distribution. (b) Cumulative distribution function for Chi-square distribution.

variables. For instance, return of a portfolio is expressed in terms of sum of return of individual asset in the portfolio. If $X_1, X_2, \ldots, X_N$ are N random variables with, say normal distribution of possibly different parameters and not necessarily independent, then the sum $Y_N = X_1 + X_2 + \ldots + X_N$ also has a normal distribution. The precise parameters of the distribution depend on the parameters of each of X_i's and their correlations.

The distributions we are now turning to are sums of other random variables, since more often than not the sum of random variables does not follow the same distribution. We will consider the sum of square of normally distributed random variables. As seen above, the sum of normally distributed random variables is normal. However, the sum of square of independent (standard) normally distributed random variables follows the Chi-square distribution. Chi-square distribution is stated with its degrees of freedom k, as $\chi^2(k)$, which is the number of normally distributed random variables added to produce the Chi-square distribution. This definition of Chi-square distribution implies that the sum of independent Chi-square random variables also has a Chi-square distribution, with degrees of freedom added. The probability density for Chi-square distribution is given by,

$$
\begin{aligned}
f(x; k) &= \frac{1}{2^{k/2}\Gamma(k/2)} x^{\frac{k}{2}-1} e^{-\frac{x}{2}}, \text{ for } 0 \leq x < \infty, \quad (1.29) \\
&= 0, \text{ for } x < 0.
\end{aligned}
$$

The mean and standard deviation of Chi-square distribution is k and $\sqrt{2k}$, respectively. We display probability density and distribution functions for Chi-square distribution in Figure 1.10. From its nature of construction, Chi-square distribution is positive valued, and its PDF takes on different shapes for increasing value of degrees of freedom.

Chi-square distribution is extensively used in inferential statistics, such as

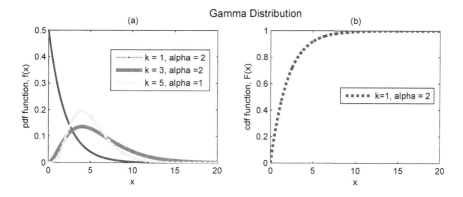

FIGURE 1.11: (a) Probability density function for gamma distribution. (b) Cumulative distribution function for gamma distribution.

in goodness-of-fit tests. If the normal random variables being used to construct a Chi-square random variable do not have a zero mean, then the sum of square of normal random variables has non-central Chi-square distribution. Chi-square and non-central Chi-square distribution are also used for the analysis of key financial risk factors.

1.2.2.10 Gamma Distribution

While the list of distributions we can potentially explore is long, our intention here is not to attempt to be exhaustive. The intent is to provide a feel for the basic examples of models of risk, and some extensions of them, in support of our future modeling developments and applications in the book. With this in mind, the last distribution we consider is the sum of exponentially distributed random variables. Sum of exponentials is not exponential, instead it makes the Gamma distribution. Gamma distribution is denoted as $\Gamma(x; \alpha, k)$, where α is the scale parameter and k gives the distribution its shape. The probability density is given by,

$$
\begin{aligned}
f(x; \alpha, k) &= x^{k-1} \frac{e^{x/\alpha}}{\alpha^k \Gamma(k)}, \text{ for } 0 \le x < \infty, \qquad (1.30) \\
&= 0, \text{ for } x < 0.
\end{aligned}
$$

The mean and variance of the Gamma random variable are $k\alpha$ and $k\alpha^2$, respectively. We display some plots of probability density and distribution in Figure 1.11. As a sum of exponential random variables, it is naturally used in reliability models, waiting times between defects. The model is also useful in modeling insurance claims, loan defaults, and other crucial risk factors.

1.3 MATLAB® Tools for Distributions

MATLAB mathematical software has a vast array of functions for working with probability distributions in its Statistics Toolbox. We list a few of these functions here. The reader is advised to look up the extensive help documentation available with MATLAB to see the details of these and other related functions. At the bottom of each function description in MATLAB help documentation, look for 'See Also' to explore other related functions. Resources such as MATLAB Primer [20] are also useful.

Normal distribution: `normpdf`, `normcdf`

Uniform distribution: `unifpdf`, `unidpdf`, `unifcdf`, `unidcdf`

Binomial distribution: `binopdf`, `binocdf`

Poisson distribution: `poisspdf`, `poisscdf`

Exponential distribution: `exppdf`, `expcdf`

Weibull distribution: `wblpdf`, `wblcdf`

Lognormal distribution: `lognpdf`, `logncdf`

Chi-square distribution: `chi2pdf`, `chi2cdf`, `ncx2pdf`, `ncx2cdf`

Gamma distribution: `gampdf`, `gamcdf`

Other: `hist`, `histfit`

1.4 Summary

Risk is ubiquitous, therefore for living a better life in the personal or professional realm, one needs to understand the meaning and implication of risk. This understanding should translate to a smarter management of risk. In this chapter, we defined risk and developed a typology of risk. The classification of risk helps isolate properties of risk that will help determine the appropriate tools for managing the risk. After the classification of risk, formalization for the quantitative definition of risk was developed. Definition of random variables and the properties of probability measure developed in this chapter are fundamental to the models for risk management developed in later chapters. Finally, we introduced certain specific models of risk and studied their properties, as examples of the general concept of probability distribution. These specific models of risk will be useful in various risk management applications.

1.5 Questions and Exercises

Review Questions

1. What is risk? Why is it important to manage risk?

2. Discuss the differences between risk and uncertainty.

3. How are objective probabilities different from subjective probabilities? Why is this distinction important?

4. How can a classification of risks by a meaningful taxonomy aid in risk management?

5. What is the difference between pure and speculative risk? Give specific examples of pure and speculative risks.

6. What are the sub-classifications of pure risk? How do these differ from one another.

7. What are the sub-classifications of speculative risk?

8. What is market risk? What are the major sub-classifications of market risk? Give specific examples of each.

9. Getting too embedded in the risk classification is dangerous from the point of view of missing their interactions. Discuss how some of the sub-classes of speculative risk may interact, and what the implication of this interaction may be.

10. Why is the development of models for risks important?

11. Define the following terms for models of risk:

    (a) Sample space
    (b) Events
    (c) σ-algebra
    (d) Random variable
    (e) Probability measure
    (f) Probability density function
    (g) Cumulative distribution function
    (h) Mean
    (i) Standard deviation

12. What is the meaning of independence of events? When are random variables independent?

13. Define the following terms for models of several risks:

    (a) Joint distribution functions
    (b) Marginal density function
    (c) Covariance
    (d) Correlation
    (e) Conditional probability

14. What is the definition of n-order central moment of a random variable?

15. What are skewness and kurtosis of a random variable? What is the skewness of a normal distribution?

16. What are the statement and significance of the Central Limit Theorem?

17. What is the relation between Bernoulli random variable model and the binomial random variable model?

18. How are the normal and the lognormal distribution related?

19. The Poisson distribution is said to be closely related to the exponential distribution. Elaborate the role of the parameter, λ, in this relation.

20. How is the Weibull random variable model an extension of the exponential model? What risks is the model used for?

21. How are the Chi-square and the Gamma random variable models constructed from the simpler distributions? What risks are the models used for?

Exercises

1. In a hurricane prone area, the hurricane season is graded on a discrete 0-6 scale, where '0' represents a mild season and '6' represents the most severe season. Historical data on hurricanes in this area have been used to assign probabilities for these grades of the hurricane season, summarized as: $\{0.1, 0.2, 0.1, 0.3, 0.14, 0.11, 0.05\}$. This pure risk causes loss to property in amounts (in millions of USD) from $\{0, \$5, \$80, \$140, \$200, \$500, \$2000\}$, corresponding to the grades of the hurricane season. Use MATLAB to perform the following computations.

    (a) What is the probability of extremes in a hurricane season?
    (b) What is the probability of greater than a '3' grade hurricane season?
    (c) If a season is predicted to be greater than a '3' grade hurricane season, what is the expected loss?
    (d) What is the mean, median, mode, and standard deviation of losses due to hurricanes in a hurricane season?

2. Annual demand forecast is being developed for a new product launch. The estimated probability density constructed based on survey data is given as,

$$
\begin{aligned}
f(x; p, \lambda) &= \lambda p e^{\lambda(x-300)}, \text{ for } x < 300, \qquad (1.31)\\
&= 0, \quad \text{for } x = 300\\
&= \lambda(1-p)e^{-\lambda(x-300)}, \text{ for } x > 300.
\end{aligned}
$$

This is the double-exponential risk model, where we will set $p = 0.3$ and $\lambda = 0.02$. All demands are in thousand (000) units. Use MATLAB to perform the following computations.

(a) From the above probability density function, construct the cumulative distribution function for the projected annual demand for the new product.

(b) How likely is the demand for the new product to be less than 350 per year?

(c) What is the probability that the demand will at least be as high as 250 per year?

(d) What is the probability the demand will exceed 400 per year?

(e) What is the mean annual demand for the product? What is the standard deviation?

3. Two risks that co-evolve are being modeled by the following joint density function,

$$
\begin{aligned}
f(x, y) &= ce^{(-x-y)}, \text{ for } x > 0, y > 0, \qquad (1.32)\\
&= 0, \quad \text{otherwise.}
\end{aligned}
$$

(a) For what value of 'c' is this a legitimate joint density function?

(b) What is the probability of the first risk exceeding in value than the second, i.e., $P(X > Y)$?

(c) Determine $P(X + Y > 1)$.

4. The relation between two risks is summarized in the following joint density function,

$$
\begin{aligned}
f(x, y) &= ye^{-y(x+1)}, \text{ for } 0 \le x, y > \infty, \qquad (1.33)\\
&= 0, \quad \text{otherwise.}
\end{aligned}
$$

(a) What is the marginal density function of each risk, X and Y?

(b) Using the joint density, obtain the covariance of the two risks, and also their correlation.

(c) What is the Conditional Density of X given Y?

5. The price of a stock is being modeled as a Lognormal distribution, where you are given that the natural log of the price, $\ln(S)$, is normally distributed as, $N(2, 0.6)$.

   (a) What are the mean and variance of the price of the stock?

   (b) Utilize MATLAB routines to compute the probability for the stock price to exceed $10.

   (c) How likely is it for the price of the stock to fall below $7?

6. If the time to the arrival, T, of the next customer at a drive-through teller at a bank is modeled as an exponential distribution, knowing that it has been 2 minutes since the previous customer left the teller does not help. This is the 'memoryless' property of the exponential distribution, i.e., $P(T > s + t | T > s) = P(T > t)$. Show that the exponential distribution possesses the 'memoryless' property.

7. In a portfolio of 600 loans, any single loan is likely to default in the next year with $p = 0.03$ probability. If loans are expected to default independent of one another, what model of risk would you like to use to assess the total number of loans that default in the portfolio in the next year? After selecting a model, compute the following quantities using MATLAB.

   (a) What is the mean number of defaults that should be expected in the portfolio in the next year? What is the standard deviation?

   (b) How likely would be more than 20 defaults in the portfolio?

   (c) What is the probability that the number of defaults will be less than 5?

Chapter 2

Framework for Risk Management

Risk management is perhaps one of the most important activities of any enterprise or household in support of its smooth and efficient operation. This is especially true when we are ready to see the goal of mastering risk as not restricted to be on defensive terms. Risk management is about how individuals and firms actively, proactively, or sometimes reactively, select the types and levels of risk that are appropriate and beneficial to assume. Some have held a defensive view of risk management, which makes sense from a pure risk perspective, that is if the focus of risk management was solely on pure risk. However, as was studied in Chapter 1, the range of risks is wide, with pure risk being important but only a subset of the risks.

For speculative risks, the risk-reward link is so integral to the risk that approaching these risks in defensive terms seems downright inappropriate. Making risk management an integral part of the management and control process for running a firm or a household allows for sounder economic management of the enterprise. Depending on the nature and extent of risk exposures, individuals, households, and firms may employ varied degrees of effort and resources towards their risk management goals. Irrespective of the degree of effort and level of resources, integrality of risk management to the firm's or household's functioning implies keeping the consideration of risk and response at the core of awareness of the firm's employees and members of a household.

For many regulated sectors, risk management is not a luxury or a choice, but an obligation. Banking, insurance, telecommunication, transportation, oil and gas, electricity, and drugs and pharmaceuticals, are all examples of sectors that are regulated, under some version of regulatory settings in many nations. What is also important to note is that the regulatory framework under which many of these sectors have functioned has changed, and will continue to change, sometimes quite dramatically. This implies that the firms of the sector, or those of related activities, must align their risk management efforts to an evolving regulatory context.

Risk management is a continual process of corporate risk monitoring, control, reduction, and management. While it may not be an activity undertaken just to satisfy regulators, the regulatory demands for risk management can result in a major impetus behind implementation and upgrades of risk management systems. For instance, in banking, the Basel Committee on Banking Supervision (BCBS), an international extension of regulatory bodies of the major developed countries, has been evolving the guidelines for regulatory control for decades, with a new impetus infused from the challenge of financial crises. Banks across the world have had to work diligently to comply with the risk management requirements, and will have to continue to do so. Therefore, management of risks is easier said than done.

Multitudes of risks that can affect a firm, an individual, or a household, their increasing complexity, the global interconnectedness of risk factors, compounded with the burden of regulatory requirements make risk management a non-trivial endeavor. In this chapter, we will focus on a formal and rigorous framework to support the risk management efforts and activities of individuals and enterprise entities. Concepts to support deliberation of the impact of risks and making choices in light of risks will be developed, with an attempt to retain their applicability for individuals, managers, and firms.

2.1   How to Handle Risk

The first prerequisite to managing risk is to identify and understand it. In Chapter 1, we defined risk as variability that can be quantified in terms of probabilities. The definition differentiated risk from uncertainty, which refers to those future non-certain happenings whose outcomes are only known as possibilities, and ex-ante probabilities are hard to assess. What we seek to manage under a rigorous risk management framework are future non-certain variabilities that are quantifiable both in terms of their outcomes and likelihoods. However, while learning to understand and manage risk, we should attempt not to narrow down our skills too much so as to find ourselves totally helpless in face of uncertainty. Since in today's increasingly connected and complex world, there is a lot of grey region between risk and uncertainty.

If risk is the variability in future happenings that can be quantified in terms of probabilities, one more distinction is needed in terms of subjective and objective probabilities. This distinction gets to the point of understanding risk, based on which choices regarding it must be made. Each one of us, based on our beliefs, historical data, and evidence can have a subjective view of probabilities of future outcomes. Objective probabilities are those backed with observational evidence, not just subjective estimates. Unavailability of relevant or insufficient data can often result in the blurring of boundary between objective and subjective probabilities. This distinction is crucial for the discussion of choices and choice-criteria for risk management.

In any context, understanding the possible sources for risk exposures is an essential prerequisite for viable and effective risk management. A full grasp of the sources of risk is greatly helped by a meaningful taxonomy of risks. A taxonomy of risks was developed in Section 1.1, which would need to be customized and applied depending on the nature of the entity developing a risk management plan and strategy. For instance, the sources of risk from an individual's and household's perspective would be quite different from those of a financial or non-financial firm. The most important classification of risk was on the basis of pure versus speculative risks. Pure and speculative risks can in their own turn be broken down into subclassifications, which highlight the source and nature of the risk exposures.

Once risks are identified, an effort supported by a risk taxonomy, managing risks in a specific context requires understanding the context so that quantities to observe are identified in order to develop measurable quantifications of the impact of the risks. For our purposes, the impact of risk will uniformly be measured in financial and/or monetary terms. A measurable and quantifiable indicator for risk in financial and/or monetary terms, though it bears shortcomings and challenges, helps in approaching risk management in a disciplined, consistent framework.

Management of risk, depending on the type of risk and nature of entity engaged in risk management, requires discovering modes by which the risks may be managed, i.e., controlled, altered, endured or avoided. This discovery phase of modes for risk management is just as important as the discovery of risk exposures themselves. Since in the end, the mapping of risk exposures with modes by which they will be managed dictates the efficacy of a risk management strategy. Once the risks are mapped with modes for managing them, an ongoing assessment of efficacy of the mapping must be in place in terms of a set of pertinent risk measures.

2.1.1 The Risk Management Framework

In Figure 2.1, the risk management process is depicted as sequentially-aligned activities of the above discussed steps for risk management. At the top is identifying risk exposures. In the taxonomy of risks, from the range of speculative risks to the variety of pure risks, one needs to identify the risks a firm will be exposed to during the course of its activities, in the foreseeable future. At this point, a sense of degree of importance of the risk exposures may also emerge for the enterprise and its divisions, which can then be verified by the quantifiable risk measures.

Identification of risks must be supported by a significant effort to have access to ample data and observational evidence for the anticipated risk exposures. In many cases, data may be hard to come by, hence a significant effort may be necessary for this step. Data and observational evidence from sources outside of internal resources may have to be acquired. This can come at a significant cost, which will need to be justified for the use the data must be put to. Models for risk discussed in Chapter 1, developed based on data and observational evidence, as well as those that will be developed in later chapters, will be needed to conduct the rest of the analysis to support risk management. Some models of risk must also help indicate how the risks will evolve in time; for this purpose a dynamic model for risk must be developed. These are developed in Chapters 5 and 6.

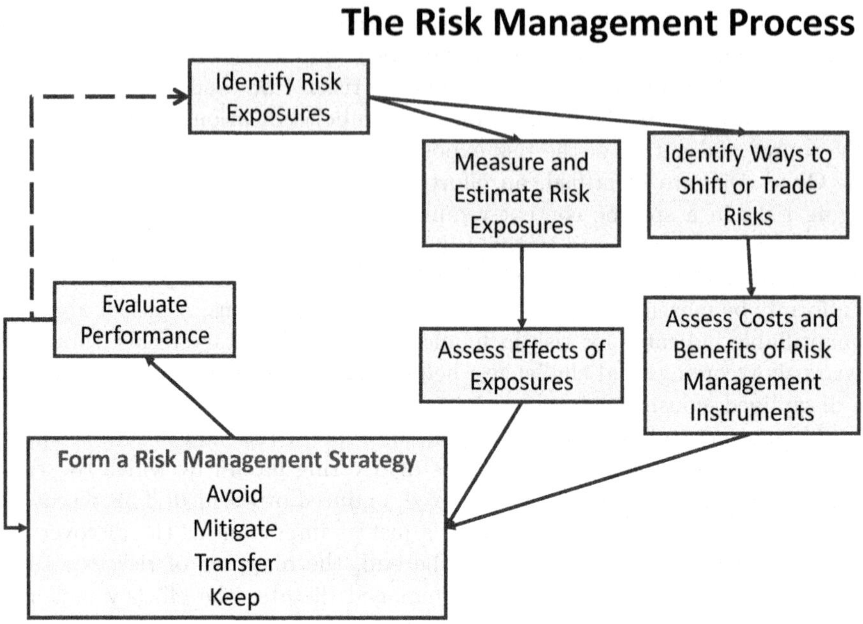

The Risk Management Process

FIGURE 2.1: The overall flowchart for the Risk Management Process.

After the identification of risks, as we follow the path of risk management process in Figure 2.1, there is a bifurcation of activities. In the first line of activity, we design appropriate measures for quantifying the impact of risks, their interactions, and their eventual effect on the bottom-line for the firm or household. Bottom-line here refers to quantities that in the end matter to the firm or household. In the second branch of activities, a process of discovery must be undertaken to identify mechanisms by which risk may be controlled, modified, reduced or transferred. This includes the exact modes, methods, contracts, terms, and conditions by which risk may be managed, along with costs incurred and benefits received from undertaking these approaches.

The two bifurcations in the process merge again in the core step of the risk management process. In this stage, the risk exposures, their quantifiable measures of individual risks, as well as their interaction and joint impact, must be juxtaposed with the mechanisms discovered and available for the management of risk. The theme followed here is summarized in the 'avoid-mitigate-transfer-keep' shorthand. We will discuss each of these terms in detail in Section 2.1.5. They refer, in summary, to how the mechanisms for risk management discovered in one branch can be optimally matched with risk exposures and their impact measured in the second branch so that the benefit of risk management is maximized. The dependencies between lines of business, integration and interactions among their risk exposures are also critical.

Evaluation of effectiveness of risk management efforts must be an ongoing activity. For this purpose, not only the appropriate and quantifiable overview measures and guidelines must be in place, but also an organizational structure should be established in a firm or a household that facilitates communication and allows familiarizing all concerned of the risk management strategy and its effectiveness. The organizational structure, the quantifiable overview measures and guidelines are all ways by which the risk management strategy is continually assessed. When the strategy is judged to be losing effectiveness, it should trigger going back to the 'avoid-mitigate-transfer-keep' drawing board. The loss of effectiveness may be triggered by any changes in specific risk types, sources of exposure, division, or activity of the organization.

Finally, in the dynamic, changing world we live in, there is only one thing we can be certain of. That is change. Over time and after some specific regime shifts, change can be so significant that the entire risk management strategy will require a thorough assessment for relevance and effectiveness. This may indeed require one to go back to the top of the process, i.e., the top box in Figure 2.1, to generate the new best response. This is captured in the final dashed arrow in Figure 2.1, which completes the circle of the risk management process, indicating it to be a continual process.

Therefore, the only way a risk management framework has value to an enterprise is if it fundamentally affects its way to do business. It must be integral in influencing decisions and actions of the agents - members of household, employees and managers of firms; otherwise, it is a mere decoration, an artifact of display with no real ability to add value. In a firm's or household's organization and functioning, risk management should not be a mere 'check-the-box' activity, since if used well, it can help an enterprise achieve and sustain optimal long-term performance and goals.

In summary, for effective management of risks, an enterprise needs to understand and assess its risks, and equally importantly needs to embrace the culture of active consideration of risks. This consideration should guide the establishing of short and longer-term strategy, organizational goals and objectives, and assign responsibilities and tasks for minimizing the likelihood and adverse effect of risks. Everyday decisions must be in concurrence and accordance to the short- and long-term strategy. With changing times, the strategy, enterprise's goals, and responses to risks must be adjusted.

It is very easy to fall into the trap of identifying and assessing risks every quarter, talking in terms of a high level risk response (e.g., accept the risk, or hedge it using xyz instrument) rather than actually managing the risks day-to-day. It is clearly not enough to understand risks for daily decisions, without paying active attention to managing them. Application of a risk management framework should not be a periodic exercise, it should be a way of life for a firm or a household.

2.1.2   Risk Preference vs. Risk Aversion

Comprehending and managing risk is all about making the right trade-off between risk and rewards. This trade-off is the byproduct of the appetite for risk an individual or a firm holds. Making choices in the presence of risk, in order to make the best risk-reward trade-off, can benefit from the concepts developed in the economic theory of choice. Theory of choice develops the framework by which individuals make choices when faced with alternatives. The simplest version of the theory addresses choices when there are alternatives, but the outcome of the choices is not risky. We will begin with the discussion of the non-risky case, then move on to choices made in the presence of risk.

An individual or a firm faced with a decision chooses from a set of alternatives. Let X denote this set of alternatives. Among all elements of X the decision-maker weighs the merit and de-merits of each option against others and attempts to select the option that suits best. This requires a more precise definition of 'best.' Therefore, in more rigorous terms, we need to define an ordering for the set of alternatives that makes it possible to compare every element of the set with every other element. We denote this ordering by '$\succeq$,' then if $x, y \in X$, $x \succeq y$ will imply that the decision-maker prefers x to y, and $x \succ y$ would imply x is strictly preferred to y.

When a decision-maker faces a choice among a number of risky alternatives, often referred to as *lotteries*, the possible outcomes of these alternatives could be N possible levels of monetary payoffs. Therefore, a risky alternative is characterized by the vector of probabilities of the outcomes in that alternative, and the decision-maker is assumed to know the probability of each outcome in each alternative. The alternatives are depicted by $(\tilde{x}, \mathbf{p})$ pair, where $\tilde{x}$ constitutes the possible outcomes of an alternative and $\mathbf{p}$ are the corresponding probabilities of the outcomes. For instance, I may try to (unsuccessfully) trick my smart six-year-old nephew to get \$10 pocket-money when the red-six face shows up on the roll of a die, versus getting \$5 for heads on the toss of a coin. The two alternatives will be summarized as, $x = (\{10, 0, 0, 0, 0, 0\}, \{1/6, 1/6, 1/6, 1/6, 1/6, 1/6\})$ and $y = (\{5, 0\}, \{1/2, 1/2\})$.

The preference relation, for both non-risky and risky alternatives, defined above is said to be rational if it satisfies certain *axioms of choice*. These axioms are reflexivity, completeness, and transitivity, and they are formally defined as follows (see Varian [89], MasColell [60]):

Reflexivity For all $x \in X$ we have $x \succeq x$.

Completeness For all $x, y \in X$ we have either $x \succeq y$, $y \succeq x$ or both.

Transitivity For all $x, y, z \in X$, if $x \succeq y$, $y \succeq z$ then $x \succeq z$.

Additional properties imposed on the preference relation allows summarizing the preference relation in terms of a utility function, $U(x)$. These properties are as follows:

Continuity For any $x, y, z \in X$ such that $x \succeq y \succeq z$, then there is an $\alpha \in [0, 1]$ such that $y \sim \alpha x + (1 - \alpha)z$. The symbol $\sim$ designates indifference between two alternatives.

Independence For any $x, y, z \in X$ and $\alpha \in (0, 1)$, we have $x \succeq y$ if and only if $\alpha x + (1 - \alpha)z \succeq \alpha y + (1 - \alpha)z$.

A utility function summarizes the preference relation, so that when $x \succeq y$, $U(x) \geq U(y)$. The utility derived from a risky alternative is called an expected utility, if the utility function is the expected value of utility derived from each outcome of the alternative. This is defined in the following,

$$U(\tilde{x}) = \sum_{i=1}^{N} u_i p_i, \tag{2.1}$$

where u_i is the utility derived from each realized outcome x_i of the alternative $\tilde{x}$ with its corresponding probability, p_i. It is worth noting that while the outcomes (and their probabilities) of an alternative are objective, the utility derived from each outcome is subjective. Therefore, this construct helps define the risk preference versus aversion of the decision-maker. In cases where the outcomes are a continuum of values, such as wealth outcomes ranging from 0 to ∞, the summation in Eqn. (2.1) must be replaced by an integral, $U(\tilde{x}) = \int_0^\infty u(x)f(x)dx$, where $u(x)$ is the utility derived from the outcome x.

We define **risk aversion** as a view towards risk that favors 'equivalent' non-risky alternatives over the risky ones. More rigorously, we define risk aversion as,

$$U(\tilde{x}) = \int_0^\infty u(x)f(x)dx \leq u\left(\int_0^\infty xf(x)dx\right), \tag{2.2}$$

for any probability density for the outcomes, $f(x)$. Therefore, expected utility of a risky alternative is less than utility of the expected value of the alternative. If the inequality in Eqn. (2.2) is strict, we call it risk aversion. However, if the relation is satisfied with equality, which will happen if the utility function, $u(x)$, is a linear function, then we would call the decision-maker **risk neutral**. A risk-preferring or risk-loving decision-maker will prefer a risky alternative over the equivalent non-risky alternative, therefore we have,

$$U(\tilde{x}) = \int_0^\infty u(x)f(x)dx > u\left(\int_0^\infty xf(x)dx\right), \tag{2.3}$$

for a risk-loving decision-maker for all probability density for the outcomes, $f(x)$.

The risk-aversion, risk-neutral, and risk-loving preferences of a decision-maker translate to the concavity, linearity, and convexity of the corresponding utility function, $u(x)$. This in turn can be indicated by the sign of the second derivative of the utility function, $u''(x)$. When $u''(x) < 0$, we obtain concavity

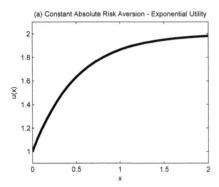
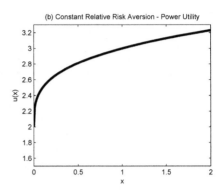

FIGURE 2.2: (a) Plot of the exponential, constant absolute risk aversion (CARA) utility function. (b) Plot of the power, constant relative risk aversion (CRRA) utility function.

of the utility function, and hence risk aversion of the preference relation. Similarly, with $u''(x) > 0$, we obtain a convex utility function, which results in a risk-loving preference relation. This is obtained from the well-known Jensen's inequality, discussed in Chapter 5.

The Arrow-Pratt measures for risk aversion [40] provide a single number indication of the nature of risk preference in a preference relation. The Arrow-Pratt coefficient of absolute risk aversion is defined as, $\frac{-u''(x)}{u'(x)}$. Given $u'(x) > 0$ for a preference relation, which implies that the decision-maker derives a higher utility from a larger monetary outcome, the coefficient of absolute risk aversion is larger when the marginal utility of wealth declines faster. The degree of absolute risk aversion of the decision-maker measures the curvature of the utility function, but is also invariant for linear transformations of the utility function, $\alpha + \beta u(x)$. A linear transformation of a utility function doesn't change the essential characteristics of the utility function. The higher the coefficient of absolute risk aversion, the more risk averse the individual is.

A decision-maker with a decreasing absolute risk aversion is willing to accept more risky alternatives as his wealth level increases. For a decision-maker with constant absolute risk aversion, the willingness to accept risky alternatives remains the same for all levels of wealth. We may wish to define a measure for risk aversion that is independent of the decision-maker's current wealth level. This is the Arrow-Pratt coefficient of relative risk aversion, defined as, $-x\frac{u''(x)}{u'(x)}$. If a decision-maker has a constant relative risk aversion, he would be willing to invest the same proportion of wealth in a risky asset independent of the current level of wealth.

In Figure 2.2(a), we provide a display of a constant absolute risk aversion utility function, $u(x) = a - be^{-\mu x}$, where μ is the coefficient of absolute risk aversion. For this utility function, popularly called the exponential utility function, $u'(x) = b\mu e^{-\mu x}$ and $u''(x) = -b\mu^2 e^{\mu x}$, therefore $u'(x) > 0$ (increasing) and $u''(x) < 0$ (concave) for all x. In Figure 2.2(b), we provide a display of a constant relative risk aversion utility function, $u(x) = a + bx^{1-\gamma}$, where γ is the coefficient of relative risk aversion. This is popularly called the power utility function, while the utility function with a constant relative risk aversion of 1, $u(x) = a + b\ln(x)$, is the logarithmic utility function.

Although at first sight the axioms of choice, and the related constructs of utility function, may look quite reasonable, it is worth taking a closer look. Imposition of these axioms has enormous implications in practical terms. Reflexivity is the weakest and most acceptable axiom. The completeness axiom implies that the decision-maker has done the investigation and introspection of all possible alternatives, however far removed they may be from the realm of common experience. It will at the least need serious work of reflection on the individual's or firm's preferences. Considering the transitivity axiom, on the other hand, implies that in a sequence of pairwise choices, there is no possibility of cycles, no matter how the options are framed or presented. Under any circumstance, the decision-maker is able to rank all of them in an order that contains no cycles. In practice, from a variety of real-world contexts, behaviorists find significant deviations from these assumptions.

2.1.2.1 Normative vs. Behavioral Choice

The axioms of choice listed in the previous section form the normative theory of choice. A decision-maker who satisfies these axioms is said to be rational. It is possible to represent the preferences in terms of a utility function, which is a map from the set of alternatives to the real line. Further assumptions of continuity, monotonicity and convexity of the preference relation imply that the utility function obtained is a continuous, increasing and concave function. Much research in economics and finance has developed elegant and elaborate theory on the basis of assumption that the decision-makers' preferences satisfy the normative axioms of choice.

As was observed in the previous section, it may be a difficult proposition for an average decision-maker to satisfy all the axioms of choice. Following this thought, many behavioral scientists and economists have attempted to understand the decision-making processes among ordinary people through experiments and studies of various kinds (see Wright [91], Bell [8]). These attempts also try to quantify the choice making process, but try to be parsimonious in their assumptions. They constitute what is called the descriptive theory of choice. The five major phenomena of choice that violate the standard model of normative theory are listed as: framing effects, nonlinear preferences, source dependence, risk seeking, and loss aversion.

These phenomena of choice have been confirmed in a number of experiments, with both real and hypothetical payoffs. These phenomena are defined as follows.

Framing Effect: Lack of description invariance implying that variations in the framing of alternatives yield systematically different preferences.

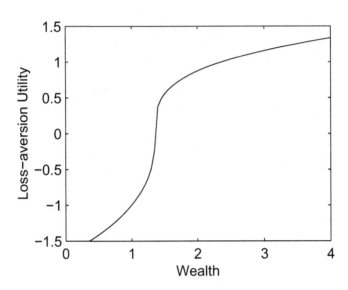

FIGURE 2.3: Plot of the loss-aversion utility, an example of behavioral utility function.

Nonlinear Preference: The expectation principle of utility theory states that utility of risky prospects is linear in outcome probabilities. However, experimental evidence indicates that people tend to transform probabilities nonlinearly, overweighing small probabilities and underweighing moderate and high ones.

Source Dependence: Willingness to bet on a risky alternative depends not only on the degree of risk, but also on its source.

Risk Seeking: As opposed to the generally assumed risk aversion in economic analysis, in certain situations people prefer more risk to less. For instance, people prefer a small probability of winning a large prize over the expected value of that prospect.

Loss Aversion: Carriers of value are gains and losses defined relative to a reference point. The losses loom larger than the gains, that is, an amount of loss elicits more "unhappiness" than the same amount of gain elicits "happiness."

In Figure 2.3, we display a behavioral utility function, the loss-aversion utility from the prospect theory. Tversky and Kahnemann [88] developed an alternative theory to the expected utility theory of normative choice, called the prospect theory. The salient features of this alternative descriptive theory are:

Reference dependence: The carriers of utility are gains and losses defined relative to a reference point.

Loss aversion: The utility function is steeper in the negative than in the positive domain, relative to the reference point, implying that losses loom larger than corresponding gains.

Diminishing sensitivity: The marginal utility of both gains and losses decreases with their size.

The utility function, as seen in Figure 2.3, is continuous and increasing, but the above properties give it an asymmetric S-shape. The utility is convex and much steeper below the reference point, and concave and less steep above it, hence the derivatives don't match at the reference point.

Behavioral finance is a fast growing field of research. In this framework, researchers have made an attempt to understand and explain the investment style, response to risky alternatives and preferences, as well as other market characteristics from a descriptive standpoint. In any risk management context, it is essential to be aware of this alternate preference structure as one implements the risk management framework, making amends to the 'rational' normative approach where necessary.

2.1.3   Risk Measures

Preference relations summarize the risk preference of an individual or a firm towards a single or several risk factors. However, various enterprises can be complex entities, with each concerned decision-maker motivated by different incentives and objectives. Therefore, in many contexts of risk management, appropriate criteria for guiding the goals of risk management must be constructed. For instance, the aggregate of the risk factors and their combined impact on a firm's solvency, an investment portfolio's profitability, or a project's viability must be understood. Risk measures are designed to provide a single-number indicator for the joint impact of multiple risk factors in potentially complex contexts.

A risk measure is a real-valued mapping, ρ, on a set of risk factors. If $\tilde{X}$ is a set of risk factors (random variables), $\rho : \tilde{X} \to R$, so that it gives a numeric summary of the combined impact of the risk factors. Given that the risk measure must be computed based on historical observations of the risk factors, a risk measure can also be described as a *statistical measure* that summarizes the aggregate portfolio-level, project-level, business-unit level, or firm-level risk due to the risk factors.

Some examples of risk measures include variance or standard deviation of portfolio returns, which are functions of risk factors affecting assets in the portfolio and the portfolio weight for each asset. Beta (β) is a risk measure used as an indicator of the impact of systematic risk due to market or a benchmark index portfolio. Alpha (α) and R-squared (R^2) are also risk measures relative to the market or a benchmark index portfolio. Sharpe ratio $(\frac{\bar{r}-r_f}{\sigma})$ is a risk measure that assesses risk relative to the reward or expected return of a portfolio, as an indicator of the level of risk exposure yielding the level of reward obtained. Each of these risk measures is unique in how it measures risk, and hence how it would be applicable in specific contexts. We will discuss these in detail later.

Risk measures are also defined to determine the amount of capital or cash required in reserve in order to withstand adverse impact of risk factors. The purpose of this reserve can be motivated by regulatory objectives to ensure that the risk exposures of a financial institution, such as a bank or an insurance company, do not threaten its stability and solvency, or may be guided internally in a firm towards the same goal. Examples of risk measures geared towards this objective are Value-at-Risk (VaR), Conditional Value-at-Risk (CVaR). We will discuss these risk measures in detail later.

Risk factors are often added or removed from the set of risks relevant in a context of risk management. Therefore, we demand properties of risk measures that keep them consistent as risk factors are added or removed from the set of relevant risks. These properties define risk measures to be convex and/or coherent risk measures.

Coherent Risk Measures

A risk measure, as defined earlier, is a mapping from a set of risk factors (random variables) to the real numbers, $\rho : \tilde{X} \to R$. Therefore, the general notation for a risk measure is $\rho(\tilde{X})$, for the risk factors, $\tilde{X}$. Let $\tilde{X}_1$ and $\tilde{X}_2$ be two sets of risk factors, we define properties of risk measures as follows (Artzner et. al [6]).

Translation Invariance: This property of a risk measure implies if risk-free cashflow is added to the risk factors, it reduces the risk as measured by the risk measure by the exact same amount. Mathematically, we can summarize this as follows for any risk-free cashflow, c,

$$\rho(\tilde{X}_1 + c) = \rho(\tilde{X}_1) - c. \tag{2.4}$$

Monotonicity: The risk measure should be such that it assigns a higher value for a more risky set of risk factors. That is, if $\tilde{X}_1 \geq \tilde{X}_2$ for almost all realizations of the risk factors, this implies that $\tilde{X}_1$ is less risky than $\tilde{X}_2$, therefore we have,

$$\rho(\tilde{X}_1) \leq \rho(\tilde{X}_2). \tag{2.5}$$

Convexity: The risk measure favors diversification, i.e., risk is reduced when it is spread over more risk factors/assets.

$$\rho(\alpha\tilde{X}_1 + (1 - \alpha)\tilde{X}_2) \leq \alpha\rho(\tilde{X}_1) + (1 - \alpha)\rho(\tilde{X}_2), \tag{2.6}$$

for all $\alpha \in [0, 1]$.

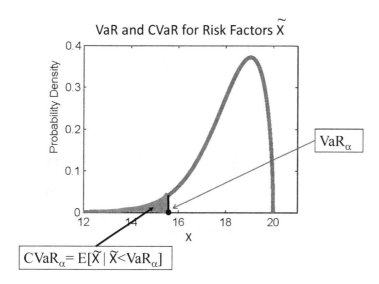

FIGURE 2.4: Display of Value-at-Risk and Conditional Value-at-Risk.

Positive Homogeneity: The risk measure scales linearly in the degree of risk exposure to risk factors $\tilde{X}_1$.

$$\rho(\alpha \tilde{X}_1) = \alpha \rho(\tilde{X}_1), \tag{2.7}$$

for positive values of α.

A risk measure that satisfies the top three properties listed above is called a **convex risk measure**, while a risk measure that satisfies all the four properties is called a **coherent risk measure**. By these properties, we can show that variance (or standard deviation) as a risk measure is neither a convex nor a coherent risk measure. This is because it does not satisfy the translation invariance or monotonicity properties.

A risk measure made popular through its use in regulatory guidelines in the banking sector is the Value-at-Risk (VaR) measure. Value-at-risk measure reports how bad things can get at a certain confidence level as a quantile of risk factor outcomes. We can define Value-at-Risk (VaR_α) at confidence level of α by,

$$VaR_\alpha = \inf\{q \in R | P(\tilde{X} \le q) \le 1 - \alpha\}. \tag{2.8}$$

Therefore, VaR_α does not tell us what the maximum loss level might be. As so defined, Value-at-Risk is not a convex or coherent measure in general. It does not universally satisfy the convexity property, and thus does not always favor diversification.

The Conditional Value-at-Risk ($CVaR_\alpha$) measure is a related risk measure to the VaR_α measure, also stated in terms of a confidence level, α. It coincides with the expected shortfall or tail conditional expectation measures, and is defined as follows.

$$CVaR_\alpha = E[\tilde{X}|\tilde{X} \leq VaR_\alpha]. \tag{2.9}$$

It can be shown that $CVaR_\alpha$ is a coherent risk measure. In Figure 2.4, we display the VaR_α and $CVaR_\alpha$ risk measures for the probability density of the risk factors, $\tilde{X}$.

Risk measures can also be constructed based on a utility function, $U : R \to R$, which is a strictly increasing and concave function. The utility-based risk measure is defined for a risk factor $\tilde{X}$ as follows,

$$\rho_U(\tilde{X}) = \inf\{u \in R | E[U(\tilde{X} + u)] \geq U(0)\}. \tag{2.10}$$

For instance, if $U(x) = ax + b$ is a linear utility function, $\rho_U(\tilde{X})$ is the smallest value u such that $E[\tilde{X}] + u \geq 0$, since $a > 0$. A utility-based risk measure can be shown to be a convex risk measure.

2.1.4 Risk Management

Why do firms engage in risk management? In fact, it is also appropriate to ask, should firms engage in risk management? Academicians have addressed this question, and in doing so, have provided valuable insight to this question. As mentioned in Chapter 1, among the Nobel laureates whose significant contributions have advanced our abilities of comprehension of risk, and its management, are Franco Modigliani and Merton H. Miller. Modigliani and Miller (M&M) are credited with propositions that address the question of value added by managers' financial decision making for a firm [64]. This includes decisions managers may make for risk management. The Modigliani and Miller propositions conclude that managers cannot increase a firm's value solely by financial transactions.

As is true for any theoretical analysis, it must be made under certain idealizations of the context. M&M propositions are made under the assumption that firms and investors make their decisions and choices in a perfect financial market. Financial markets are assumed to be highly competitive, there are no transaction costs, informational asymmetry, or taxes. Under these stringent assumptions, managers cannot increase value of a firm by choosing a certain financing structure for the firm or by engaging in financial transactions for risk management. In competitive, liquid financial markets with no informational asymmetry, investors and shareholders can themselves make their investment or risk management choices.

The M&M proposition does not include decisions managers must make to manage business, strategic or operational risk. Therefore, even under the assumptions of the M&M world, risk management of these risks is important for a firm to create value for its shareholders. In fact, no management of risks is a type of risk management, by adopting the 'keep' strategy of risk management from Figure 2.1. A firm can choose to hedge the price of a commodity in order to create synergies for its operations. For example, airline companies engage in hedging jet fuel price risk by trading in crude oil futures. Failing to do so can erode their profits entirely, and threaten their solvency. Hedging the price of a commodity essential to its production process or service delivery can stabilize a firm's costs. This can translate to pricing policies that make the firm more competitive in the marketplace. Performing due diligence for capital budgeting regarding large-scale strategic investments is also an exercise in risk management.

Firms also transfer a range of operational risks by taking out insurance policies for these risk exposures. The policy offers indemnification in case of events that may cause damage to the property or other assets of a firm. The indemnification can also safeguard against disruption in the production process and in the delivery of services. A number of different liability insurance products are also utilized, or are required for firms to hold, in order to minimize their own risk of loss or loss to the public due to negligence. Engaging in these risk transfers is also a key risk management activity.

The M&M propositions apply specifically to financial contracts, i.e., derivatives transactions, put in place for the purpose of managing financial risks in the balance sheet of a firm. But even this must be questioned due to the stringent assumptions under which the propositions are constructed. In practice, firms do pay taxes based on a complex taxation system. All segments of markets are not always highly competitive and there is ample informational asymmetry between managers and investors. Moreover, managers are often biased by their objectives for the firm and for their self-interest. Therefore, firms do engage in financial risk management in a variety of ways.

Informational asymmetry between managers and investors (or shareholders) of a firm is in terms of a significantly higher level of knowledge managers have of the business and financial activities of the firm. Managers are, therefore, disinclined to give a variable signal to investors regarding the financial health of the firm. Variability can also be interpreted as arising due to managerial incompetence. A firm seen to have less variable income and sound financial health also has higher debt capacity and easier access to financing. A lower variability in income also assures that funds will be available when attractive investment opportunities come by for a firm. Moreover, financial distress is very costly for a firm, hence managed variability helps in lowering the possibility and probability of a firm landing in a financially distressed state. A nonlinear corporate tax structure also encourages making taxable income more uniform through quarters and years. Lastly, but quite importantly, managers' personal risk aversion may play a significant role in dictating which risk they consider managing actively.

Empirical evidence of risk management of financial positions of a firm is ample [83][66] for currency, interest-rate, commodities risk, etc. Sometimes firms don't need to take explicit positions to create the hedge, since there is a natural hedge in the firm's activities. For instance, if a firm has a supplier in a country, as well as sales in the same country, a significant amount of demand risk and currency risk can be naturally hedged. Despite market imperfections, it is not obvious that all firms should manage risks in their financial positions. In the case of small and medium-sized firms, resources may be already stretched that allocating them to dedicated risk management (using hedges) may not be feasible. Hedging financial risk based on half-baked knowledge and half-clear intent can be more harmful than helpful, irrespective of the size of the firm.

Besides following the flowchart in Figure 2.1 for performing risk management, a firm must define its risk appetite, that is applicable at the firm-level as well as at the different business-unit levels. Risk appetite expressed in terms of a variety of risk measures aids more robust risk control and oversight. A firm should aspire to develop an organizational structure to support and implement risk management strategies and inculcate a culture for responsible participation by managers and employees. An organizational structure that permeates top-down vision and bottom-up involvement creates the necessary engagement to deliver the desired goals of risk management.

Performance evaluation measures utilized for business units and for firm-wide assessment of risk management framework help determine the value added from these efforts and strategies. Risk-adjusted Return on Capital (RAROC) and Shareholder Value Added (SVA) are commonly used measures. Firms may consider static as well as dynamic strategies for different risk exposures. The static approach makes a one-time response, implements it and allows the world to respond. Adjustments to the strategy, if needed, are made once the outcomes of the strategy are known, and if the outcomes are found to not meet the objectives of risk management. Dynamic strategies respond to the world's responses on the fly; as a sequence of risks unfold, the strategy changes and adapts the response. Clearly the dynamic strategies are more complex than the static ones, and must be utilized for the clear advantage they may offer. Sophistication of strategies developed must be commensurate to skills of both developers of the strategy and their implementers.

2.1.5 Elements of the Framework

In Section 2.1.1, the risk management framework was presented as a flowchart of a logical sequence of activities. The end goal of these activities was to have a firm-wide or a business-unit specific strategy for the management of its risks. Built into the framework was an assessment of the strategy, as well as assessment of the entire implementation of the framework. These are crucial for the efficacy of the risk management framework, as the identified risk exposures change and as new risk exposures appear. The most important component of the flowchart in Figure 2.1 is at the bottom of the flowchart, where 'avoid-mitigate-transfer-keep' decisions must be made for each risk exposure. We discuss each action component of a risk management strategy in detail next.

2.1.5.1 Avoid

Individuals and firms knowingly or unknowingly avoid many risk exposures. The purpose of including 'avoid' as a risk management action is to make the decision of avoiding a risk exposure a conscious one. A conscious decision makes the decision-maker weigh the pros and cons of the risk exposure and evaluate its impact on the individual or the firm. It also makes the decision-maker explore if there were possible ways by which the harmful effects of the risk can be reduced, without giving up the beneficial consequences of the risk exposure. There can also be related risks that get ignored if the first risk is not carefully evaluated, where the related risks can offer possibilities of higher gains. Therefore, 'avoid' is not a decision one arrives at without careful thought. However, after careful consideration, it may turn out that for a risk the cons significantly outweigh the pros, in which case, the risk exposure is rightly avoided.

A mining company could be scoping a mining project for the viability of its copper deposits, and as such it may not appear to be a profitable project. However, if the firm considers the prospect of recovering significant amounts of gold and other precious metals along with the copper from the mining project, it may have to reconsider its 'avoid' decision. Similarly, a pharmaceutical company could be evaluating a research and development project for a drug. As such even if the project doesn't look attractive for its profitability, and the firm may be inclined to 'avoid' the risks of the project, the R&D for the drug may have the potential to lead to other drug discoveries.

2.1.5.2 Mitigate

Mitigate is a risk management action of risk reduction by taking on multiple risks. Consider two risks, R_1 and R_2, that a firm can select to be exposed to for the rewards they can offer. The firm can choose either of the risks by allocating all its resources to that single risky project, or it can choose to allocate a fraction of available resources to both the risky projects. If the firm uses standard deviation of the outcomes of the risky projects to be the risk measure, and mean value to be the measure for reward, the 'mitigate' action dictates how mixing the two risks can improve the firm's prospects of these risk exposures.

In Figure 2.5, we plot the mean and standard deviation for the combination of two risks for a range of allocation, $(w, 1 - w)$, in the two risky projects. The mean and standard deviation of the combined risk is given by,

$$\mu_\pi = w\mu_1 + (1 - w)\mu_2, \tag{2.11}$$

$$\sigma_\pi = \sqrt{w^2\sigma_1^2 + (1 - w)^2\sigma_2^2 + 2\rho w(1 - w)\sigma_1\sigma_2}, \tag{2.12}$$

where μ_1, μ_2 are mean of the two risks, respectively, σ_1, σ_2 are the standard deviations, and ρ is the correlation between the two risks. In Figure 2.5(a), the first risk has a mean of $\mu_1 = 10\%$ and a standard deviation of $\sigma_1 = 15\%$, while

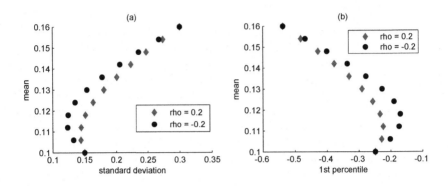

FIGURE 2.5: (a) Plot of mean and standard deviation of combined risk for a range of weights on two individual risks. (b) Plot of mean and first percentile of combined risk for a range of weights on two individual risks, assuming normal distribution of combined risk.

the second risk has a mean of $\mu_2 = 16\%$ and standard deviation of $\sigma_1 = 30\%$. As Figure 2.5(a) shows, with changing allocation, w, in the two risks, the firm can benefit by exploring other risk-reward options. In some cases, it is possible to reduce the risk below either of the options, while improve on the reward relative to at least one of the two options. This is the benefit of the 'mitigate' action of risk management.

When the correlation between risky projects is positive, there is a possibility to reduce risk by choosing multiple risky projects, as is seen in Figure 2.5(a) for $\rho = 20\%$. However, when the correlation is negative, the possibility of risk reduction can be significantly further improved. In Figure 2.5(b), a plot of mean and first percentile of the combined risk is plotted for a range of allocations to the two risks, under the assumption that the combined risk has a normal distribution. The first percentile is chosen as a measure of downside risk, therefore the higher first percentile value is the better. We again see in this figure that combining risks makes it possible to explore options with lowered downside risk.

Benefits of the mitigate action extend beyond the example of two risky projects discussed above. One key assumption made in the results is that there are no transaction costs, and taking on two projects doesn't change the reward and risk characteristics of each risky project. Therefore, while mitigation is an important mode for risk management, its benefits must be cost-effectively achievable.

In the case of operational risk, or more generally for pure risk, the 'mitigate' action is not necessarily applied by taking on multiple risks. Instead, in this case, mitigation is achieved by educating the stakeholders for correct behaviors, appropriately securing property assets, and making processes and systems robust so that the likelihood of occurrence of loss events is reduced.

2.1.5.3 Transfer

Risks that are not avoided and cannot be controlled by mitigation can be considered for a transfer. Transferring of risk entails passing a specific portion of the risk and its impact to another entity, firm or enterprise. Risk transfer strategy has to be carefully constructed since the transfer is effected at a cost. In the case of operational risk, and in general for pure risk, the insurance mechanism provides the opportunity to transfer risk. For different types of market risks, the vast variety of derivative instruments can enable risk transfers. Strategic and business risk management can benefit from transfer of risk in terms of joint ventures in strategic projects, partnerships, and subcontracting and outsourcing opportunities.

A variety of pure risk can be transferred by individuals and firms by purchasing insurance contracts. However, insurance is not available for all kinds of pure risk, since many of the pure risks are not insurable. As discussed in Section 1.1.1, particular pure risk affects a single individual or a small group of individuals, while fundamental pure risk has much more macro-level effects. Therefore, private insurance has a higher likelihood of being able to profitably offer insurance for particular pure risk, while governments often have to provide support for fundamental pure risk. When insurance is available for a specific pure risk, it should be available at an affordable premium. Premium is dependent on the type of coverage and the extent of risk transfer sought, therefore determination of exact risk transfer strategy through insurance contracts is a carefully constructed decision.

Derivative instruments are created to allow the transfer of market risk and credit risk. Market price risk of equity, interest rates, currency and various commodities is transferred by utilizing these derivative instruments. Derivatives can be broadly classified into option contracts, forwards/futures contracts, and swaps. Some of the contracts are traded on exchanges, while others are set up as bilateral agreements, also called 'over-the-counter' contracts. Derivatives for credit risk are mostly designed as swaps or options. Options are so named since they give the buyer the right to perform an action or exercise the option. Swaps, forwards/futures don't offer a right, instead they are obligatory. Depending on the risk exposures and the objectives of risk management, choice of derivatives and the strategy for taking positions in them must be determined for the optimal risk transfer.

A forward contract allows buying or selling a security, commodity, or instrument at a future time for a set price. The set price of settling the contract in future is called the 'forward price'. Therefore, a forward contract helps set the future price of buying or selling the underlying asset. A manufacturer or a producer of a raw material or commodity has interest in being able to sell its product or produce at a reasonable price, and thus eliminate large fluctuations in its revenues. A manufacturer or producer that utilizes these raw materials or commodities also benefits from reducing fluctuations in its costs by utilizing forward contracts. The optimum level of forward contract utilization for risk management depends on the degree of exposure to the underlying risk and the firm's chosen measure of risk.

Forward contracts are designed so that no premium is charged upfront. However, option contracts require payment of a premium. In an option contract, the buyer of the option gets the privilege, while the seller (or writer) of the option takes over the obligation to fulfill the terms of the contract. For this privilege, the buyer must pay a premium upfront. A call option is designed to allow buying the underlying asset in the future at a set exercise price or strike price, while a put option is designed for selling the underlying asset. For instance, a firm that would want to sell the underlying asset in the future, but wants to eliminate the downside risk, can take a position in put options. Determination of the exercise price and extent of position in put options depends on the level of exposure to the underlying risk, cost of put options, risk measure used for the risk management objectives of the firm, as well as the firm's risk aversion.

2.1.5.4 Keep

Finally, after the avoid, mitigate, and transfer considerations are explored, and included in the risk management strategy, what is left behind is essentially the 'keep' action. However, it should not be assumed that the keep action is a default and passive action. The risk kept must be continually monitored, measured and evaluated. The goal is as time progresses, the kept risk has the same characteristics as was intended originally. If the properties of the kept risk change beyond an acceptable level, the strategy for risk management must be reassessed and reconstructed. Additionally when adverse outcomes are realized of the kept risks, the firm should be prepared to cushion the impact of the adverse outcome to ensure solvency of the firm. When firms operate under a regulatory structure, the ability to cushion losses from kept risk is not an option, but a requirement.

2.2 Example Contexts to Apply the Framework

The risk characteristics of the multitudes of risk exposures of a firm must be summarized in meaningful ways to aid the development of a risk management strategy. The avoid-mitigate-transfer-keep actions produce a response to specific risk characteristics, guided by the objectives of risk management. In this section, we elaborate on some of these specific characteristics that are of greatest importance.

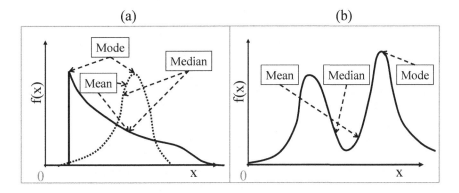

FIGURE 2.6: (a) Unimodal distribution of risk and its central tendencies. (b) Bimodal distribution of risk.

2.2.1 Analysis Using Central Measures

A measure of central tendency provides the typical value of the risk. Mean, median, and mode of a distribution are summaries of the risk that provide this information. In this regard, it is worth noting that these summaries provide information about the central tendency if the distribution is unimodal. For a bimodal or a multi-modal distribution, there are multiple tendencies, therefore mean and median may not strictly summarize the central tendency of the distribution. In a unimodal distribution, if the density function is increasing or decreasing, only mean and median provide the information of central tendency. In Figure 2.6, these central tendencies are marked for a unimodal and a bimodal distribution of risk. In case of bimodal, the mean and median are not indicative of the actual values the risk will realize with high likelihood.

Dispersion from a measure of central tendency is a risk measure. The most common risk measure of this type, one which we have already looked at, is variance or standard deviation. This measure was formally defined in Chapter 1 as, $\sigma^2 = E[(X - \mu)^2]$, where X is the risk and μ is its mean. In the consideration of mitigation, the commonly used measure of central tendency and dispersion are mean and variance, respectively. We discussed formulation of objective for risk management via risk mitigation in terms of these measures in Section 2.1.5.2.

Dispersion or spread from the central tendency can also be measured as a mean-absolute deviation, defined as $E[|X - \mu|]$. This measure can also be used for optimally mitigating risk by allocating resources among several risky projects, as discussed in Section 2.1.5.2. Mean-absolute deviation, as a measure of spread or dispersion, does not square the distance from the mean, therefore the influence of extreme observations is less than it is in the definition of variance or standard deviation measures. Mean-absolute deviation as a measure of risk has been used for risk mitigation [80, 53, 52], which has some implications. One of the implications is that the risk mitigation problem can be formulated as a linear programming problem. The formulation would not require information regarding covariance (or correlation) between risky projects under consideration. Moreover, mean-absolute deviation risk measure doesn't distinguish positive deviations from negative ones, similar to variance, which may be a disadvantage.

Median absolute deviation is a modification on mean absolute deviation, where mean is replaced by median in accounting for deviations of the risk. Median, $p_{1/2}$ or the 50th-percentile of the risk, is known to be a more robust measure of central tendency, since compared to the mean value of the risk it is much less affected by the extreme values of the risk. Median absolute deviation is a robust estimator of dispersion of risk for the same reason, however it is a somewhat more complex risk measure to work with for the development of an optimum risk mitigation strategy.

2.2.2 Tail Analysis

In Section 2.1.4, we discussed the motivations and reasons for firms to engage in risk management. One reason was that they don't have a choice, since the regulatory structure for their industry, such as in banking and insurance, requires them to actively engage in risk management. The motivation for the regulatory structure is to prevent financial weakness of one firm causing a chain-effect, commonly referred to as a domino-effect, resulting in many more firms in the industry and beyond to suffer. Hence, their focus is on the downside risk or the risk of financial distress of the firms in the industry. Systemic risk arising from many firms in these industries simultaneously suffering distress can be very detrimental to the nation's economy, with possible spillover effects on the global economy. Even if a firm is not affected by a regulatory structure dictating its risk management goals, as stated in Section 2.1.4, the firm's risk management goals may include controlling the possibility of financial distress. Assessing downside risk requires zooming into the tails of probability distributions.

Risks and their probability distributions can have a variety of tail characteristics, and since by definition events of the tail are less likely to occur, the data to support statistical analysis of tail risk is meager. Depending on the context, characteristics of either the left, right, or both tails of a distribution may be important from a risk management perspective. Tails can also be thin (light) or fat (heavy). Heaviness of a tail can be roughly defined as when probability of extreme value does not decay rapidly enough as the value gets increasingly extreme. Mathematically, this can be summarized as, there exists a positive parameter $\alpha > 0$ such that,

$$P(X > x) \sim x^{-\alpha}, \quad x \to \infty. \tag{2.13}$$

Examples of single fat-tailed (or one-tailed) distribution include Pareto

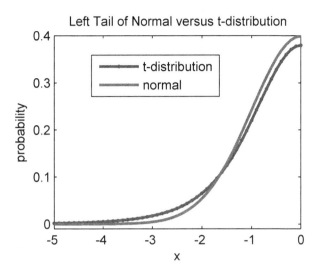

FIGURE 2.7: Probability density plot displaying light-tail and heavy-tail.

distribution, lognormal distribution, Weibull distribution with shape parameter less than 1, while examples of two-fat-tailed distributions include Cauchy distribution and t-distribution. Figure 2.7 displays the tails of a light and a heavy-tailed distribution. Although both curves go down to zero as x decreases to -5, the probability density for t-distribution decays much more slowly. For the range $(-5$ to $-1.5)$, the density curve of the t-distribution is significantly higher than that of the normal distribution. This is the fatness of the tail of t-distribution, resulting in higher probability of occurrence of extreme events.

For a set of observations, detecting the heaviness of the tail of the underlying distribution is the first investigation of tail analysis. A simple probability plot is revealing in this investigation. In Figure 2.8, probability plots for two datasets are given. Figure 2.8(a) is a sample drawn from a normal distribution, hence a normal probability plot yields an undeviating straight line. While in Figure 2.8(b) the sample is drawn from a t-distribution, but is plotted on a normal probability plot to indicate how the tails deviate from the straight line. Both the extreme ends of data deviate significantly from the straight line, indicating that the underlying distribution of the data has left and right fatter tails than the normal distribution. Goodness of fit tests, such as the Kolmogorov-Smirnov (KS) test, the Berk-Jones test and the score test [51], can be applied to quantitatively verify fatness of tail of the risk distribution.

A typical example of tail risk measure is the Value-at-Risk (VaR) risk measure defined and discussed in Section 2.1.3. Other measures discussed

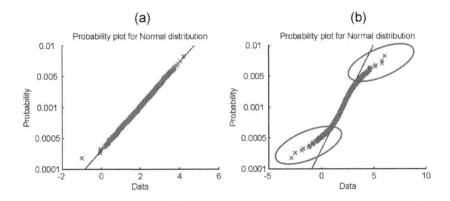

FIGURE 2.8: (a) Probability plot for a dataset that matches the light-tailed normal distribution model. (b) Probability plot for a dataset that displays heavy-tail deviations from the normal distribution model.

in that section that apply to tails of a distribution were Conditional Value-at-Risk (CVaR) and expected shortfall. We will explore these risk measures further in later chapters in the context of specific risk types.

2.2.3 Scenario Analysis

Risk monitoring and assessment is an important component of risk management. This is captured in the top left branch after identification of risk exposures in Figure 2.1. It is also utilized at the bottom loop of the flowchart where a strategy for management of risks on the avoid-mitigate-transfer-keep theme is created and implemented, but a continuous performance evaluation of the strategy is required.

Scenario analysis conducts a specific kind of risk assessment; it is a process of estimating the expected impact of risk exposures over a given period of time, assuming that specifically identified outcomes of risks are realized. To obtain valuable insights from scenario analysis, the specifically identified outcomes for the risks must be carefully picked. To that goal, scenario analysis commonly focuses on estimating the impact of risks in the unfavorable outcomes, or the 'worst-case' scenarios. One method could be to apply two to three standard deviation outcomes of the risks to see their combined impact on the performance of the firm or portfolio. In this way, the range of impact of risks is ascertained for the given time period.

In general, however, there are many different ways to approach scenario analysis. In fact, scenario analysis does not have to rely on historical data, and does not need to assume that past observations must be the only ones valid for the future. Instead, scenario analysis can be developed to try and consider possible developments, modifications, and evolutions by which future outcomes may transpire, which may be disconnected from the past.

Scenario analysis is commonly used for business and strategic risk analysis, where little is known about probabilities associated with risks due to uniqueness of a firm's experience with these risks. Scenarios may be created as a 'normal' case, 'pessimistic,' or 'optimistic' case to examine the range of performance the project may have, and economic soundness of taking it up. Scenarios may also be constructed based on secondary or tertiary risk factors not considered in the core analysis to develop an improved understanding of the impact of these broader sets of risks on the project. For instance, a trucking company may consider price of fuel, demand for its services, risk of labor cost fluctuation, and equipment failure risk as its primary risks, but would benefit from a secondary analysis of considering scenarios for weather and hurricane risks for its service geography.

2.2.4 Stress Testing

Some would consider scenario analysis as a tool within stress testing, while others may treat stress testing as a special case of scenario analysis. In essence, they are both attempting to do the same thing - that is, without providing the entire probability distribution, picking outcomes of risks and assessing their impact on the firm, project or portfolio. Even if the distinction between scenario analysis and stress testing is not agreed upon, the fact remains that they are both indispensable for risk assessment, monitoring, and management.

Stress testing can be considered to have at least one distinct component from scenario analysis. Stress testing can imply physically applying stress or force on a physical system. It is a form of testing where the strength and stability of a physical system or its prototype is determined by testing it beyond its normal operational capacity, often to a breaking point, in order to observe the results. For example, in reliability engineering, systems may be put through extreme stress to determine their modes of failure and to ensure stability when they are used in a normal environment.

Alternately, stress testing can be applied to a non-physical object, such as a software or a model, with the goal of assessing the object's strength, integrity and soundness. Stress testing a model evaluates the model with extreme choice of inputs to determine what the response of the model might be. If under these extreme settings the model can produce behavior expected of the system it is an abstraction of, this clearly instills greater confidence in the model.

The word 'stress' in stress testing suggests consideration of impact of extreme outcomes of risk. Therefore, the objective of stress testing is to complement tail analysis conducted on the basis of VaR or CVaR measures of risk. Estimates of these tail risk measures must rely on historical data, under an assumption of stationarity and repeatability in the future of these past observations. However, this could be far from reality, since a changing and evolving world changes the risk exposures and their characteristics in fundamental ways. By not demanding probability assessments, and focusing only on possibility of outcomes, stress testing relieves itself of relying heavily on past observations. In doing so, it provides the opportunity to look deep in the tail, beyond what VaR or CVaR can report about tail risk based on past observations.

Stress testing is important and valuable for market risk, but is also useful to obtain insights about tail risk in credit, operational, liquidity, business and strategic risks. It is of great significance to note that under stressed economic conditions, the usual silos of risks break down. New interactions of risks, not seen in normal economic conditions, emerge that require special attention. Stress testing can explore the modalities by which such interactions may emerge, and assess their impact on the firm or portfolio.

Once the risk exposures are identified, in the most simplistic way, stress testing can be applied by developing stress shocks for each risk exposure. A stress shock is a significant level of change in parameters that define the risk characteristics of a risk exposure. Examples could be interest rates applied to a bond, probability of default of loans, mean return of a stock, correlation between assets, or volatility of a commodity. These shocks can be applied one at a time on the performance measure for the firm or the portfolio, in order to see the impact of the stressed level, as well as to assess the degree of their influence on the firm.

In reality, stress shocks don't arrive in a piecemeal manner. Therefore, a more advanced approach of stress testing would require combining the stress shocks of all the risk exposures in meaningful ways to construct stress envelopes. Meaningfulness of the stress envelope can be guided by events of the past, an intuitive understanding of the context, and a stretch of imagination. This is where we must venture into the domain of uncertainty. Since the demand of assigning objective probabilities to the occurrence of the stress envelopes is removed, exploration and assessment of uncertainty is possible for the much needed insight. Moreover, inspired by past experience of a stressed sequence of events, market crashes, and failures, stress envelopes capturing the dynamic evolution of circumstances should also be constructed and evaluated for their impact. The sequentiality of events is also very important to understanding risk beyond looking at stress envelopes as static, one-period events.

2.3 MATLAB Tools for Risk Measures

MATLAB mathematical software has an array of functions for working with statistical and risk measures in its Statistics and Finance Toolbox. We list a few of these functions here. The reader is advised to look up the extensive help documentation available within MATLAB to see the details of these and other related functions. At the bottom of each function description in MATLAB help documentation, look for 'See Also' to explore other related functions. Resources such as MATLAB Primer [20] are also useful.

Heavy-tail test: `kstest`

Central Tendencies: `mean`, `median`, `mode`

Measures of Dispersion: `var`, `mad` (used for both mean/median absolute deviation), `range`

2.4 Summary

In this chapter, we developed the necessary constructs to be able to assess the exposure and management of risk. The constructs include a risk management framework, in which all risk management problems can be broken down into stages, and each stage is facilitated by appropriate tools. We developed a detailed discussion of each stage of the risk management framework and appropriate tools for each stage, including risk discovery, models for risk preference, definition and properties of various risk measures. Finally, we applied the risk management framework to some example contexts. Each example context is developed with a certain risk management objective, and is applicable to broad risk types discussed in Chapter 1. In later chapters, we will develop these example contexts for specific risk types.

2.5 Questions and Exercises

Review Questions

1. Why is it important to construct a framework for risk management?

2. What are the steps adopted for implementing a risk management framework?

3. What are the major actions that constitute a risk management strategy?

4. How does an appropriate organizational structure support effective risk management?

5. What are axioms of choice? What purpose do they serve?

6. When is a decision-maker considered risk-seeking, risk-averse, and risk-neutral?

7. What are the Arrow-Pratt measures of risk aversion? How does the coefficient of absolute risk aversion differ from the coefficient of relative risk aversion?

8. What is the descriptive theory of choice? How does it differ from the normative theory?

9. What is a risk measure? How can risk measures be useful?

10. What are the properties of a convex risk measure? Coherent risk measure?

11. What are the Value-at-Risk (VaR) and Conditional Value-at-Risk (CVaR) measures? For a risk distribution, which of the two will be higher?

12. Why do firms engage in risk management?

13. How can the discrepancy between the practice of risk management and the Modigliani-Miller (M&M) propositions be resolved?

14. Why are financial institutions, for instance, banks and insurance companies, required to do risk management?

15. Why should the 'avoid' decision in a risk management strategy be an active decision?

16. What is the goal of 'mitigate' decision of a risk management strategy? In the case of speculative risk, how is this goal achieved?

17. In the case of pure risk, how is the 'mitigate' response to risk achieved?

18. How is transfer of risk accomplished in case of pure risks and speculative risks?

19. Why is there a premium requirement for some derivative instruments designed to transfer risk, while no premium requirement for others?

20. How is the 'keep' response to risk not a default and passive response?

21. What is the central tendency of risk? What is central measure of risk? Give two examples of both.

22. What is a bimodal distribution? How does it differ from a unimodal distribution of risk, in terms of central tendency of risk?

23. Why is tail analysis of risk important? Why is it challenging?

24. How is heaviness of tail of a risk distribution detected?

25. What is scenario analysis? How is it different from stress testing?

26. How can scenario analysis and stress testing be used to make decisions under uncertainty?

Exercises

1. What may be the implication of the descriptive theory of choice on risk management?

2. Find some examples from recent history of failures of risk management. Discuss the causes for these failures, and how they could have been avoided.

3. Determine the coefficient of absolute risk aversion and coefficient of relative risk aversion for the following utility functions:

   (a) Exponential utility function: $u(x) = a - be^{-\mu x}$
   (b) Power utility function: $u(x) = a + bx^{1-\gamma}$
   (c) Logarithmic utility function: $u(x) = a + b\ln(x)$

4. Demonstrate that variance as a risk measure does not satisfy the translation invariance property of risk measures. Demonstrate the same for the monotonicity properties.

5. Demonstrate that the Value-at-Risk (VaR_α) in general is not a convex or coherent risk measure.

6. Demonstrate that the Conditional Value-at-Risk ($CVaR_\alpha$) is a coherent risk measure. What advantages does Conditional Value-at-Risk have over Value-at-Risk risk measure?

7. What is the advantage of the influence of extreme observations being less in the definition of a risk measure?

8. Compute the mean-absolute deviation and median-absolute deviation of the following distributions, and compare them with the standard deviation of the distributions.

   (a) Normal distribution
   (b) Uniform distribution
   (c) Lognormal distribution

9. How is it a disadvantage that the mean-absolute deviation or the variance risk measures don't distinguish between positive versus negative deviations from the central tendency of a risk?

10. Which risk management responses, among 'avoid-mitigate-transfer-keep', will heaviness of the tail of a risk distribution affect more? Explain how.

11. For the following example contexts, after identifying potential relevant risk exposures, develop stress shocks, stress envelopes, and scenarios to conduct a meaningful stress testing and scenario analysis.

    (a) A well-diversified equity fund

    (b) A multinational automobile manufacturer

    (c) A property and casualty insurance company

Chapter 3

Regulations and Risk Management

Everyone seems to agree that some form of government control of business is necessary, however there are differences of opinion regarding the exact purpose and mechanism of control. One camp believes the government's principal role is to maintain competition, supported by the view that an efficient market and maintaining competition will generally produce benefits to the society. The other camp doesn't place as much trust on the markets; it believes that there are ample lessons from history to suggest that governments should focus on preventing abuse of consumers, and regulations should be imposed to prevent market failures. As a result, government's control of business takes two forms, antitrust focuses on maintaining competition, while regulation applies certain performance standards to the firms of an industry.

The focus of antitrust is to curtail monopoly power from forming by preventing collusion and opposing mergers that may lead to excessive concentration. The hope is if government prevents monopoly and unfair competition, it will lead to healthy competition in an industry resulting in public welfare. The classical view is that competition serves consumers by forcing inefficient firms out of the market, that failure of some firms from time to time leading to 'survival of the fittest' is a wholesome phenomenon. There is a flaw in this view, especially when it pertains to industries like banking and insurance, which makes regulation a more important channel for government control for these industries.

The rationale for regulation in banking and insurance is that these industries are vested in the public interest. They are pervasive in their influence, therefore a failure of one firm in these industries can affect persons other than those directly involved with the failing firm. Contracts in these industries are long-term in nature, for instance, individuals purchase insurance to protect themselves against financial loss at a later time, therefore it is important that insurers promising to indemnify the insured for future losses keep their promise. Unless there is a way to seamlessly transfer all the liabilities of a bank or insurer to another institution, letting 'unfit' banks or insurers fail in favor of competition is not a solution. This is where regulation comes in.

Regulation is a more direct involvement of government in the affairs of a business, to ensure that the enterprise runs robustly. A strong enterprise can in turn create a healthy industry. The firms in banking and insurance sectors hold vast sums of money in trust for the public, therefore should be subject to government regulation due to their fiduciary nature. Beyond the solvency aspect of the rationale for regulation, these industries conduct business with their customers through varying degrees of complex contracts. Regulation is deemed necessary to ensure that the contracts offered are fair and are fairly priced.

The linkages between regulations and activities of risk management are fundamental. Even when risk management is not considered a high priority activity in an organization, the regulatory framework can force the organization to conduct at least the minimally necessary level of active risk management, monitoring and reporting. However, regulations are neither costless, nor perfect. In this chapter, we provide an overview of evolution of regulatory bodies and their roles, and of the regulatory framework affecting and influencing risk management in the finance sector. The intent here is not to be comprehensive in addressing this topic, which is clearly not achievable in a book of this nature. The intent here is to develop a regulatory context for risk management, from which the practical issues underlying the rest of the chapters can be better appreciated.

3.1   Regulations Overview

So how much regulation is enough, and how should the regulatory structure of the industries that require regulation for robust functioning be organized? These are non-trivial questions, ones we don't have crisp, clear and conclusive answers for. The bigger issue is, even if we knew the answers to these questions, since the development of regulatory structures may have happened in a somewhat organic fashion, making a complete revamp may be impossible. Countries that started later in their regulatory efforts get the opportunity to learn from the successes and failures of others. One thing is however definite, regulatory environment is always on the move, sometimes at a rather fast pace, and at others not rapidly enough. Sometimes in retrospect it even seems to have proceeded with some trial-and-error or with misguided objectives. In this section, we provide an overview of changes over the years in the regulatory environment for banking, securities and insurance sectors, primarily from the US perspective.

Figure 3.1 shows a summary of regulatory bodies for banking, investment banking and insurance for the G8 countries. Group of Eight or G8 is a forum of governments of the world's eight largest economies, which originated in 1975 with six original members (G6), namely France, Germany, Italy, Japan, the UK, and the US. Soon after, Canada was included to form the Group of Seven or G7 countries, and in 1997, Russia's addition created the current G8 countries. The EU is represented within the G8, but cannot host or chair summits, while China and Brazil, which are among the top 8 economies by nominal GDP, are not included. The G8 + 5, however, includes the heads of

Country/Sector	Banking	Securities (Investment Banking)	Insurance
US	Federal Reserve System; State Regulators; OCC; FDIC	SEC; CFTC; FINRA; State Securities Regulators; State Attorneys General	State Regulators; NAIC
UK	FSA	FSA; FRC	FSA
Japan	JFSA; BOJ	JFSA; SESC	JFSA
Germany	BaFin	BaFin	BaFin
Italy	Banca d'Italia	CONSOB	ISVAP
Canada	OSFI; CDIC	CSA; IIROC; Provincial Securities Commissions	OSFI
France	Commission Bancaire (Banque de France)	AMF	CCA
Russia	FFMS	FFMS	FFMS

FIGURE 3.1: The table shows the main regulatory agencies for the Group of Eight (G8) countries for banking, investment banking, and insurance industry, as of 2012.

state of the five leading emerging economies - Brazil, China, India, Mexico, and South Africa. The G8, or expanded less formal groups, coordinate or discuss future efforts for coordination on several themes, including those pertaining to their respective finance ministries. There have been other efforts underway for international coordination of regulatory activities, which we will touch upon in later sections.

3.1.1 Regulatory Evolution for Banking

Alexander Hamilton, on becoming the first Secretary of the Treasury, received approval from the US Congress to create the Bank of the United States (BUS) in 1791, using the Bank of England as the model. However, at that time, BUS was not the only bank, and not the only bank issuing notes. Between 1782 and 1837, 700 banks sprang up in the US, while there still was no national currency. Given the number of active banks, some states started some rudimentary banking regulation, with Massachusetts, New Hampshire and New York leading the way. As BUS was growing as a premier bank with broad presence across the US, it was seen as a threat to the state banks, and the Congress refused to renew its charter in 1811!

Promoting competition to create a robust banking industry was not the priority at this time, since a second effort to revive a Second BUS failed purely due to political forces in the 1812-1832 period, after which an era of state banking prevailed. As a response to the aftermath of the Second BUS shutting its doors in 1836 and the ensuing events leading up to the panic of 1837, New York created the first bank supervisory authority and formal banking commission, which imposed reserve requirements. Leading up to the US Civil War, there still was no common US currency, and the bank notes issued by hundreds of state banks served as currency, with serious issues regarding circulation of counterfeit notes.

After the US Civil War, the 'greenback' was created, and a national banking system was established as well, backed by the National Banking Act of 1863. The state banks' notes were gradually taxed out of existence. The national banks being regulated by the federal government with the aid of the Office of Comptroller of the Currency (OCC) resulted in a dual banking regulation system. By 1899, the Secretary of the Treasury was using national banks, and their clearinghouse certificates, as a means to stabilize money markets during times of uncertainty.

As regulations for commercial banking took strength, trust companies picked up in popularity as a mechanism for avoiding functional regulation being imposed on commercial banking. In 1907, the Knickerbocker Trust collapsed with deposits in excess of $60 million, sending ripples of panic, while the federal government appeared helpless in dealing with this crisis. An individual, J.P. Morgan, served as a 'one-man Federal Reserve Bank' through this crisis. The panic of 1907 resulted in the enactment of the Federal Reserve Act of 1913, which in fact further fragmented banking regulation. Besides the state regulation, the federal bank regulatory authority was split into the OCC at the Treasury and the Federal Reserve System. OCC retained the responsibility for examining and regulating national banks, while the Fed became responsible for monetary issues.

The Fed found itself struggling with monetary policy from its inception. Moreover it was marred with power struggles within the Federal Reserve system, specifically the New York Federal Reserve Bank's efforts to gain prominence over the other Reserve Banks in the system. There were grave consequences of this fight for power in the events that led to the stock market crash of 1929. Disputes with the New York Federal Reserve Bank, and specifically dealings of Charlie Mitchell with National City Bank (now Citibank), the then director of New York Fed, largely paralyzed monetary policy during almost the entire year of 1929.

Meanwhile the McFadden Act of 1926 had allowed national banks to establish branches under conditions similar to those permitted by state banks. Restrictions on branch banking were resulting in a large number of very weak banks that would be unable to cope with a serious economic downturn. Bank failures reached epidemic proportions after the stock market crash of 1929. In 1923, there were 91 national and 580 state banks that had a total of over 2000 branches. By 1932, one in four banks in the US had failed. During Franklin Roosevelt's US presidency, new legislation was enacted to strengthen the banking system. The Federal Deposit Insurance Corporation (FDIC) was created to protect customer bank deposits to avoid bank runs. The Glass-Steagall Act of 1933 sought complete divorcement of commercial and investment banking. After the World War II decree of the Treasury to keep interest rates at artificially low levels, the Treasury-Federal Reserve Accord of 1951 strengthened the role of the Fed in more independently managing federal monetary policy.

The 1960s marked the beginning of an era in which financial services firms sought to expand and diversify their business across regulatory boundaries. Banks began looking for loopholes to expand business and avoid banking regulations. The Bank Holding Company Act of 1956 restricted the ability of bank holding companies to enter in other lines of business or to purchase other banks. Such activities required Fed approval, but the BHC Act of 1956 did not apply to one-bank holding companies. The number of one-bank holding companies grew rapidly as this loophole was exploited. In response Congress acted to close the one-bank holding company exception through the Bank Holding Company Act Amendments of 1970.

Increasing competitive pressures, for instance by money market funds becoming alternatives for customer deposits, made customer deposits become a smaller factor in banking. Banks were exploring new markets, including international markets, such as in Latin America, which caused huge losses to the industry later. On the other hand, banks in other countries were getting stronger, and the World War I domination of US banking in world finance was dwindling with only 4 of the top 20 banks in the world being American. During the middle of the 1980s, over 40 US banks failed each year. In 1984, the FDIC had a list of over 500 'problem' banks, which doubled to 1000 institutions by 1986. Artificial regulatory restraints on their business and disintermediation was the root cause of these problems.

When the congress lifted the interest rate ceilings for deposits, there were major interest rate mismatch issues between bank assets and liabilities. Major S&L (Thrift) crises emerged after the Depository Institutions Deregulation and Monetary Control Act of 1980, when Thrifts began exploring speculative, perhaps questionable, investment opportunities, building up to the 1987-1988 banking debacle, when 700 banks and over 1000 S&Ls were closed down, with cost to taxpayers predicted to range from $500 billion to $1 trillion. All pointing to less than adequate risk management practices used at banks.

The rapid evolution of the banking industry continued, with 1980s seeing a change from relationship banking to transactional banking. Securitization of assets made banks into conduits for loans, as their underwriters and distributors. Banks were more engaged in 'mezzanine' finance - funding in the middle of corporate capital structure. Banks became less and less like traditional banks and more like financial services firms. First they were edging their way into securities, followed by beginning to provide insurance products, such as life insurance, annuities, etc. The bottom line was, banks could not depend on deposits business as their prime source for generating revenue. In response to these changes, the SEC attempted to adopt a rule in 1985 that would have required banks to register with it if the banks engaged in securities business.

The Senate didn't support this on grounds that the SEC was extending beyond its power. While in the early 1990s, only six states allowed interstate bank branches, by the middle of 1990s every state was permitting multi-office banking. The Riegle-Neal Interstate Banking and Branching Efficiency Act of 1994 opened doors widely to interstate banking, and the continuing waves of mergers and acquisitions changed the industry terrain.

When in October 1999, the Gramm-Leach-Bliley Act of 1999 finally repealed the Glass-Steagall Act of 1933, many argued that the Glass-Steagall Act was already 'dead.' The Gramm-Leach-Bliley Act of 1999 or the Financial Services Modernization Act of 1999 authorized the creation of financial holding companies that could engage in a broad array of financial services, including commercial and investment banking, securities and insurance. The historical functional regulatory system was, however, maintained as is, as the basis for regulating the expanded activities of the banks and their holding company structures. This meant traditional commercial banking activities would be regulated by bank regulators, securities by the SEC and state securities commissions, futures and options by CFTC, and insurance by multiple state insurance regulators!

The current answers to how much regulation is enough, and how should the regulatory structure for banking be organized will appear very suboptimal even to the most naive eyes. The differences between products of banks and non-bank financial firms have become increasingly blurred. The emergence of similar products by different firms operating under different regulatory regimes results in complicated competitive and regulatory issues. Some of these issues are being addressed, however functional regulation does not reflect current realities. Banking evolution has continued; with increasing popularity of Internet banking and credit cards, cash has become an increasing anachronism. Adaptability of the national banking system is increasingly important as advances in technology and telecommunications accelerate the rate of changes. Attention is needed for the realities of the marketplace.

The financial crises of 2007-2008 shook US banking in unprecedented ways and magnitude. In 2012, the aftereffects are still felt in almost all parts of the globe. Although there were many culprits to the creation of the crisis, the 'too big to fail' financial institutions became the centerpiece of everyone's headache and a $700 billion to trillions of dollars of cost to the US taxpayer, depending on the way one looks at it. Regulatory reforms in response to the crises are taking shape all over the world. The Dodd-Frank Wall Street Reform and Consumer Protection Act of 2010 was signed into law by President Barack Obama on July 21, 2010. The Act implements financial regulatory reforms passed as a response to the 2008 financial crisis and ensuing recession. It is said to be the most significant financial regulation change in the US since the regulatory reform after the Great Depression, which significantly affects all federal financial regulatory agencies and almost every aspect of the financial services industry [7].

3.1.2 Regulatory Evolution for Investment Banking

Investment banks are intermediaries. Unlike commercial banks that accept deposits from customers and make consumer or commercial loans, investment banks don't accept deposits, they sell investments. The securities originated by an investment bank are typically not held by the bank, but are instead sold to third parties, investors. Investment banking practices have existed for centuries, such as through the merchant banks that helped finance foreign trade, overseas voyages and investments. In the US, investment banking received a boost during the Civil War. Around this period, banking houses were syndicated to meet the federal government's need for money to fund its war efforts. Jay Cooke launched the first mass securities selling operation in US history, selling $830 million worth of government bonds to a wide group of investors, followed by war bonds worth $1.5 billion to the general public. This marked the first mass-market securities sales operation in the US, a practice that continued later in the 1800s to finance the expansion of the transcontinental railroads.

Through the late nineteenth century, the investment banking industry was dominated by two distinct groups, the German-Jewish immigrant bankers and the so-called 'Yankee houses', both of which had ties with their respective merchant banking operations in London. One exception of Kuhn, Loeb had ties with European sources of capital through the German investment banking community. The 1800s saw the emergence of some of the most famous firms in investment banking, including JP Morgan (1871) and Goldman Sachs (1869). Goldman Sachs was among the pioneers of the initial public offering (IPO), and managed one of the largest IPOs at that time, for Sears, Roebuck and Company in 1906. We have already discussed the role of J.P. Morgan through the panic of 1907.

In the early twentieth century, there was no legal requirement to separate the operations of commercial and investment banks. Therefore, deposits from the commercial banking side of the business served as an in-house supply of capital that could be used to fund the underwriting business of the investment banking side. The investment banking industry was highly concentrated and was dominated by an oligopoly of firms. In this period, investment banking expanded dramatically. One reason was an increase in the number of individuals who owned stocks, as a result of the prosperous years after the First World War.

In 1913, the Pujo Committee unanimously determined that a small syndicate of financiers had gained consolidated control of numerous industries through the abuse of public trust in the US. The chair of the House Committee on Banking and Currency, Representative Arsne Pujo, had convened a special committee to investigate a 'money trust', whose report found that the officers of JP Morgan & Co. sat on the boards of directors, a total of 341 directorships, of 112 corporations, thus had gained control of major manufacturing, transportation, mining, telecommunications, and the financial markets of the

US. The report revealed that a handful of men held manipulative control of the New York Stock Exchange and attempted to evade interstate trade laws. These findings generated support for the passage of the Federal Reserve Act in 1913 and the passage of the Clayton Antitrust Act in 1914, which built on the Sherman Antitrust Act of 1890.

The post World War I run-up in stock prices created an unsustainable bubble that finally collapsed in 1929. The US plunged into one of the worst depressions in history. By the beginning of 1933, the banking system in the United States had effectively ceased to function. More than 11,000 banks failed or merged, and a quarter of the population was out of work. We have already discussed in the context of banking, the reform steps taken by the incoming President Franklin Roosevelt administration, and the incoming Congress in response to the Great Depression, including the Glass-Steagall Act of 1933 for the separation of commercial and investment banking. In order to comply with this new regulation, most large banks split into separate entities. For example, J.P. Morgan split into three entities: J.P. Morgan, which continued to operate as a commercial bank, Morgan Stanley, which became an investment bank, and Morgan Grenfell, which operated as a British merchant bank.

The Securities Act of 1933 was legislated pursuant to the interstate commerce clause of the Constitution, and required that any offer or sale of securities using the means and instrumentalities of interstate commerce should be registered. This 1933 Act was the first major federal legislation to regulate the offer and sale of securities, prior to which regulation of securities was chiefly governed by state laws. The Securities Exchange Act of 1934 governed the secondary trading of securities, such as stocks, bonds, and debentures, in the US. It was a sweeping piece of legislation, which formed the basis of regulation of the financial markets and their participants in the US. The 1934 Act also established the Securities and Exchange Commission (SEC), the agency primarily responsible for enforcement of US federal securities law. These Acts required full disclosure of accurate information for publicly offered securities and a prospectus filed with the SEC.

After the 1933-34 reforms, major Wall Street investment banks focused on deal-making, serving as advisers to corporations on mergers and acquisitions, as well as public offerings of securities. In the 1980s, with the advances in computing technologies and use of sophisticated mathematical models, investment banks began developing and executing trading strategies, and the emphasis shifted from deal-making to trading. Prominent investment bank firms, such as Salomon Brothers, Merrill Lynch and Drexel Burnham Lambert, earned an increasing amount of their profits from trading for their own account. The high frequency and large volume of trades enabled them to generate a profit by taking advantage of small changes in market conditions. Financial innovations, such as popularized use of high yield debt, also called junk bonds, got used in corporate finance, especially in mergers and acquisitions. This new source of capital sparked investment banks' contribution in an explosive increase in leveraged buyouts (LBOs) and hostile takeovers.

As discussed in the context of banking, the 1999 repeal of the Glass-Steagall Act by the enactment of the Gramm-Leach-Bliley Act of 1999 removed the separation that was created in 1933 between investment banks and commercial banks. The subprime crisis of 2007 fundamentally involved investment banks, leading to the largest US investment banks being the center stage of the 2008 financial crisis. Investment banks Lehman Brothers and Bear Stearns, over 80-100 years old, collapsed. Merrill Lynch was acquired by Bank of America, which remained in trouble. Goldman Sachs and Morgan Stanley converted themselves into a traditional bank holding company so they could be eligible to receive emergency taxpayer-funded assistance. Initially, banks received part of a $700 billion Troubled Asset Relief Program (TARP) funds, which was intended to stabilize the economy and thaw the frozen credit markets. Eventually, taxpayer assistance to banks reached many trillions of dollars. As discussed earlier, the post-crisis regulatory reforms under the Dodd-Frank Wall Street Reform and Consumer Protection Act of 2010 will have wide ranging impacts, also on issuance and trading of securities.

3.1.3 Regulatory Evolution for Insurance

The regulation of insurance companies in the US began around the US Civil War (1861-1865), with New Hampshire being the first state to establish its insurance commission in 1851, followed by New York in 1859. Within ten years, in the Paul vs. Virginia case of 1869, the US Supreme Court ruled that insurance was not an interstate commerce, and that states rather than the federal government had the right to regulate the insurance industry. This decision applied for the next 75 years, when in 1944, in a price fixing case against the South-Eastern Underwriters Association (SEUA) in violation of the Sherman Antitrust Act of 1890, SEUA was found guilty. In this landmark case, the US Supreme Court ruled that insurance was, in fact, interstate commerce when conducted across state lines, and therefore, was subject to federal regulation. This ruling led to significant turmoil in the industry and among state regulators. It was the conflict between the regulation and the antitrust roles of government control, with the antitrust component being a federal issue, while state regulators had had a strong hold on insurance regulation for decades.

In 1945, the McCarthy-Ferguson Act was passed by the US Congress, which made regulation and taxation of the insurance industry a state responsibility. It also stated that federal antitrust laws apply to insurance only to the extent that the insurance industry is not regulated by state law. The Sherman Antitrust Act of 1890 is still applicable to insurers. Since the passage of the McCarthy-Ferguson Act, or the Public Law 15, states have had the primary responsibility of regulating insurance. This is done through the states' Commissioner of Insurance and the National Association of Insurance Commissioners (NAIC), founded in 1871. Quick summary of regulatory entities for insurance in the US and seven other G8 countries is given in Figure 3.1.

There has been a continued dialogue of state versus federal regulation of insurance in the interim years, with advantages and disadvantages of both being weighed in, as well as allegations of state regulators not being able to meet consumer interests satisfactorily. After the Financial Modernization Act of 1999, from convergence of financial services, state regulators have experienced an increasing challenge of regulating activities in the insurance markets. Commercial banks started offering insurance products, and vice versa, and product innovations have made the importance of coordination and cooperation between different regulators of financial institutions a necessity.

The International Association of Insurance Supervisors (IAIS), which was formed in 1994, is an international effort on the same lines as the Basel Committee for banking. In the US, since insurance is under state regulators, and NAIC is a voluntary federation of 50 separate Commissions, it isn't equipped to offer global leadership. Europe is at the forefront of this effort, defining the Solvency II framework for life and non-life insurers and reinsurers. Solvency II enforcement should begin in 2014.

3.2 Regulations and Banking

In the previous section, we provided a quick historical overview of evolution of regulatory changes in the US banking, investment banking and insurance industries. The repeal of the Glass-Steagall Act of 1933 by the Gramm-Leach-Bliley Act or the Financial Services Modernization Act of 1999 erased the distinction between commercial banks, investment banks and insurance companies. There were many regulatory entities before the 2007 subprime crisis to oversee different segments of the financial services industry. A summary of regulatory authorities for G8 countries was provided in Figure 3.1. The reforms post 2007-2008 crisis by the Dodd-Frank Wall Street Reform and Consumer Protection Act of 2010 have, in fact, increased the number of regulatory authorities. In this section, we provide a brief snapshot description of the banking-related regulatory authorities and their roles.

US Federal Reserve System

The Federal Reserve System is the central bank of the US, and was founded in 1913, as stated earlier. Similar to the functions of central banks of most developed countries, the role of the Federal Reserve is to provide a safer, more flexible, and stable monetary and financial system. Over the years, its role in banking and the economy has expanded, with today its four general roles being:

1. Conducting the nation's monetary policy by influencing the monetary and credit conditions in the economy in pursuit of maximum employment, stable prices, and moderate long-term interest rates.

2. Supervising and regulating banking institutions to ensure the safety and soundness of the nation's banking and financial system and to protect the credit rights of consumers.

3. Maintaining the stability of the financial system and containing systemic risk that may arise in financial markets.

4. Providing financial services to depository institutions, the US government, and foreign official institutions, including playing a major role in operating the nation's payments system.

The Fed is composed of a central, governmental agency - the Board of Governors - in Washington, DC, and twelve regional Federal Reserve Banks. The Board and the Reserve Banks share the above responsibilities of the Fed. A major component of the System is the Federal Open Market Committee (FOMC), which is made up of the members of the Board of Governors, the president of the Federal Reserve Bank of New York, and presidents of four other Federal Reserve Banks, who serve on a rotating basis. The FOMC oversees open market operations, which is the main tool used by the Federal Reserve to influence overall monetary and credit conditions in the economy.

The Federal Reserve implements monetary policy through its control over the federal funds rate - the rate at which depository institutions trade balances at the Federal Reserve. It exercises this control by influencing the demand for and supply of these balances through the following means:

- Open market operations - the purchase or sale of securities, primarily US Treasury securities, in the open market to influence the level of balances that depository institutions hold at the Federal Reserve Banks

- Reserve requirements - requirements regarding the percentage of certain deposits that depository institutions must hold in reserve in the form of cash or in an account at a Federal Reserve Bank

- Contractual clearing balances - an amount that a depository institution agrees to hold at its Federal Reserve Bank in addition to any required reserve balance

- Discount window lending - extensions of credit to depository institutions made through the primary, secondary or seasonal lending programs

The goal of the Fed's monetary policy is to promote effectively the goals of maximum employment, stable prices and moderate long-term interest rates. Stable prices in the long run are a precondition for maximum sustainable output growth and employment as well as moderate long-term interest rates.

Office of Comptroller of the Currency

The primary mission of the Office of Comptroller of the Currency (OCC) is to charter, regulate, and supervise all national banks and federal savings associations. OCC also supervises the federal branches and agencies of foreign banks. OCC's goal in supervising banks and federal savings associations is to ensure that they operate in a safe and sound manner and in compliance with laws requiring fair treatment of their customers and fair access to credit and financial products. As discussed in Section 3.1.1, the OCC predates the Fed, since the OCC was established in 1863 as an independent bureau of the US Department of the Treasury. The President, with the advice and consent of the US Senate, appoints the Comptroller to head the agency for a five-year term.

The OCC is headquartered in Washington, DC, with four district offices plus an office in London to supervise the international activities of national banks. The OCC's nationwide staff of bank examiners conducts on-site reviews of national banks and federal savings associations (or federal thrifts) and provides sustained supervision of these institutions' operations. Examiners analyze loan and investment portfolios, funds management, capital, earnings, liquidity, sensitivity to market risk, and compliance with consumer banking laws for all national banks and federal thrifts. They review internal controls, internal and external audit, and compliance with law, as well as evaluate management's ability to identify and control risk. The OCC functions with four objectives in support of the agency's mission to ensure a stable and competitive national system of banks and savings associations:

- Ensure the safety and soundness of the national system of banks and savings associations.

- Foster competition by allowing banks to offer new products and services.

- Improve the efficiency and effectiveness of OCC supervision, including reducing regulatory burden.

- Ensure fair and equal access to financial services for all Americans.

For serving its mission, the OCC has the power to regulate the national banks and federal thrifts as follows:

- Examine the national banks and federal thrifts.

- Approve or deny applications for new charters, branches, capital, or other changes in corporate or banking structure.

- Take supervisory actions against national banks and federal thrifts that do not comply with laws and regulations or that otherwise engage in unsound practices. Remove officers and directors, negotiate agreements to change banking practices, and issue cease and desist orders as well as civil money penalties.

- Issue rules and regulations, legal interpretations, and corporate decisions governing investments, lending, and other practices.

The OCC's operations are funded primarily by assessments on national banks and federal savings associations. National banks and federal thrifts pay for their examinations, and they pay for the OCC's processing of their corporate applications. The OCC also receives revenue from its investment income, primarily from US Treasury securities. By law, the OCC is prohibited from releasing information from its safety and soundness examinations to the public. National banks and federal savings associations must, however, submit regular reports of their condition and income to the FDIC, available on its Web site.

Federal Deposit Insurance Corporation

The Federal Deposit Insurance Corporation (FDIC) is an independent agency created by the Congress to maintain stability and public confidence in the nation's financial system by insuring deposits, examining and supervising financial institutions for safety and soundness and consumer protection, and managing receiverships. The FDIC is a recognized leader in promoting sound public policies, addressing risks in the nation's financial system, and carrying out its insurance, supervisory, consumer protection, and receivership management responsibilities. As discussed in Section 3.1.1, the FDIC was created in 1933 in response to the thousands of bank failures that occurred in the 1920s and early 1930s, as an independent agency of the federal government. Since its inception on January 1, 1934, no depositor has lost a single cent of insured funds as a result of a failure. The FDIC insures more than $7 trillion of deposits in U.S. banks and thrifts; deposits in virtually every bank and thrift in the country.

The FDIC insures deposits only. It does not insure securities, mutual funds or similar types of investments that banks and thrift institutions may offer. The FDIC directly examines and supervises more than 4,900 banks and savings banks for operational safety and soundness, more than half of the institutions in the banking system. Banks can be chartered by the states or by the federal government. Banks chartered by states also have the choice of whether to join the Federal Reserve System. The FDIC is the primary federal regulator of banks that are chartered by the states that do not join the Federal Reserve System. In addition, the FDIC is the back-up supervisor for the remaining insured banks and thrift institutions.

The FDIC also examines banks for compliance with consumer protection laws, including the Fair Credit Billing Act, the Fair Credit Reporting Act, the Truth-In-Lending Act, and the Fair Debt Collection Practices Act, to name a few. Finally, the FDIC examines banks for compliance with the Community Reinvestment Act (CRA) which requires banks to help meet the credit needs of the communities they were chartered to serve.

To protect insured depositors, the FDIC responds immediately when a bank or thrift institution fails. Institutions are generally closed by their chartering authority, which may be a state regulator or the Office of the Comptroller of the Currency. The FDIC has several options for resolving institution failures, but the one most used is to sell deposits and loans of the failed institution to another institution. Customers of the failed institution automatically become customers of the assuming institution. Most of the time, the transition is seamless from the customer's point of view.

The FDIC is headquartered in Washington, D.C., but conducts much of its business in six regional offices, three temporary satellite offices and in field offices around the country. It is managed by a five-person Board of Directors, all of whom are appointed by the President and confirmed by the Senate, with no more than three being from the same political party. The FDIC receives no Congressional appropriations, it is funded by premiums that banks and thrift institutions pay for deposit insurance coverage and from earnings on investments in U.S. Treasury securities.

Regulatory Expansion from Dodd-Frank Act of 2010

The Dodd-Frank Act of 2010 has neither simplified supervision nor decreased the number of regulators [56]. There are new agencies created, which include the Financial Stability Oversight Council (FSOC), the Office of Financial Research (OFR), and the Consumer Financial Protection Bureau (CFPB). Financial Stability Oversight Council performs comprehensive monitoring to ensure the stability of the US financial system. The Council is charged with identifying threats to the financial stability of the US; promoting market discipline; and responding to emerging risks to the stability of the US financial system. The Council consists of 10 voting members and 5 nonvoting members and brings together the expertise of federal financial regulators, state regulators, and an insurance expert appointed by the President.

The Office of Financial Research will produce, promote, and sponsor financial research aimed at developing the analytical tools needed to assess threats to financial stability. This research will support the work of the Financial Stability Oversight Council in assessing potential risks to the financial system. The OFR will also establish forums and networks to bring together experts within and outside the regulatory system in research on these issues. The OFR will work with academia and the private sector to promote best practices in risk management through publications and forums.

The Consumer Financial Protection Bureau (CFPB) will be working to give consumers the information they need to understand the terms of their agreements with financial companies. Congress established the CFPB to protect consumers by carrying out the federal consumer financial laws. Among other things, CFPB conducts rule-making, supervision, and enforcement for federal consumer financial protection laws, restricts unfair, deceptive, or abusive acts or practices, takes consumer complaints and promotes financial education. It is engaged in research of consumer behavior, monitors financial markets for new risks to consumers, and enforces laws that outlaw discrimination and other unfair treatment in consumer finance.

Bank for International Settlements

International efforts have been underway for decades to coordinate regulation of banking, especially given the increasingly global footprint of US and foreign banks. The Bank for International Settlements (BIS) was established in 1930, making it the world's oldest international financial organization. The mission of the BIS is to serve central banks in their pursuit of monetary and financial stability, to foster international cooperation in those areas and to act as a bank for central banks. As a broad outline, the BIS has been pursuing its mission by:

- Promoting discussion and facilitating collaboration among central banks.

- Supporting dialogue with other authorities that are responsible for promoting financial stability.

- Conducting research on policy issues confronting central banks and financial supervisory authorities.

- Acting as a prime counterparty for central banks in their financial transactions.

- Serving as an agent or trustee in connection with international financial operations.

Headquartered in Basel, Switzerland, BIS has two other representative offices in the Hong Kong Special Administrative Region of the People's Republic of China and in Mexico City. Since central banks and international organizations are its customers, the BIS does not accept deposits from, or provide financial services to, private individuals or corporate entities.

Basel Committee on Banking Supervision, under the auspices of the BIS, has constructed a series of accords for defining sound regulation of banks. Basel I, which was the first round of deliberations by central bankers from around the world, in 1988 published a set of minimum capital requirements for banks. This was also known as the 1988 Basel Accord, and was enforced by law in the G10 countries in 1992. The world changed with financial innovations and new financial conglomerates. Therefore, a more comprehensive set of guidelines, known as Basel II, were developed. Basel II focussed on three pillars.

Pillar I: The first pillar deals with maintenance of regulatory capital calculated for three major components of risk that a bank faces - credit risk, operational risk and market risk. The credit risk component can be calculated in three different ways of varying degree of sophistication, namely standardized approach, Foundation Internal Rating-Based Approach (IRB) and Advanced IRB. For operational risk, there are three different approaches - basic indicator approach (BIA), standardized approach and the internal measurement approach (IMA). An advanced form of IMA is the advanced measurement approach (AMA). For market risk the preferred approach is Value-at-Risk (VaR).

Pillar II: The second pillar deals with the regulatory response to the first pillar, giving regulators much improved 'tools' over those available to them under Basel I. It also provides a framework for dealing with all the other risks a bank may face, such as systemic risk, concentration risk, strategic risk, reputational risk, liquidity risk and legal risk, which the accord combines under the title of residual risk. It gives banks a power to review their risk management system.

Pillar III: The third pillar aims to complement the minimum capital requirements and supervisory review process by developing a set of disclosure requirements which will allow the market participants to gauge the capital adequacy of an institution.

Before Basel II could be fully developed and adopted, the financial crisis of 2007-2008 occurred. The crisis has highlighted the importance of liquidity risk and stress testing. Basel III strengthens bank capital requirements and introduces new regulatory requirements on bank liquidity and bank leverage.

The Financial Stability Forum (FSF) was established in 1999 as a group of major national financial authorities such as finance ministries, central bankers, and international financial bodies. The Forum was to promote international financial stability, facilitate discussion and co-operation on the supervision and surveillance of financial institutions, transactions and events. FSF was managed by a small secretariat housed at the Bank for International Settlements in Basel, Switzerland. The 2009 G-20 London summit decided to establish a successor to the FSF, the Financial Stability Board (FSB). The FSB includes members of the G20, who were not all members of the FSF.

3.3   Regulations and Investment Banking

The world of investing is fascinating and complex, but with the right approach to navigate, it can be very fruitful. Unlike the banking world, where deposits are guaranteed by the federal government, stocks, bonds and other securities can lose value. Our discussion of the evolution of banking and securities in Section 3.1 talked of crashes, failures, and panics, many of which were results of speculation and issues related with financial intermediaries that support the investment activities of the society. Even though the Financial Services Modernization Act of 1999 removed the separation of commercial banking from investment banking, the regulatory authorities have maintained a functional regulation framework. We now look at regulatory authorities that regulate securities.

Securities and Exchange Commission

The mission of the US Securities and Exchange Commission (SEC) is to protect investors, maintain fair, orderly, and efficient markets, and facilitate capital formation. SEC recognizes the need for individual investors when it states, 'as more and more first-time investors turn to the markets to help secure their futures, pay for homes, and send children to college, our investor protection mission is more compelling than ever.' On the other hand, as the securities exchanges have evolved into global for-profit competitors, there is even greater need for sound market regulation. The SEC also defines the significance of its actions with an eye toward promoting capital formation that is necessary to sustain economic growth.

The SEC oversees the key participants in the securities world, including securities exchanges, securities brokers and dealers, investment advisors, and mutual funds, where the SEC is concerned primarily with promoting the disclosure of important market-related information, maintaining fair dealing, and protecting against fraud. Crucial to the SEC's effectiveness in each of these areas is its enforcement authority. Each year the SEC brings hundreds of civil enforcement actions against individuals and companies for violation of the securities laws. Typical infractions include insider trading, accounting fraud, and providing false or misleading information about securities and the companies that issue them.

All investors, whether large institutions or private individuals, should have access to certain basic facts about an investment prior to buying it, and so long as they hold it. To achieve this, the SEC requires public companies to disclose meaningful financial and other information to the public. This provides a common pool of knowledge for all investors to use to judge for themselves whether to buy, sell, or hold a particular security. Only through the steady flow of timely, comprehensive, and accurate information can people make sound investment decisions. To help support investor education, the SEC offers the public a wealth of educational information on its website, which also includes the EDGAR database of disclosure documents that public companies are required to file with the commission.

The Securities Act of 1933 and the Securities Exchange Act of 1934, which created the SEC, were designed to restore investor confidence in US capital markets after the 1929 stock market crash and the Great Depression, by providing investors and the markets with more reliable information and clear rules of honest dealing. The main purposes of these laws can be reduced to two common-sense notions, first that companies publicly offering securities for investment dollars must tell the public the truth about their businesses, the securities they are selling, and the risks involved in investing. Second, people who sell and trade securities - brokers, dealers, and exchanges - must treat investors fairly and honestly, putting investors' interests first.

The Division of Risk, Strategy, and Financial Innovation at SEC was established in September 2009 to help further identify developing risks and trends in the financial markets. This new division is providing the commission with sophisticated analysis that integrates economic, financial, and legal disciplines. The division's responsibilities cover three broad areas: risk and economic analysis; strategic research; and financial innovation. The emergence of derivatives, hedge funds, new technology, and other factors have transformed both capital markets and corporate governance.

The division is working to advise the commission through an interdisciplinary approach that is informed by law and modern finance and economics, as well as developments in real world products and practices on Wall Street and Main Street. Among the functions being performed by the division are: (1) strategic and long-term analysis; (2) identifying new developments and trends in financial markets and systemic risk; (3) making recommendations as to how these new developments and trends affect the commission's regulatory activities; (4) conducting research and analysis in furtherance and support of the functions of the commission and its divisions and offices; and (5) providing training on new developments and trends and other matters.

Commodities and Futures Trading Commission (CFTC)

US Congress created the Commodity Futures Trading Commission (CFTC) in 1974 as an independent agency with the mandate to regulate commodity futures and option markets in the US. The agency's mandate has been renewed and expanded several times since then, most recently by the Dodd-Frank Wall Street Reform and Consumer Protection Act of 2010. In 1974 the majority of futures trading took place in the agricultural sector. The CFTC's history demonstrates, among other things, how the futures industry has become increasingly varied over time and today encompasses a vast array of highly complex financial futures contracts.

Today, the CFTC assures the economic utility of the futures markets by encouraging their competitiveness and efficiency, protecting market participants against fraud, manipulation, and abusive trading practices, and by ensuring the financial integrity of the clearing process. Through effective oversight, the CFTC enables the futures markets to serve the important function of providing a means for price discovery and offsetting price risk.

Title VII of the Dodd-Frank Wall Street Reform and Consumer Protection Act of 2010 amends the Commodity Exchange Act to establish a comprehensive new regulatory framework for swaps and security-based swaps [79]. The legislation is enacted to reduce risk, increase transparency, and promote market integrity within the financial system by, among other things: (1) providing for the registration and comprehensive regulation of swap dealers and major swap participants; (2) imposing clearing and trade execution requirements on standardized derivative products; (3) creating robust record-keeping and real-time reporting regimes; and (4) enhancing the SEC's rule-making and enforcement authorities with respect to, among others, all registered entities and intermediaries subject to the SEC's oversight.

Financial Industry Regulatory Authority (FINRA)

The Financial Industry Regulatory Authority (FINRA) is the largest independent regulator for all securities firms doing business in the US. FINRA's mission is to protect investors by making sure the securities industry operates fairly and honestly. FINRA oversees about 4,400 brokerage firms, about 162,930 branch offices and approximately 630,020 registered securities representatives. It has approximately 3,200 employees and operates from Washington, DC and New York, NY, with 20 regional offices around the country.

FINRA touches virtually every aspect of the securities business, from registering and educating industry participants to examining securities firms; writing rules; enforcing those rules and the federal securities laws; informing and educating the investing public; providing trade reporting and other industry utilities; and administering the largest dispute resolution forum for investors and registered firms. When rules are broken, FINRA can fine, suspend or expel firms or individual brokers from the business. FINRA can require firms to return money to investors who have been harmed.

Every investor deserves a fundamental protection when investing in the stock market, whether they are investing in a 401(k) or other thrift, a savings or employee benefit plan, a mutual fund, an exchange-traded-fund (ETF) or a variable annuity, FINRA works to ensure that:

- Anyone who sells a securities product has been officially tested, qualified and licensed.

- Every securities product advertisement used is truthful, and not misleading.

- Any securities product promoted or sold to an investor is suitable for that investor's needs.

- Investors receive complete disclosure about the investment product before purchase.

On the lines of international coordination in banking through Bank for International Settlements (BIS), International Organization of Securities Commissions (IOSCO) has been addressing cross-border issues faced by securities regulators. The role of IOSCO became more significant after many cross-border consolidations in exchanges, which has now resulted in securities markets being dominated by a small number of global intermediaries. IOSCO grew out of an association of stock exchange commissions of the Americas set up in 1974. By 1983, it had become a global organization. Since then, effectively all securities regulators have become members, including second regulators in some cases (CFTC of the US).

IOSCO's role is to assist its members to promote high standards of regulation and to act as a forum for national regulators to cooperate with each other and other international organizations. Its main objectives can be summarized as follows.

- Cooperate together to promote high standards of regulation in order to maintain just, efficient and sound markets.

- Exchange information on their respective experiences in order to promote the development of domestic markets.

- Unite their efforts to establish standards and an effective surveillance of international securities transactions.

- Provide mutual assistance to promote the integrity of the markets by a rigorous application of the standards and by effective enforcement against offenses.

3.4   Regulations and Insurance

Even though we motivated the need for regulation of insurance from the fiduciary responsibility of insurers toward the customers in terms of insurance contracts, their fair definition and pricing, the scope of regulation of the insurance industry goes further. It addresses entry into market by licensing of insurance companies, investment practices of insurers, and similar aspect of insurer solvency. Complexity of insurance products requires regulatory scrutiny and licensing of practitioners to ensure their competence.

Although regulation in many sectors aims at enforcing competition and preventing artificially high prices, insurance regulations were designed for the reverse objective. They were designed to prevent excessive competition, hence the regulation of entry into market and licensing. It was feared that excessive competition will lead insurers into providing products at unsustainable prices, which is a serious risk in insurance, since cost of production in insurance often is not known until the contract has completed full term. Moreover, as insurers vie for customers in an excessively competitive environment, issues related with adverse selection can significantly damage the health of the industry. The original goals of regulation of insurance, solvency, and equity, are expanded to availability and affordability of insurance. One may sense an inkling of contradiction of these new additional goals, especially with the original goal of solvency.

In the US, regulation of insurers is by state regulators through the state's commissioner of insurance, who is appointed by the governor of the state. The commissioner of insurance is responsible for administration of insurance laws and general supervision of the insurers' business in the state. The National Association of Insurance Commissioners (NAIC) is the medium for exchange of information and coordination of regulatory activities among the states. Since 1989, in order to establish minimum standards for financial regulation of insurers by states, NAIC created the Financial Regulation Standards Accreditation Program (FRSAP). The accreditation program includes 16 model laws and rules affecting regulation of insurers for managing general agents, credit for insurance, examination processes, liquidation proceedings, reinsurance intermediaries, and risk retention. By 1995, 45 states and the District of Columbia were accredited by this program, which once accredited requires an annual evaluation process and recertification review every five years.

We now provide an overview of the scope of regulatory authority the state's commissioner of insurance must conduct. The power to license insurers is perhaps the most important activity. The license is a certification of the company with regards to its financial stability and soundness, therefore to qualify an insurer should have a certain amount of capital and surplus. Regulatory requirement dictates insurers' accounting standards, which significantly differ from the generally accepted accounting principles (GAAP). The set of accounting procedures used by insurers is called statutory accounting, which emphasizes the insurer's ability to fulfill its obligations under the contracts it issues. Statutory accounting is mostly ultraconservative. For instance, in their balance sheets insurance companies recognize only those assets that are readily convertible into cash.

In insurance, the premium is collected in advance of the delivery of the service, therefore insurance laws require specific recognition of the insurer's fiduciary obligations by maintaining policy reserves and risk-based capital. Insurers maintain unearned premium reserve and loss reserves to respect the difference in timing of premium payment and service delivery, or loss event and indemnification. The accumulated premium must be invested by the insurer, and there are also restrictions for investment. Investment is allowed in US and Canadian government bonds, mortgage loans, certain high-grade corporate bonds, and to some extent in preferred or common stock. Property and liability have lesser investment restrictions than life insurers, where the latter can invest only a small percentage of their assets in common stocks.

Regulators also oversee the insurance rates, where the requirement is that rates must be adequate, not excessive, and not unfairly discriminatory. The unfairly discriminatory implies insurers may not charge significantly different premiums for two customers with approximately similar risk profiles. Any variation of rates must have an actuarial basis. In rates and in sales practices, regulators check for unfair practices. For a failing insurer, the regulators take steps to rehabilitate the company; this may involve reinsuring a substantial portion of the firm's business or a merger with a stronger insurer. Finally, every licensed insurer active in a state, both foreign and domestic, must submit an annual report to the state's commissioner of insurance. The report details the insurer's asset and liabilities, its investments, income loss payments, expenses, and additional information as required by the commissioner.

3.5 Summary

Risk management is a crucial activity in all firms and all households, but in certain firms and institutions it is fundamentally important. This well-recognized fact has led to the development of a regulatory environment in various segments of the financial sector. The regulatory environment is an integral and evolving feature of the financial sector. In this chapter, we presented an overview of regulatory systems in place and their historical evolution for some key segments of the US financial sector. The intention here was not to give a comprehensive description of the regulatory environment, but to present some salient features of it to demonstrate the importance of the contexts of risk management problems, issues, and models developed in the rest of the book.

3.6 Questions and Exercises

Review Questions

1. Why is government control of business necessary?

2. What is the antitrust objective of government control of business?

3. How does competition among firms benefit society?

4. What is the role of regulation as a government control of business?

5. Why is competition not always considered good for the banking and insurance industry?

6. What is the link between regulations and risk management in financial services firms?

7. What are the regulatory agencies active in US and Canada?

8. Why did the first and the second attempt at a Bank of United States fail?

9. When was the first federal regulatory authority for banking created that is currently active?

10. What was the significance of the Knickerbocker Trust collapse in 1907?

11. What role did Charlie Mitchell play in the Federal Reserve System, and how did it affect the Federal Reserve's management of events of 1929?

12. What were the major developments in the regulation of the financial services sector after the Great Depression?

13. How was the banking environment changing in the 1960-1980 period?

14. What was the impact of the ceiling for depositary interest rates? What was the impact of lifting this ceiling?

15. What were the magnitude and impact of the 1987-88 S&L crisis?

16. What were the regulatory implications of the repeal of the Glass-Steagall Act of 1933 by the Financial Services Modernization Act of 1999?

17. What were 'too-big-to-fail' financial institutions through the 2007-2008 financial crises?

18. What is the scope of the Dodd-Frank Wall Street Reform and Consumer Protection Act of 2010?

19. What were the findings of the 1913 Pujo committee, and what were the implications of these findings?

20. How did J.P. Morgan split in response to the Glass-Steagall Act of 1933?

21. What were the major federal legislations and regulatory authority created for securities in the post Great Depression era?

22. What was the impact of the 2008 financial crisis on US investment banks?

23. What was the vacillation in US regulation of insurance regarding insurance being interstate commerce or not?

24. After the Financial Modernization Act of 1999, what are the challenges for the state insurance regulators?

25. What is the International Association of Insurance Supervisors (IAIS)?

26. When and how is the US Federal Reserve System organized?

27. What are the roles of the Fed?

28. How does the Federal Reserve implement its monetary policy?

29. What is the primary role of the OCC? How is the OCC funded?

30. What power does the OCC have in regulating banks and federal thrifts?

31. What is the Federal Deposit Insurance Corporation? How expansive is its role in regulating banks?

32. What are the new regulatory agencies created under the Dodd-Frank Act of 2010? What are their roles?

33. What is the Bank for International Settlements? What is its mission?

34. What is the progression of the Basel Accords from Basel I to Basel III?

35. What is the SEC and what is the scope of its activities?

36. What is the SEC's EDGAR system?

37. What is the role of the new Division of Risk, Strategy, and Financial Innovation at the SEC?

38. What role does CFTC fulfill in the regulatory needs of the financial sector?

39. How does the Dodd-Frank Act of 2010 change the role of CFTC?

40. What is the International Organization of Securities Commissions (IOSCO)? What are the objectives of IOSCO?

41. Why were regulations in insurance built to prevent excessive competition?

42. What is the scope of activities of a state's Commissioner of Insurance in the US?

43. What is NAIC's Financial Regulation Standards Accreditation Program (FRSAP)?

44. How do accounting standards in insurance differ from those of GAAP?

45. What reserves and investment restrictions do insurers maintain/observe?

Exercises

1. Write an essay on the role competition has played in the history of banking in the US.

2. What lessons can the rest of the world learn from the historical evolution of regulations in banking, securities, and insurance in the US?

3. Write an essay on the proactive versus reactive nature of regulations in the financial services sector.

4. What were the reasons for the 2007-2008 global financial crises?

5. Read the cited article by E.A. Ludwig [56] and discuss the strengths and weaknesses of the Dodd-Frank Wall Street Reform and Consumer Protection Act of 2010.

Part II

Modeling and Simulation
of Risk

Chapter 4

Principles of Simulation and Generating Random Variates

Making decisions in the face of risk, however significant its impact, can be quite challenging. The greater the impact of a risk on a person, property or wealth, the higher the level of care necessary to make choices in order to ensure the outcomes are as desirable as possible. The risk management framework of Chapter 2 formulated a guideline to develop a strategy for handling risk; however, the actual implementation of the framework requires quantifying the relevant risks and their impacts. We developed models for the quantification of risk in Chapter 1. To assess the impact of risk, as is fundamentally needed for the implementation of a risk management framework, in today's complex financial markets and institutions requires a well-developed and versatile tool-set. The regulatory environment in place for various segments of the financial sector described in Chapter 3 makes assessment and management of risk under stipulated constraints not only a desired goal, but a mandatory activity of the institutions.

4.1 Principles of Simulation

One key tool in the tool-set to assess, measure, and manage risk is simulation. Simulation modeling and analysis constitutes a set of quantitative tools and methodologies, and indeed software systems, to facilitate delving into the role risk plays in an environment. The intentional use of a general reference to an 'environment' is to establish the fact that simulation is a versatile tool, relevant and widely used in many sectors and areas of study beyond risk management and finance. In fact, the modeling techniques underlying simulation are broad, not all of which incorporate randomness; some 'simply' describe how a complex system evolves deterministically over time. Models built using complex systems of ordinary or partial differential equations and solved by numerical techniques fall in this category. For us, however, simulation will be used to model and analyze risk in order to answer important risk management questions. The rest of the chapter will define and lay-out the basics of how to use simulation for this purpose.

4.1.1   What Is Simulation?

The word 'simulation' is derived from the Latin word *simulatus*, past participle of *simulare*, which means to copy, represent. Simulation is an attempt to duplicate the operation of a system or the behavior of a quantity of interest without incurring the expense and expending the effort to build or operate the system, or generate observations for the quantity of interest by natural means. More operationally, simulation involves creating an experimental set-up to study the dynamics of the system or the quantity of interest. Instead of 'solving' the system, a simulation analyst operates a 'model' of the system in the experimental set-up under different conditions to assess the system behavior.

The model of the system almost always is an abstraction of the relatively complex reality, where a simplified description is constructed for the purpose of gaining insight into the system behavior. Modeling involves not only creating abstractions, but also coding it in an appropriate computing environment, which is used for running experiments. There are dedicated commercial products to facilitate building simulation models for different industry segments - manufacturing, communications networks, civil infrastructure, such as roadways, railways and air-traffic control, medical and health care services, and indeed, financial services. The objectives of a simulation study can range from asset pricing, performance analysis, capability analysis, comparison studies, sensitivity analysis, optimization study, or constraint analysis.

Models are abstractions of the real system, where a system is viewed as a group of objects or quantities joined together in some regular interaction or interdependence. Models developed for facilitating simulation are constructed with a view of using them to conduct specific experiments, predict events, or determine a future course of action that would in some sense be best. In short, models are built with a definite use in mind, hence models themselves are not the focus, the use they would be put to is. Models are a means to an end. Identifying the objective of a simulation study helps determine the level of scope and detail appropriate in the model of a system for the set goal. Abstraction in model building is justified since it brings perspective to the need for detail, introducing a level of detail that improves understanding of system behavior and the necessary modifications of the system. Controlling details allows the analyst to have more control over variations to the system studied, and to organize beliefs and empirical observations.

Conducting a simulation analysis of a problem can be organized in logical stages, laid out in Figure 4.1, as the structure a simulation study. Once the objective of the study is stated, including the measures for evaluating the outcomes, appropriate data must be acquired, and equations and algorithms need to be defined to describe the model. It is generally useful to run a simplification of the model or just a portion of the model for testing the model's validity. After reviewing the usefulness and robustness of the model, translating the mathematical model into the desired computational environment can

The Structure of a Simulation Study

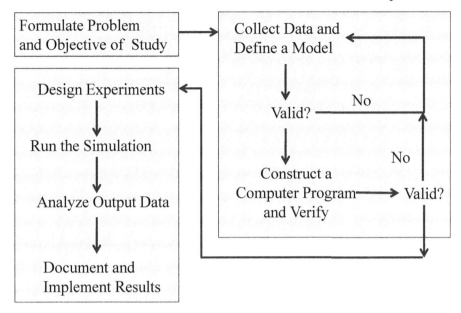

FIGURE 4.1: The guideline for how to structure a simulation study.

be initiated. Developing portions of the model separately and putting these modules together makes for a more robust strategy. A well-tested computer model should be the basis of running the necessary designed experiments, and the corresponding simulation runs. The output data needs to be analyzed to achieve the goals of the simulation study.

It is often tempting to jump into writing up the computer model to perform the simulation runs and experiments, and generate the output results, cutting short the due diligence on the conceptual model development. This can lead to erroneous assumptions or implementation for the intended problem and objectives. Figure 4.2 emphasizes the two stages for a simulation study. The first stage is where the conceptual model building and assessment in natural language and mathematical formalism takes place. The second stage would then translate the conceptual model into a computer representation of the model in a chosen simulation software using appropriate programming structures.

We have looked at models of risk in Chapter 1. In the next chapter, we will advance the modeling of risk so that its evolution in time may be described. All subsequent chapters will continue to advance the modeling approaches appropriate for specific contexts of risk management. Therefore, proficiency in conceptual and mathematical modeling of risk for the objectives of risk management will be developed throughout the book. In the rest of this chapter, we will focus on the core of a simulation model – generating random variates for

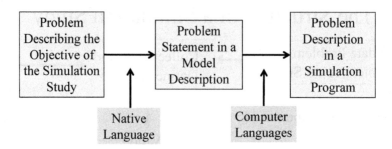

FIGURE 4.2: The stages to build the simulation model.

the various models of risk. We will also discuss methods for testing, validating, and running simulation models.

4.2   Random Number Generation

Simulation is to copy, therefore for simulating risk, the inherent nature of risk - generating random outcomes - needs to be copied. Before digital computers came into existence, random numbers were directly obtained from actual random processes: for example, by rolling a die, tossing a coin, electronically by the noisy output of a valve. Obviously, such numbers were not statistically reliable, and a particular sequence of random numbers could not be reproduced for comparative studies.

Today digital computers allow the implementation of simple deterministic algorithms to generate sequences of random variables, quickly and reproducibly. Such numbers are, strictly speaking, not truly random, but with sufficient care they can be made to resemble random numbers by most of their properties. Hence, these numbers are called pseudo-random numbers. Actually, pseudo-random numbers is a term reserved for random numbers corresponding to the simplest of models of risk - the uniform random variable, $U(0,1)$. Random numbers generated for other models of risk, such as, normal, lognormal, Weibull, etc., are called random variates.

We will first describe methods for generating pseudo-random numbers, followed by methods for generating other random variates. This order of presentation is particularly meaningful, since pseudo-random numbers are the building block for all other random variates. Two basic deterministic, recursive methods for creating a sequence of pseudo-random numbers follow.

4.2.1 Linear Congruential Generator

The Linear Congruential Generator (LCG), a basic iterative method for generating pseudo-random numbers, is based on the primary relation as follows,

$$X_{n+1} = aX_n + b \ (mod \ c), \tag{4.1}$$

where a, c (> 0) are positive integers, and b (≥ 0) is a non-negative integer. The recursion needs to start from a number, X_0, which is called the 'Seed.' The 'mod' operator is short for 'modulo,' representing the remainder after division of a number by another number. For instance, $7(mod \ 3) = 1$, $9(mod \ 5) = 4$, etc. Therefore, the LCG recursion generates a sequence of numbers taking integer values from 0 to $c - 1$; the remainders when $aX_n + b$ is divided by c. When a, b, and c are picked appropriately,

$$U_n = X_n/c, \tag{4.2}$$

seem uniformly distributed in the unit interval $[0, 1]$. For example, $a = 16,807 = 7^5$, $b = 0$, $c = 2^{31} - 1$ (prime) is a relatively carefully selected set of values for these parameters.

The congruential algorithm is widely used, but there is one problem with it. It displays a looping characteristic, captured by the 'period' of a random number generator. Period of a random number generator is defined as the smallest positive integer p which satisfies $X_{i+p} = X_i$ for all $i > k$, for some $k \geq 0$. In a large-scale simulation on a supercomputer, this becomes a major weakness, for example, a 32-bit LCG will have a period less than 2^{32}. For a supercomputer this sequence will get exhausted in just a few minutes. Therefore, a, b, and c need to be chosen with care, with not only long period in mind, but also good statistical properties, such as apparent independence, computational and storage efficiency, reproducibility (same seed!), and facilities for separate streams (change the seed).

There are alternatives to random number generation than using the LCG. These alternatives are also based on iterative procedures, developed with the goal of achieving the desirable properties listed above. One of them is the lagged Fibonacci generator.

4.2.2 Lagged Fibonacci Generator

An alternative random number generator to the Linear Congruential generator is the lagged Fibonacci generator. The recurrence relation for the lagged Fibonacci generator is

$$X_n = X_{n-r} \ op \ X_{n-s}, \tag{4.3}$$

where s and r are the lags with $0 < s < r$ and $n \geq r$ and op is a binary operator. For example, the binary operator could be 'addition (mod c)' or 'subtraction (mod c).' If *op* is 'addition (mod c),' the lagged Fibonacci generator will become, $X_n = X_{n-r} + X_{n-s}(mod\ c)$. It should be noted that the lagged Fibonacci generator iterations begin from $n \geq r$, therefore the first r iterates, $\{X_k; k = 1 \ldots r\}$, have to be obtained from some other scheme. The linear congruential generator can be used for initialization of the lagged Fibonacci generator.

A good property of the lagged Fibonacci generators is that extremely long periods are possible with these generators, and several have been shown to exhibit good global properties if the parameters are chosen carefully. Excellent references for more detailed information on random number generators are books by Law and Kelton [55], Glasserman [30], and Knuth [50]. These books offer a more rigorous discussion of properties and implementation issues for various random number generators.

While performing a simulation-based analysis, it is good to know where the random numbers are coming from. The particular random number generator in use may have some artifacts that can sometimes result in misleading conclusions. Hence, it may even be advisable to test run the simulation model with two different random number generators to make sure the results are genuine, and not artifacts of the properties of the generator.

4.3   Generation of Discrete Random Variates

Pseudo-random numbers are the building block for generating all other random variates. Therefore, the basis for generating random variates for all other models of risk with good properties is getting good pseudo uniform random numbers. We will now describe methods for generating other models of risk, starting with simple discrete models of risks. The common aspect of these methods for generating other models of risk from pseudo-random numbers is an appropriately designed transformation of the latter. We begin with a very simple risk model, a random variable that has a finite number of outcomes, two or more.

4.3.1   n-Outcome Random Variate

Two-point random variable, X, is one that takes two values $x_1 < x_2$ with probabilities p_1 and $p_2 (= 1 - p_1)$, respectively. Even though this is a very simple model of risk, it is popularly used in describing time-dependent models of risk, such as the binomial tree model, the random walk model, etc. The transformation of pseudo-random numbers, $U(0, 1)$, that can be used to generate

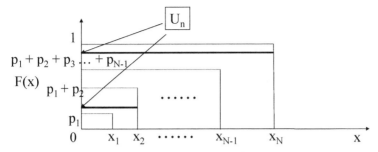

FIGURE 4.3: N-outcome discrete random variate generation.

random variates for a two-point random variable is,

$$X_n = x_1 \text{ if } 0 \leq U_n \leq p_1, \tag{4.4}$$
$$= x_2 \text{ if } p_1 < U_n \leq 1, \tag{4.5}$$

for $n \geq 1$. Therefore, for every pseudo-random number generated, a random variate for the two-point random variable model is produced. This is not always the case for other methods and for other models of risk.

The method for the two-outcome random variable can be easily extended to N-state random variable taking values $x_1 < x_2 < \ldots < x_N$, with non-zero probabilities $p_1, p_2, \ldots, p_N$, respectively. From the rules for probability distributions, we have that $\sum_{i=1}^{N} p_i = 1$. We will define $s_0 = 0$, $s_j = \sum_{i=1}^{j} p_i$ for $j = 1, \ldots, N$, and using these quantities generate the N-point random variates as follows,

$$X_n = x_{j+1} \text{ if } s_j \leq U_n \leq s_{j+1} \quad j = 0, \ldots, N, \tag{4.6}$$

for $n \geq 1$. Figure 4.3 is a pictorial depiction of the above algorithm, which shows that the above algorithm is effectively an inversion of the cumulative mass function.

Not all discrete random variables have a finite number of outcomes; some can have an infinite number of outcomes. If a discrete random variable has infinite outcomes, the above transformation has to be modified into an iterative procedure. We describe this next in the context of the Poisson random variate.

4.3.2 Poisson Random Variate

Poisson random variate is a counting discrete random variate with an infinite number of outcomes starting from $n = 0$. Poisson risk model was one of the example risk models studied in Chapter 1. Examples of risks that can be modeled using the Poisson model are number of customers arriving at a ticket counter per unit time, the number of defects per unit square centimeter of a semiconductor, or number of breakdowns of a server per month. The possible outcomes of a Poisson random variable are $0, 1, 2, \ldots$. Given the parameter

λ, the probability that there will be j number of occurrences of event of interest by this model of risk are given by $p_j = e^{-\lambda} \lambda^j / j!$ (the probability mass function of the Poisson distribution). The following algorithm generalizes the transformation for finite outcome discrete random variable to generate random variates by the infinite outcome Poisson distribution.

Step 1: Initialize Set $s_1 = 0$, $s_2 = p_1$, $j = 1$ and generate a uniform random number U_n;

Step 2: Check If $s_1 \leq U_n \leq s_2$, then $X_n = j$.
And exit.

Step 3: Else Update $s_1 = s_2$, $s_2 = s_2 + p_{j+1}$, $j = j + 1$;
Go to Step 2.

Therefore, the method transforms every pseudo-random number U_n generated into a Poisson random variate, X_n, for all $n \geq 1$. However, the generation of random variates now needs more work than was needed in the N-outcomes random variate case.

4.4   Generation of Continuous Random Variates

We now move from discrete random variables to considering continuous random variables. We have, in fact, already seen one example of a continuous random variate generation, the uniform distribution, $U(0, 1)$. The uniform random variates generated by the random number generators were key to generating the discrete random variates. They will continue to be so for continuous random variates. Let X be a continuous random variable with a probability distribution function, $F_X(x)$. Therefore, by the property of probability distribution functions, $F_X : R \rightarrow [0, 1]$. This basic property of all probability distribution functions is exploited in the first method presented, the inverse transform method.

4.4.1   Inverse Transform Method

The principle behind the inverse transform method for generating random variates for a random variable, X, is as follows. Let U $(0 < U < 1)$ be a uniform random variate generated by a random number generator. If we can find an X such that $X(U) = F_X^{-1}(U)$ for every uniform random variate, U, generated, then X will be the desired random variate. Here F^{-1} is the inverse of the probability distribution function of the random variable, X, assuming it exists and can be computed with sufficient ease. In general, if the cumulative distribution function of the random variable, X, is not continuous, we will

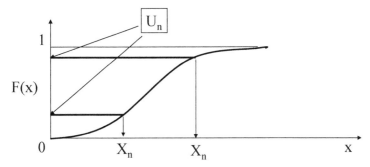

FIGURE 4.4: A pictorial depiction of the principle behind the inverse transform method.

define $X(U) = \inf\{x : U <= F_X(x)\}$. Figure 4.4 shows the principle behind the inverse transform method pictorially.

The inverse transform method is very effective for generation of random variates for the exponential random variable model. In Chapter 1, the probability distribution function for the exponential random variable was given to be, $F_X(x) = 1 - e^{-\lambda x}$, for the parameter $\lambda > 0$. Applying the inverse transform method gives,

$$X_n(U_n) = F_X^{-1}(U_n) = -(ln(1 - U_n))/\lambda = -ln(U_n)/\lambda \quad \text{for} \ \ 0 < U_n < 1, (4.7)$$

where $n \geq 1$. The last equality in Eqn. (4.7) is true since, if U_n is $U(0,1)$, then so is $1 - U_n$. Therefore, given a sequence of random numbers, U_n, generated by a random number generator, $X_n = -ln(U_n)/\lambda$ is a sequence of exponentially distributed random variates. This is a very simple and efficient transformation.

In principle the inverse transform method should be all one should need to know about continuous random variates generation. But, this is mostly not efficient, since computing the inverse of many probability distribution functions is not easy. Consider the normal distribution as an example. As was discussed in Chapter 1, in the case of normal distribution, the integrals of the probability density function to compute the distribution function must be evaluated numerically. Therefore, inverting the probability distribution function is not an efficient option. In such cases other techniques are needed. We develop some methods to generate other continuous random variates.

4.4.2   Acceptance-Rejection Method

The acceptance-rejection method, as the name suggests, builds an iterative procedure by which in every iteration the output is either accepted as a valid random variate outcome or is rejected. The method is useful when the inverse transform method or any other simple method is not available for a random variable model. For a random variable, X, with a probability density function, $f_X(x)$, the method relies on a second random variable, Y, with a probability density function, $g_Y(y)$. The key determinant for the selection of the second

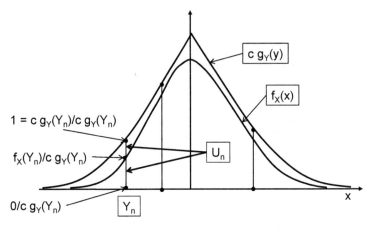

FIGURE 4.5: A pictorial depiction of the principle behind the acceptance-rejection method.

random variable, Y, is that a method for generating random variates for Y should be available. The acceptance-rejection method will use this fact as a basis for generating random variates for X by the probability density, $f_X(x)$.

The method requires picking a constant, c, such that $\frac{f_X(y)}{g_Y(y)} \leq c$; for all y. We have assumed for the definition of the constant c that the two random variables, X and Y, are defined on the same sample space, say Ω. The algorithm behind the acceptance-rejection method is as follows.

The Acceptance-Rejection Algorithm

Step 1: Generate Y_n with probability density $g_Y(y)$.

Step 2: Generate a random number U_n.

Step 3: If $U_n \leq \frac{f_X(Y_n)}{(cg_Y(Y_n))}$, set $X_m = Y_n$ (Accept); $m \leftarrow m + 1$.
Otherwise Reject, and **return to Step 1**.

In Figure 4.5, we give a pictorial description of the acceptance-rejection method. Every time a Y_n is generated by its probability density, $g_Y(y)$, for which we have an easier method for generating random variates, we also generate a uniform random variate, U_n. The probability that $U_n \leq \frac{f_X(Y_n)}{(cg_Y(Y_n))}$ is $\frac{f_X(Y_n)}{(cg_Y(Y_n))}$, which is the probability that the Y_n generated could also be a realization of X generated by $f_X(x)$. Hence, it is accepted in this scenario, but rejected otherwise. Therefore, for every random variate generated for density, $f_X(x)$, there are at least two uniform random numbers needed, one to generate Y_n and the other for the acceptance-rejection decision.

While one may be concerned about the efficiency of this method on the grounds of frequency of rejection versus acceptance, it is reassuring that the following result can be proven for the acceptance-rejection method.

Theorem: The random variates generated by the acceptance-rejection method, in fact, have the density $f_X(x)$. The number of iterations of the algorithm that are needed to create a desired random variate is a geometric random variable with mean c.

For the proof of this result, the reader may refer to Nelson [67].

Therefore, the acceptance-rejection is in contrast to the inverse transform method, where every uniform random number generated produces a random variate by the desired distribution. In the acceptance-rejection method, number of iterations before obtaining the next random variate by the desired distribution is not deterministic. The lower the value of c, fewer may be the iterations needed before successfully generating a random variate for X. The trick, therefore, is to find the smallest constant, c, such that $\frac{f_X(y)}{g_Y(y)} \leq c$.

Let's consider an example to demonstrate the application of the acceptance-rejection method. Let $g_Y(y) = \exp(-|y|)/2$ defined on $(-\infty, \infty)$, which is the double-exponential density. The inverse transform method can be easily modified to generate random variates for the double-exponential distribution, as follows.

Step 1: Generate two random numbers, U_1 and U_2.

Step 2: Let $Y_n = -ln(U_1)$.

Step 3: If $U_2 \leq \frac{1}{2}$, then set $Y_n \leftarrow -Y_n$.
Otherwise return Y_n.

If $f_X(x) = \frac{\exp(-x^2/2)}{\sqrt{2\pi}}$, which is the probability density function for the standard normal distribution model. As stated earlier, for the normal distribution the integrals of the probability density function to compute the distribution function must be evaluated numerically. Therefore, inverting the probability distribution function to apply the inverse transform method is not an efficient option. In order to apply the acceptance-rejection method, we need to find a constant c, such that $\frac{f_X(y)}{g_Y(y)} \leq c$ for all y. It can be shown that $\frac{f_X(y)}{g_Y(y)} \leq 1.3155$ in this case, hence we can take $c = 1.3155$.

The acceptance-rejection method is a good method to generate random variates for the normal distribution; however, the normal distribution is such a popular model of risk that specialized methods have been developed to generate random variates for it. We study some of these next.

4.4.3   Normal Random Variate

As we will study in Chapters 5 and 6, the normal random variable model of risk is a very popular and a frequently used model of risk. In Chapter 5, we will be using the normal distribution to describe the evolution of risk over time, while in Chapter 6 we will use simulation to solve these models. Due to the frequent use of the normal distribution model, we will describe two specialized methods for generation of normal random variates.

4.4.3.1 Box-Muller Method

The Box-Muller method avoids using the probability distribution function for generation of normal random variates. It instead uses the fact that if U_1 and U_2 are two independent $U(0, 1)$ uniformly distributed random variables, then G_1 and G_2 defined by the following transformation are two independent standard normal (Gaussian) random variates. As described in Chapter 1, by standard normal random variates, symbolically $N(0, 1)$, we mean outcomes for a normal random variable with a mean of 0 and standard deviation of 1.

$$G_1 = \sqrt{-2\ln(U_1)}\cos(2\pi U_2) \tag{4.8}$$
$$G_2 = \sqrt{-2\ln(U_1)}\sin(2\pi U_2) \tag{4.9}$$

Now, if the two uniform random numbers, U_1 and U_2, are truly independent, we would have generated two independent standard normal variates. However, if the two uniform random numbers are two successive random numbers from a congruential random number generator, this will not be the case; the two normal random variates, (G_1, G_2), will make a spiral structure in the $R \times R$ space. Therefore, to be able to use both the normal random variates generated in one calculation, where independent normal random variates are necessary, the two uniform random numbers must come from different linear congruential streams - corresponding to different seeds.

The transformation underlying G_1 and G_2, i.e., Eqns. (4.8) and (4.9), are simple, but there is one disadvantage of the Box-Muller method. The transformations require computing trigonometric functions (sin, cos), which are somewhat computationally demanding, thus resulting in inefficiency in normal random variates generation. An alternate method that bypasses this computational burden is the Polar-Marsaglia method.

4.4.3.2 Polar-Marsaglia Method

The Polar-Marsaglia method is based on the fact that if U is a uniformly distributed random variable, $U(0, 1)$, then $V = 2U - 1$ is $U(-1, 1)$, i.e., V is uniformly distributed on the interval $(-1, 1)$. If V_1 and V_2 are independent and distributed as U(-1,1), obtained by transforming as above two independent standard uniform random variables, U_1 and U_2, then we define $W = V_1^2 + V_2^2$, when $V_1^2 + V_2^2 \leq 1$. Therefore, W lies in the unit circle.

It can be shown that W so defined is uniformly distributed as $U(0, 1)$. This can be visualized by considering the pair (V_1, V_2) as points on the $R \times R$ plane. Since the points (V_1, V_2) are equally likely to fall anywhere in the unit circle inscribed with the box, $[-1, 1] \times [-1, 1]$, their distance from the origin, $(0, 0)$, is equally likely to be between 0 and 1. Similarly, the angle made by (V_1, V_2) on the positive x-axis, θ, is uniformly distributed as $U(0, 2\pi)$, and is

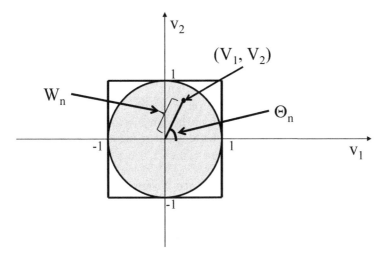

FIGURE 4.6: A pictorial depiction of the construction of the Polar-Marsaglia method.

independent of W. See Figure 4.6 for a pictorial representation of the (W, θ) pair constructed.

The square defined by the values (V_1, V_2) has an area of 4, whereas the area of the circle inscribed by W is $\pi r^2 = \pi$. Therefore, the area of the circle is $\pi/4$ fraction of the area of the square in Figure 4.6. The point (V_1, V_2) will fall inside the circle with a probability of $\pi/4$. When that happens, $W = V_1^2 + V_2^2 \leq 1$, and we can use the random numbers, V_1, V_2, to make the following computations,

$$\cos \theta = V_1/\sqrt{W}, \tag{4.10}$$
$$\sin \theta = V_2/\sqrt{W}. \tag{4.11}$$

This allows rewriting the Box-Muller method by replacing the sin and cos functions as follows,

$$G_1 = V_1\sqrt{-2\ln W/W} \tag{4.12}$$
$$G_2 = V_2\sqrt{-2\ln W/W} \tag{4.13}$$

In the above modification of the Box-Muller method, θ replaces $2\pi U_2$, since it has the same distribution as $2\pi U_2$. W plays the role of U_1 in the Box-Muller method, again due to distributional similarity.

The Polar-Marsaglia method can also be seen as an example of the *acceptance-rejection method*. As seen in the previous section, there are other acceptance-rejection methods for generation of normal random variates, where acceptance/rejection is the central theme of the method. Here the acceptance/rejection was implemented depending on whether the (V_1, V_2) pair fell inside the unit circle or not. If it fell inside the circle, the algorithm could continue and generate a pair of independent normal random variates, otherwise not. Therefore, although a proportion of the generated uniformly distributed random numbers are discarded, this method is often computationally more efficient than the Box-Muller method, especially when a large quantity of normal random variates are to be generated.

In simulation experiments, since often a very large number of random variates need to be generated to serve the purpose of the simulation study, the efficiency of the random variates generator algorithms is of significant concern.

4.4.3.3   Generation of Multi-Variate Normal

The Box-Muller and Polar Marsaglia methods give independent normal (Gaussian) random variates with zero mean and standard deviation of 1, when uniform random numbers U_1 and U_2 used are independent and uniform, $U(0,1)$. In practice, often a pair, (X_1, X_2), or an n-dimensional set, $(X_1, \ldots, X_n)$, of correlated Gaussian random variates are required. Let X_1, X_2 be jointly Gaussian with mean μ_1, μ_2, and variance σ_1^2, σ_2^2, respectively, and covariance of $\rho\sigma_1\sigma_2$. We now describe how the independent standard normal random variates can be transformed to be a correlated pair of normal variates, of a given correlation structure.

To begin with, assume μ_1, $\mu_2 = 0$ and σ_1, $\sigma_2 = 1$. We first generate Y_1, Y_2, and Y_3 random variates that are independent, standard normal, $N(0,1)$. Using these three independent standard normal random variates, we define the following,

$$X_1 \;=\; \sqrt{1 - |\rho|}\,Y_1 + \sqrt{|\rho|}\,Y_3, \qquad\qquad (4.14)$$

$$X_2 \;=\; \sqrt{1 - |\rho|}\,Y_2 \pm \sqrt{|\rho|}\,Y_3, \qquad\qquad (4.15)$$

where in Eqn. (4.15), '+' is used when $\rho \geq 0$ and '−' for $\rho < 0$. The (X_1, X_2) pair thus obtained is jointly standard normal, with desired correlation structure. In order to get random variates with the general mean and standard deviation characteristics, the following transformations are applied, $X_1 \leftarrow \mu_1 + \sigma_1 X_1$ and $X_2 \leftarrow \mu_2 + \sigma_2 X_2$.

The method described above for bi-variate normal random variates with general correlation structure can be extended for n-dimensional $(n > 2)$ correlated normal random variates. This requires using the Cholesky factorization of the correlation matrix, given as $Corr = RR^T$. Once an n-dimensional standard, independent normal variates, Y are produced, the desired correlation structure is introduced as, $X = RY$. Following this, as done for 2-dimensional case, required mean and variance can be introduced for each element of X.

4.4.4 Chi-Square and Other Random Variates

In Chapter 1, several additional specific models of risk were presented and discussed. Some of these, such as the lognormal distribution and Weibull distribution, are extensions of more basic models of risk. Therefore, if a method is known for generating random variates for the simpler model, the method can be extended to create random variates for the extended models of risk. For instance, we discussed a few methods for generating normal random variates in Sections 4.4.3.1 and 4.4.3.2. If a normal random variate, X_n, is produced by any of these methods, after appropriately choosing its mean and standard deviation, the desired lognormal random variate can be produced as, $V_n = \exp(X_n)$.

The Weibull distribution was presented as a more general distribution than the exponential model of risk in Chapter 1. The inverse transform method is efficiently utilized for generating random variates for the exponential model of risk. The same can be applied to generate Weibull random variates. Similarly, the gamma risk model, $\Gamma(x; \alpha, k)$, was discussed as a sum of exponential random variables in Chapter 1. This relation can be utilized for generating gamma random variates.

Finally, we consider the chi-square, χ_d^2, distribution, where d is a positive integer denoting the degrees of freedom. If $Y_1, \ldots, Y_d$ are independent standard normal random variates ($N(0,1)$), then $Y_1^2 + \ldots + Y_d^2$ has a χ_d^2 distribution. Moreover, for constants $\alpha_1, \ldots, \alpha_d$, the distribution of $\sum_{i=1}^{d}(Y_i + \alpha_i)^2$ is noncentral chi-square with d degrees of freedom and noncentrality parameter, $\nu = \sum_{i=1}^{d} \alpha_i^2$, $\chi_d^2(\nu)$. This relationship between chi-square and normal models of risk can be utilized for generating chi-square random variates. In general, however, for more general values of parameters, such as degrees of freedom, d, in the case of chi-square and scale-shape (α,k) parameters in the case of the gamma model, more specialized methods for random variates generation would be needed.

We have so far seen some basic methods for random variates generation, both for discrete and continuous models of risk. The intention of this introductory view was to give the reader a sense of the building blocks behind the sophisticated simulation software available today, and the underlying assumptions and issues. We next move to discussing testing of random variates for their quality and accuracy.

4.5 Testing Random Variates

Once appropriate methods for random variates generation are created, they may be used to perform the required experiments for the simulation study. The quality and reliability of the experiments, however, will rely on the correctness

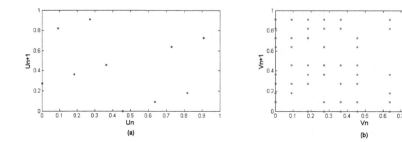

FIGURE 4.7: Display of output from a linear congruential generator. (a) 1000 numbers generated lie on three parallel lines. (b) The 1000 numbers after implementing shuffling.

and quality of random variates generated. Therefore, testing the outputs of the random number generators and the random variates generating algorithms is an important step for assuring the validity of a simulation analysis. Since the random numbers are the basic building block for all other random variates, it is necessary to have independence wherever this is a requirement. We begin with testing the independence of the uniform random numbers.

4.5.1 Testing for Independence of Random Numbers

Statistical independence is usually a difficult property to test. No single test for it is totally satisfactory and fully reliable. Among the random number generators, a generator like the linear congruential generator (LCG) is an important case to consider to test independence of its output, since in the linear congruential iterations each random number is determined only by its immediate predecessor. If the generator fails one of the independence tests, a remedial strategy may be developed.

Different plotting techniques is one approach for testing independence. Autocorrelation plot and scatter plots are some of the plotting techniques that may be used. A simple test for independence involves plotting the successive pairs (U_n, U_{n+1}) for $n = 1, 2, 3, \ldots$ as points in the unit square of the $R \times R$ plane, with the U_n on the x-coordinate and U_{n+1} on the y-coordinate. If the random numbers generated by the random number generator are independent, the plots will scatter about with no apparent patterns.

Applying this plotting technique to the output from a linear congruential generator reveals that the points lie on one of c different straight lines of slope a/c. We display this for a simple generator to enhance the pattern, with $a = 6, b = 3, c = 11$, in the left panel of Figure 4.7. For a more sophisticated LCG with parameters suggested in Section 4.2.1, a large number of random numbers generated should fairly evenly fill the unit square. The presence of patches without any of these points is an indication of bias in the generator.

These patterns can be eliminated by applying additional remedial procedures to the numbers produced by a generator.

4.5.1.1 Shuffling Procedure

A shuffling procedure attempts to introduce greater 'randomness' in the deterministic output from a random number generator. Sample steps of a shuffling procedure can be as follows,

1. Generate 20 or more random numbers from a random number stream: $\{U_1, ..., U_{20}\}$.

2. Pick one of these 20 numbers with equal probability, $1/20$. This will require generating another random number from a different stream, V_1. The randomly generated index ranging from $1 \ldots 20$ is obtained as, $j = \text{floor}(20 * V_1) + 1$. Assign the random number from the $U(1:20)$ stream corresponding to the randomly picked index, say U_j, to the shuffled random number stream W, i.e., $W_1 = U_j$.

3. Replace the random number picked from the U stream, say U_j, with the next random number in the U sequence, i.e., U_{21}.

4. Repeat Steps 2-3. This results in a shuffled sequence of random numbers, W, of the initial sequence, U.

Note that this procedure requires two-times more random numbers generated than an unshuffled sequence of the same length. This is because a random number stream also needs to be generated to determine the index of a randomly picked number from a set of 20 or so numbers. Shuffling procedures have been found to be effective in reducing patchiness in poor generators. They also provide a possibility to lengthen the periods when using linear congruential generators.

In the right panel of Figure 4.7, we apply a shuffling procedure to the output of the left panel from a simple linear congruential generator. The shuffling improves the output, although it doesn't make this generator attractive for practical use. Shuffling improves the independence properties of a random number generator, but how about its *reproducibility*? Remember experiments being reproducible is a desired property for a simulation environment. As long as a randomly picking index in the shuffling procedure can be reproduced, the shuffling procedure retains the random number generator's reproducibility.

Besides plotting strategies to test independence, other quantitative hypothesis tests can also be conducted to test the independence of the output of a random number generator. The most useful hypothesis tests for this purpose are the Runs Tests, such as, Runs above and below the Median, Runs Up, Runs Down, etc. Besides independence, for the uniform distribution and other general distributions, we will also need to test if the random variates generated are truly representative of the desired distribution.

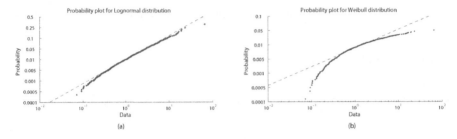

FIGURE 4.8: Display of Probability Plots. (a) Lognormal probability plot. (b) Weibull probability plot.

4.5.2   Testing for Correctness of Distribution

Testing accuracy of the distribution of random variates generated is not any different from statistical inference for testing the distribution of data acquired from real-world experiments. As in standard statistical inference for real-world data, one simple way to test correctness of distribution is to plot a histogram for the output of a random variates generator and compare it visually with the graph of the true density function it is supposed to simulate. This is the simplest assessment possible, and remains a subjective evaluation.

A less subjective evaluation is to produce a probability plot for the simulated data. For instance, if the random variates are supposed to be representing a normal probability distribution, one can plot the data in a normal probability plot. If the points fall effectively on a straight line, the hypothesis that the data represent a normal probability model cannot be rejected. In Figure 4.8, we plot lognormal and Weibull probability plots in the left and right panel, respectively, for lognormal data generated using the Polar-Marsaglia method. The lognormal random variates are obtained by taking the exponential of the output of the Polar-Marsaglia method, i.e., $\exp(G_1)$, where G_1 is defined in Eqn. (4.12). The probability plots are constructed using the *probplot* function in MATLAB [61]. The lognormal probability plot is arguably a straight line, and would pass the so-called '*fat-pencil*' test, i.e., if a fat-pencil were put on the points in the plot, it would hide all the points. However, the Weibull probability plot, which was chosen to create a contrast, is by no means a straight line.

For more rigorous quantitative evaluation, we would set up standard statistical tests for testing validity of a hypothesized distribution for the simulated random variates data. The tests are developed based on quantifying the 'distance' between the histogram of the simulated data and the true probability density. The first test is more suitable for discrete models of risk.

4.5.2.1   The χ^2 Goodness of Fit Test

Using the designed random variates generator, we generate a large number, N, of independent and identically distributed random variates. We form a cumulative frequency histogram $F_N(x)$ for these random variates, which we wish to compare with the true cumulative mass function, $F(x)$. The procedure requires subdividing the random variates data into $k + 1$ mutually exclusive categories and counting the numbers, $N_1, N_2, \ldots, N_{k+1}$, of them falling into each of these categories. Clearly, $N = N_1 + N_2 + \ldots + N_{k+1}$.

Breaking the data into $k + 1$ categories works easily for discrete random variables, with possibly each discrete outcome of the random variable making a category. For continuous random variables, one needs to set arbitrarily chosen break-points. Following the categorization of the data, the true probability for each of the categories is computed, say $p_1, p_2, \ldots, p_{k+1}$. In the discrete case, this would essentially be the probability mass function of the random variable. Given the true probability for each category, the (true) expected number of values falling in each category should be $Np_1, Np_2, \ldots, Np_{k+1}$.

The Pearson statistic is designed to test the hypothesis whether the simulated data represents the hypothesized distribution. We measure the 'distance' between the true expected number of observations in each category and the observed number of observations in each category, as follows,

$$\chi^2 = \sum_{j=1}^{k+1} \frac{(N_j - Np_j)^2}{Np_j}. \tag{4.16}$$

If the hypothesis is supported, in other words, if the random variates are generated from the desired distribution, the statistics should have a small value. The Pearson statistic is asymptotically (as N goes larger) distributed according to the χ^2-distribution with k degrees of freedom. The expected value of χ^2 with k degrees of freedom is, $E[\chi^2] = k$, and $Var(\chi^2) = 2k$, as discussed in Section 1.2.2.9. Therefore, to complete the hypothesis test, we pick a significance level $100\alpha\%$ and determine a value of $2\chi^2(1 - \alpha, k)$, such that,

$$P(\chi^2 \leq \chi^2(1 - \alpha, k)) = 1 - \alpha. \tag{4.17}$$

Restating the Eqn. (4.17) for the probability of the complementary event, $\chi^2 > \chi^2(1 - \alpha, k)$, gives,

$$P(\chi^2 > \chi^2(1 - \alpha, k)) = \alpha. \tag{4.18}$$

The event, $\chi^2 > \chi^2(1 - \alpha, k)$, is the event of Type I error, when the null hypothesis is rejected when in fact it is true. The null hypothesis in our case is that the simulated data are accurate, i.e., they represent the desired model of risk. However, the particular sample of size N is finite, and hence, there is a possibility that the distance measured in Eqn. (4.16) comes out to be large, leading to the erroneous conclusion of rejecting the null hypothesis when in fact it is true. The probability of Type I error measures the probability of making this error, rejecting the null when the null is true, in Eqn. (4.18). Therefore, $0 < \chi^2 \leq \chi^2(1 - \alpha, k)$ defines our acceptance region for the null hypothesis that $\chi^2 \sim 0$, or the 'distance' between the observed number of observations and true expected observations in each category is essentially zero. If the χ^2 value computed by Eqn. (4.16) satisfies $\chi^2 \leq \chi^2(1 - \alpha, k)$, we accept (or as is stated in Statistics texts, fail to reject) the null hypothesis at the significance level of $100\alpha\%$. If however, $\chi^2 > \chi^2(1 - \alpha, k)$, then the null is rejected, which implies that the random variates are not acceptably generated according to the desired distribution.

4.5.2.2   Kolmogorov-Smirnov Test

For a continuous random variable, the discrete categories of the χ^2 goodness of fit test are (i) artificial, (ii) subjective, and (iii) do not fully take into account the variability in the data. These disadvantages are avoided in the Kolmogorov-Smirnov test, which is based on the Glivenko-Cantelli theorem [11, 48]. If $\{X_i; i = 1 \ldots N\}$ are the N random variates generated for a continuous model of risk with a cumulative distribution function, $F(x)$, then the cumulative frequency function, $F_N(x)$, for the random variates generated can be described as,

$$F_N(x) = (\#\text{of}\,X_i's \leq x)/N. \tag{4.19}$$

Therefore, if the sample $\{X_i\}$ is sorted in increasing order to get $\{X_{(i)}\}$, then

$$F_N(X_{(i)}) = i/N, \qquad i = 1, 2, 3, \ldots, N. \tag{4.20}$$

The Glivenko-Cantelli theorem states that the cumulative frequency function, $F_N(x)$, will converge to the true cumulative distribution function, $F(x)$, as the number of random variates generated, N, becomes large. Formally, it states the following,

$$D_N = \sup_{-\infty < x < \infty} |F_N(x) - F(x)| \to 0 \quad a.s. \quad \text{as} \quad N \to \infty. \tag{4.21}$$

In order to apply the Kolmogorov-Smirnov (KS) one-sided test at $100\alpha\%$ significance level, based on the above result, for the null hypothesis: H_0 : $\sqrt{N}D_N = 0$, we follow the following steps. The test utilizes the fact that the test statistic, $\sqrt{N}D_N$, follows a Kolmogorov distribution with cumulative density function, $H(x)$. The steps involve first computing the test statistic, followed by defining the acceptance region for the chosen significance level, and finally, checking if the computed test statistic falls in the acceptance region.

1. Compute the value of the test statistic, $\sqrt{N}D_N$, from the random variates as follows.

Define $D_N^+ = \max_{1 \leq i \leq N}\{\frac{i}{N} - F(x_{(i)})\}$.

Define $D_N^- = \max_{1 \leq i \leq N}\{F(x_{(i)}) - \frac{i-1}{N}\}$.

Compute $D_N = \max\{D_N^+, D_N^-\}$.

2. Find the acceptance region $(0, x_{1-\alpha})$, where $x_{1-\alpha}$ value is such that $H(x_{1-\alpha}) = 1 - \alpha$, where H is the Kolmogorov distribution function.

3. Finally, if $\sqrt{N}D_N < x_{1-\alpha}$ then we accept (or as in Statistics texts, fail to reject) the null hypothesis at $100\alpha\%$ significance level; otherwise the null hypothesis is rejected and the random variates are concluded to not represent the desired distribution.

Being based on asymptotic results, the sample size or number of random variates generated to conduct the test, N, is important, both for the χ^2 goodness of fit test and the Kolmogorov-Smirnov test. A guideline suggests that $N > 35$ suffices for these tests, but in a simulation lot many random variates can be generated without much problem, hence sample size is not an issue.

We have spent considerable effort in describing the procedures for testing random variates' independence and accuracy. The primary motivation for this description was to understand the nuts-and-bolts behind the otherwise black-box simulation software routines, including those available in MATLAB. Just as random variates generation can be conveniently accomplished by using the packaged routines in a simulation software, most statistical software (including MATLAB) come packaged with functions and routines to conduct the above tests. For instance, in MATLAB, KS test can be conducted using the *kstest* routine, which requires giving the input random variates data, specifications for the desired distribution, and the significance level, α, to conduct the test. The χ^2 goodness of fit test can be performed using the *chi2gof* function in MATLAB.

4.6   Validation of Model

The models developed for complex systems, even after considerable abstraction and simplification, can be quite complex. Each module or component of the model can have several variables, each described by various models of risk, with complex interaction between the variables, as well as the model components, to determine the overall system behavior or performance. We have so far discussed the principles behind building a model for a simulation study and the methods underlying capturing the riskiness of various variables of the model. Besides capturing the riskiness of the variables by generating random variates by various probability distributions, we also looked at methods to test the accuracy of the models of riskiness. We now need to move to the higher level of testing the accuracy of simulation models.

Testing the accuracy of the simulation model can be broken down into two major pieces: testing the accuracy of the computer representation of the model (right-most box in Figure 4.2) and testing the accuracy of the model for its ability to capture desired characteristics of the real system (the top loop in the right box in Figure 4.1). Testing of programming accuracy is often referred to as **model verification**, and includes programming error detection and debugging, while testing the model accuracy is **model validation**. Model validation can ask fundamental questions about whether the conceptual model correctly reflects the real system, or whether the conceptual model is really capable of addressing the necessary issues about the real system. The results of a validation analysis may result in going back to the drawing board, as suggested in Figure 4.1.

Clearly, model verification and validation is a very important exercise, since without a level of confidence on the model accuracy, its recommendations cannot be trusted. Despite its importance, often a modeler can miss paying sufficient attention to this step. Moreover, a key point to note is, model validation should be a continuous, ongoing activity throughout the time the model is being used to make decisions. This is important, since no model is good for all times and under all conditions; as times and conditions change, the assumptions underlying a model may no longer hold, and hence, must be assessed. Reasons for overlooking the need for testing model accuracy can be multi-fold, ranging from good-old laziness or ignorance, to overconfidence on one's modeling capabilities, or pressures of time and budget.

The challenge behind testing models is, while model building is a fun and creative activity, model testing can be quite effortful and a drag. But, being skeptical of one's own work is a good rule to follow throughout model building and model usage. The gap between model verification and validation can be nicely summarized by, if the computer program representation of a model runs, it does not mean it is OK! As in the planning for any other project, explicitly setting aside time and resources for model testing is a good practice to combat pressures of time and budget, or as a means for instilling the discipline to overcome laziness.

4.6.1   Techniques for Model Verification

Errors creeping into model building can be classified as syntactical or semantical. Syntactical errors are unintentional addition, omission, or misplacement of notation that either prevents the model from running or causes it to run incorrectly. Misplaced decimal points or parentheses can have a dramatic impact on the outcome. Semantical errors are errors in the meaning or intention of the modeler, such as a wrong condition inserted in an if-then-else statement. Semantical errors are harder to detect, but can have a very damaging effect on the usefulness of the model.

Best practice for developing good models is that the entire development of the simulation project should be done so that it facilitates testing its accuracy. Writing spaghetti software code, or other poor organization of code, such as without good descriptive variable names, descriptive comments, good flow of code logic, that makes the code hard to understand even by its own creator, is clearly not advisable. A stitch in time does, indeed, save nine, if not more!

As stated in Section 4.1.1, the code for a simulation model should be built modularly, starting simple and gradually growing to capture the complexity of the model, with staged verification, and preferably, validation. Step-wise refinement and progressively adding complexity to the model, verifying and validating the model in each step, develops the model in several passes, and ensures the model's accuracy, along with guaranteeing that the model possesses the right level of complexity needed for the purpose of the simulation study. Use of unstructured control, such as 'goto' type statements, should be avoided, instead logic control should be structured, such as using 'if-then-else', 'do-while', etc. statements. Not only model code and its logic, but also data supporting the model should be thoroughly and clearly documented. This helps detect and remove unintentional errors in model data, logic, and construction. It also facilitates communication and collaboration for efficient, error-free model building and usage.

Performing a top-down and bottom-up model code review helps in a thorough inspection of the code and its accuracy. A top-down review begins with looking at the major module and works its way to the lower level modules, while a bottom-up review begins looking at the smallest modules and builds up the verification process towards the upper major modules. Running the model, provided it runs, to check for reasonable output can also help identify semantical errors. Plotting outputs provides a visual aid for verification, where some errors can be detected visually that may otherwise go unnoticed. The plots of outputs provide help in identifying a problem, rather than discovering the cause of a problem. For aid in locating syntactical errors or tracing the code to identify semantical errors, software tools come with debuggers, which help uncover the source of the error.

To a seasoned programmer, most of the points made here would be trivialities, or second nature to account for when building the model, but for a newbie learning the discipline they can save many hours of neck-breaking, mind-numbing debugging work.

4.6.2   Techniques for Model Validation

Model validation tests if the model is a meaningful and accurate representation of the real system for the purpose of the simulation study. Validation of models should be a continuous activity throughout the time the model is being used to make decisions. As stated earlier, this is important, since as times and conditions change, the assumptions underlying a model may no longer hold, and hence, must be assessed. Validation is not only of the model structure, but begins at the data-gathering stage to support building the model. As is said, 'garbage-in, garbage-out.' The validation looks for functional validity, namely that the model's output behavior has sufficient accuracy for the model's intended purpose. Therefore, among other things, model validation helps develop a trust that simulation results may be used for real-world decisions.

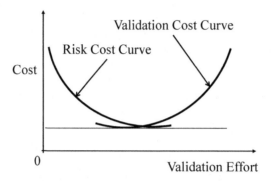

FIGURE 4.9: Display of Validation Cost vs. Risk Cost Curve.

Validation is a hard and painstaking process, but just as important, if not more, than any other activity in a simulation study. Therefore, the more novel ideas generated for performing validation, the better. We suggest some ideas here. It helps to compare the simulation results with other models, or simpler versions of the same model, for which analytical results may be available. Conducting the simulation under degenerate or extreme parameters or conditions should give anticipated degenerate or extreme output, for instance, should $X_0 = 0$ imply $X_T = 0$, or if investment weights $w_t^i = 0$ (for all i and t), what should be the anticipated portfolio performance, similarly, if the strike, $K = 0$, what should the option be worth, etc. On the other hand, reasonable values should result in reasonable outcomes. Not only that, if you show the results from the model to a knowledgeable expert, would he/she agree that the results seem reasonable and representative of the real system. In this regard, the simulation model can be subjected to a Turing test, where we ask the knowledgeable expert to discriminate between model and system outputs. If the expert cannot detect a difference, there is more evidence for model validity.

As in model verification, plots and summary statistics of output variables are useful means for model assessment. Comparing these with similar information for the real system creates a context for judging if the model is capturing the necessary system characteristics. Conducting sensitivity analysis with the model involves changing some input parameters to determine their effect on the model's behavior and output. These effects of sensitivity analysis on the model should be similar to how the real system would behave under similar changes. Current and historical data for the system is useful to support the validation.

Although validation is a continuous process, and should be performed throughout the simulation study, there is still an optimum level of effort on validation, beyond which returns may be less valuable. Overdoing is not good, although this is hardly ever the problem in practice. A balance is needed between validation costs and cost incurred due to risk of making decisions based on an invalid model. This trade-off is illustrated in Figure 4.9, where the lowest total cost point between the validation cost curve and the risk cost curve is sought. Some models meant for ongoing use need continued validation, which should be performed only in a controlled modification mode to prevent unwieldy growth of the model that hurts its robustness.

4.7 Output Analysis

A simulation model is a computer-based statistical sampling experiment, therefore appropriate statistical techniques must be used to design and analyze the simulation experiments. Once the verification and validation steps are satisfactorily accomplished, in the final box of Figure 4.1, the simulation model is used for running the required simulation experiments and simulation output analysis. Very often a great deal of time and money is spent on model development and 'programming,' while little effort is made for analyzing simulation output data appropriately. A little more care and attention is needed for designing and implementing the output analysis in a simulation study.

The level of effort or precision required in a simulation output analysis depends on several factors, such as the nature of the problem, importance of decision, validity of the input data, and availability of a verified and valid model. In some cases a rough analysis using judgmental procedures may suffice, while in others a detailed statistical analysis will be necessary. We will focus here on statistical analysis of simulation data.

As with statistical analysis of data obtained from the real system, statistical analysis of simulation output comprises descriptive and inferential statistics. The goal of descriptive statistics, as the name suggests, is to **describe** the properties of the system based on the statistical properties of the simulated data obtained from running experiments using the model. The data generated by running experiments using the model is finite, depending on the design for sample size sought, and is hoped to be representative of the population. Inferential statistics tries to make conclusions or **infer** knowledge about the population based on the sample data produced by running simulation experiments, assuming the data are representative of the population. These concepts of sample and population, and descriptive or inferential statistics, are similar to those in statistical analysis of data obtained from the real-world.

The simulation output variables of a model are in their turn random variables, potentially a complex function of the input risk factors and their interactions in the model. An experimental sample, called replications, are intended to be independent observations of the output variables obtained after every 'run' of the simulation model. The independence of observations of the output variables from different runs of the simulation model depends on the properties of the random number generator in use, such as long cycle period and ability to use different seeds. A reasonable number of replications (sample size) is a good indicator of what can be expected in any subsequent replication. Clearly, the sample size, amount of data generated from simulation experiments to make descriptive or inferential statistics, determines the accuracy and quality of inferences made from the data. Independence of observations in the simulated data is a requirement of most descriptive and inferential statistics procedures. The idea is to generate a large enough sample to draw valid inferences about the population, where more is always better, but sample generation time and computing cost are the primary constraints.

4.7.1 Descriptive Output Analysis

Let the simulation output quantity of interest or the performance measure for the simulation study be θ. Using the replications, we want to estimate the value of the performance measure. We can seek two types of estimates under a descriptive output analysis.

Point Estimator: A *formula* for a single value estimate of the performance measure, denoted by $\hat{\Theta}$.

Point Estimate: The actual value a point estimator takes when specific data values are plugged into the formula, denoted by $\hat{\theta}$.

Interval Estimate: Gives a range of values the performance measure will have with a degree of confidence. It is also called a *Confidence Interval*.

For example, the descriptive statistic of interest of an output variable could be its mean, μ. To create an estimate of the mean, we can use a point estimator or an interval estimator. Sample mean estimator, defined by

$$\bar{X} = \frac{\sum_{i=1}^{N} X_i}{N}, \tag{4.22}$$

is a point estimator, while a confidence interval based on sample mean is an interval estimator. The confidence interval estimator will be,

$$(\bar{X} - z_{\alpha/2} \frac{\sigma}{\sqrt{N}}, \bar{X} + z_{\alpha/2} \frac{\sigma}{\sqrt{N}}), \tag{4.23}$$

if the standard deviation of the performance measure, σ, is known, otherwise will be,

$$(\bar{X} - t_{N-1,\alpha/2} \frac{s}{\sqrt{N}}, \bar{X} + t_{N-1,\alpha/2} \frac{s}{\sqrt{N}}), \tag{4.24}$$

where s is a point estimator for the standard deviation of the performance measure. $z_{\alpha/2}$ is the $1 - \alpha/2$-th percentile of a standard normal distribution and $t_{N-1,\alpha/2}$ is the $1-\alpha/2$-th percentile of a standard t-distribution with $N-1$ degrees of freedom. When we substitute the values for X_i's to be data obtained from N runs of the simulation model, say x_i, we obtain point estimates and interval estimates for the mean. The interval estimates are created so that with $100(1 - \alpha)\%$ confidence the true mean, μ, lies within the confidence interval.

The narrower the confidence interval, the better the accuracy of the interval estimate. However, there is an inverse relationship between confidence level, $100(1 - \alpha)\%$, and the width of the confidence interval. If we desire a higher confidence level, the width of the confidence interval becomes bigger. On the other hand, if we require a tighter confidence interval, the confidence level will drop. For a given confidence level, getting a tighter confidence interval can be accomplished by increasing the number of observations generated. In particular, for a confidence interval with width $2e$ and a confidence level of $100(1 - \alpha)\%$, number of observations exceeding,

$$N > (\frac{z_{\alpha/2}\sigma}{e})^2, \tag{4.25}$$

will suffice. If the standard deviation of the output variable, σ, is not known, a point estimate for it is used to get an approximate number of observations needed for the desired interval estimate accuracy.

Mean of the output variable is one example of a performance measure. Often other summary descriptive statistics are required for the output variables of a simulation study. A general functional of the output variable, X, that may be defined as, $\theta = E[f(X)]$, or can be a conditional expectation, $\theta = E[f(X)|g(X)]$. Functions $f(.)$ and $g(.)$ are appropriately well-defined functions. For instance, variance of the output variable can be computed by picking function, $f(x) = (x - \mu)^2$. Similarly, semi-variance, conditional variance, percentiles, Value-at-Risk (VaR), Conditional Value-at-Risk (CVaR), etc. can be estimated. For each functional, a point estimator needs to be used to create the point estimate or the interval estimate. We will introduce these point estimators in the later chapters, where the functionals are used in specific contexts. Here our objective was to introduce the general concepts underlying simulation output analysis.

4.7.1.1   Designing Simulation Run by Properties of Estimators

Every point estimator for a functional of a simulation output random variable, $f(X)$, has properties that should be understood. These properties would guide the development of design of simulation experiments. An estimator is said to be unbiased if in expectation it gets right what it is attempting to estimate, i.e., in our earlier notation $E(\hat{\Theta}) = \theta$. Clearly, being unbiased is a good property for a point estimator to have. However, beyond bias, there is a second important property of an estimator to consider, which is the variance of the estimator, $V(\hat{\Theta})$. The higher the variance of an estimator, the poorer the estimator, since for any given simulated data, the estimate produced by the estimator can be quite off from the true population value for that functional of the output variables, $\theta = E[f(X)]$. In that, one seeks a minimum variance unbiased estimator, MVUE, for the functionals of interest.

Bias and variance of an estimator, in the simulation approach to solving problems, depends heavily on the computational effort made. The greater the computational budget, the lower the level of achievable variance of an estimator. However, computational budget is never infinite, just as compute time is rarely unlimited. Therefore, design of a simulation study must determine the trade-off between bias, variance and compute time for various quantities being estimated.

The general guideline for the design of simulation runs the decision process is as follows:

1. If the compute time for each replication to generate θ_i is fixed, say τ, and the estimator is unbiased, select number of runs to fit $V(\hat{\Theta})\tau$ within the computational budget.

2. If the compute time for each replication to generate θ_i is stochastic, say $\tilde{\tau}$, and the estimator is unbiased, select number of runs to fit $V(\hat{\Theta})E[\tilde{\tau}]$ within the computational budget.

3. If more than one estimators are available, select one with least Mean-square error $(MSE(\hat{\Theta}) = bias(\hat{\Theta})^2 + V(\hat{\Theta}))$. Follow guideline for step 1 or 2, depending on the nature of replication compute time.

4.7.2 Inferential Output Analysis

Inferential analysis of simulation output variable is in essence identical to the inferential analysis of real-world data. The analyst postulates a hypothesis regarding the value of a functional of a simulation output variable and utilizes the data generated from simulation experiments to test if the hypothesis has support or not. The test is conducted on the basis of a test statistic, which is essentially a point estimator for the hypothesized functional. For instance, the sample mean, $\bar{X}$, estimator for the population mean functional, $\mu = E[X]$.

Developing the test statistic utilizes knowledge of the *sampling distribution* of the estimator. The sampling distribution of the sample mean estimator, $\bar{X}$, is an asymptotically normal distribution, by the central limit theorem. This fact is used to construct the test statistic,

$$\frac{\bar{X} - \mu}{\sigma/\sqrt{N}}, \qquad (4.26)$$

if σ is known. And,

$$\frac{\bar{X} - \mu}{s/\sqrt{N}}, \qquad (4.27)$$

when σ is not known, and is estimated using the point estimator, s. In the σ unknown setting, the test statistic (Eqn. (4.27)) is approximately a t-distribution with $N - 1$ degrees of freedom.

Under the hypothesis that the true population mean of the output variable, $\mu = \mu_0$, an acceptance region is constructed as,

$$\left(\mu_0 - z_{\alpha/2}\frac{\sigma}{\sqrt{N}}, \mu_0 + z_{\alpha/2}\frac{\sigma}{\sqrt{N}}\right), \qquad (4.28)$$

in the case when the standard deviation of the performance measure, σ, is known. Otherwise, the acceptance region becomes,

$$\left(\mu_0 - t_{N-1,\alpha/2}\frac{s}{\sqrt{N}}, \mu_0 + t_{N-1,\alpha/2}\frac{s}{\sqrt{N}}\right), \qquad (4.29)$$

when σ is not known. If the computed sample mean, $\bar{X}$, falls in the acceptance region, the simulated data supports the hypothesis of the mean output variable level being μ_0. If the computed sample mean does not fall in the acceptance region, the hypothesized value of mean can be rejected. As before, $z_{\alpha/2}$ is the $1 - \alpha/2$-th percentile of a standard normal distribution and $t_{N-1,\alpha/2}$ is the $1 - \alpha/2$-th percentile of a standard t-distribution with $N - 1$ degrees of freedom. The value α is called the significance level, which is the probability of making an erroneous conclusion, namely rejecting the hypothesis when it is in fact true (Type I error). Clearly, we would like to minimize the probability of making an erroneous conclusion, but we cannot indefinitely reduce this probability without creating another problem, that of not being able to reject the hypothesis when it is in fact false (Type II error).

The principle applied above to the mean functional, $\mu = E[X]$, can be applied to any other functional for which inferential analysis is needed. A point estimator, its sampling distribution, and a hypothesized value of the functional of the output variable will need to be defined. Using these and a chosen significance level, an acceptance region will be constructed and the test performed. We will consider specific details of other inferential analysis in the context of specific problems in later chapters. Here our objective was to introduce the basic principles behind inferential simulation output analysis.

4.8   MATLAB Tools for Simulation

MATLAB mathematical software has a vast array of functions for simulating random variates in its Statistics Toolbox. We list a few of these functions here. The reader is advised to look up the extensive help documentation available with MATLAB to see the details of these and other related functions. At the bottom of each function description in the MATLAB help documentation, look for 'See Also' to explore other related functions. Resources such as MATLAB Primer [20] are also useful.

Random Number Generator: rand

Normal distribution: `normrnd`

Uniform distribution: `unifrnd, unidrnd`

Binomial distribution: `binornd`

Poisson distribution: `poissrnd`

Exponential distribution: `exprnd`

Weibull distribution: `wblrnd`

Lognormal distribution: `lognrnd`

Chi-square distribution: `chi2rnd, ncx2rnd`

Gamma distribution: `gamrnd`

Debugging support: `assert, echo, error, keyboard, return, warning`

Other: `probplot, kstest, chi2gof`

4.9 Summary

In this chapter, we set down the principles to follow for constructing a simulation framework. The logical steps needed to accomplish a simulation study were laid out, followed by discussing each step in detail. A simulation study can be successfully performed if each step is given its due importance, including model development, verification, validation, designing simulation experiments and conducting appropriate output analysis. In order to not leave the nuts-and-bolts of a simulation framework as a black-box, and be a more informed user of simulation software and tools, we also learned various algorithms to generate and test various random number and random variates. The principles of simulation developed in this chapter will be utilized in all future chapters to address various risk management problems.

4.10 Questions and Exercises

Review Questions

1. What is simulation? When is simulation used?

2. What are the advantages and disadvantages of simulation analysis?

3. What role do models serve when solving a problem using simulation analysis?

4. Discuss the steps that should be adopted for a successful implementation of a simulation study.

5. Why is it advised that a simulation model be developed in stages?

6. What is a random number generator? What are random variates?

7. What are the desirable properties of a random number generator?

8. What is a linear congruential generator? How does it differ from a lagged Fibonacci generator?

9. Describe the shuffling procedure applied to a linear congruential generator. Why is a shuffling applied to a random number generator?

10. In what terms can random number generators and random variates generated by various algorithms be tested for their required properties?

11. What is the χ^2 goodness of fit test? When and how is it applied?

12. What is the Kolmogorov-Smirnov test? When and how is it applied?

13. What is model verification? What is model validation? Why are these activities considered an important part of addressing problems using simulation analysis?

14. What are the techniques utilized for model verification?

15. How can model validation be effectively implemented?

16. What is the purpose of a descriptive versus an inferential simulation output analysis?

17. What is a point estimator? Point estimate? How is it related to an interval estimate?

18. How does the number of replications in a simulation experiment relate with the accuracy of an estimate of the output variables?

19. What is bias, variance and mean square error of an estimator? How can these be utilized for providing a guideline to design simulation runs?

20. What are Type I and Type II error of inferential analysis?

Exercises

1. Implement your version of the linear congruential generator (LCG) in MATLAB with parameters set as, $a = 16,807 = 7^5$, $b = 0$, $c = 2^{31} - 1$. Pick a seed, X_0, of your choice.

2. Construct a lagged Fibonacci random number generator given by, $X_n = X_{n-r} + X_{n-s}(mod\ c)$, in MATLAB with parameters, $r = 7$, $s = 3$, $c = 2^{31} - 1$. Initiate the iteration using the LCG of Exercise 1.

3. Computing software, like MATLAB, comes packaged with random number and variate generation functions. We can get to see how this is done by implementing some of these ourselves. Construct your own functions in MATLAB that generate 1000 realizations by the following:

   (a) Binomial random variates, with $n = 10$, $p = 0.7$.

   (b) Poisson random variate, with $\lambda = 5$.

4. Implement the following specialized methods in MATLAB for generating standard normal random variates.

   (a) Box-Muller method

   (b) Polar-Marsaglia method

5. Utilize your normal random variate generators from the previous exercise to produce a pair of correlated normal random variates, (X_1, X_2), with the following correlation matrix.

$$\rho = \begin{pmatrix} 1 & 0.4 \\ 0.4 & 1 \end{pmatrix} \qquad (4.30)$$

   The mean and standard deviation of the two random variables is given as, $E[X_1] = 3, E[X_2] = 10$, and $\sigma_{X1} = 2, \sigma_{X2} = 4$.

6. Construct and implement an acceptance-rejection method for generating standard normal random variates in MATLAB using the double-exponential density.

7. For a sequence of random numbers generated by the linear congruential generator (LCG), $\{U_n\}$, plot the successive pairs (U_n, U_{n+1}) for $n = 1, 2, 3, \ldots$ as points in the unit square of the $R \times R$ plane, with the U_n on the x-coordinate and U_{n+1} on the y-coordinate. Use the following simple choice of parameters for this LCG: $a = 6, b = 3, c = 11$.

8. Generate a sequence of random numbers, $\{U_n\}$, from MATLAB's built-in random number generator (**rand**) and plot the successive pairs (U_n, U_{n+1}) for $n = 1, 2, 3, \ldots$ as points in the unit square of the $R \times R$ plane, with the U_n on the x-coordinate and U_{n+1} on the y-coordinate. Compare the plot with a similar plot for the LCG of the previous exercise.

9. Apply the shuffling procedure described in Section 4.5.1.1 to a linear congruential generator with parameters, $a = 6, b = 3, c = 11$. Plot the shuffled sequence as successive pairs (U_n, U_{n+1}) for $n = 1, 2, 3, \ldots$ as points

in the unit square of the $R \times R$ plane, with the U_n on the x-coordinate and U_{n+1} on the y-coordinate. How does this shuffled sequence compare to a similar plot of unshuffled sequence.

10. Generate 1000 random variates by the MATLAB routines for the following discrete models of risk and test that the sample generated truly represents the intended distribution using the χ^2 goodness of fit test. Use an $\alpha = 0.05$.

   (a) Uniform distribution by `unidrnd`.

   (b) Binomial distribution by `binornd`.

   (c) Poisson distribution by `poissrnd`.

11. Generate 1000 random variates by the MATLAB routines for the following continuous models of risk and test that the sample generated truly represents the intended distribution using the Kolmogorov-Smirnov test. Use an $\alpha = 0.05$.

   (a) Normal distribution by `normrnd`.

   (b) Weibull distribution by `wblrnd`.

   (c) Gamma distribution by `gamrnd`.

12. Consider a portfolio of 100 loans, where the loans can be classified into three categories per their credit risk characteristics: the low, medium, and high risk loans. The portfolio has 20 high risk loans, each of principal $100K$, 30 medium risk loans, each of principal $150K$, and 50 low risk loans, each of principal $200K$. In the duration of a year, a high risk loan has 20% probability of default independent of any other loan, a medium risk loan has 10% probability of default, and a low risk loan has 5% probability of default. The recovery rate on default for low, medium, and high risk loans is modeled as, $U(0.7, 0.9)$, $U(0.6, 0.8)$, and $U(0.5, 0.7)$, respectively.

   (a) Formulate an expression (formula) for the value of the loan portfolio in one year.

   (b) Create an estimator for the expected value of the loan portfolio in one year. Similarly, create an estimator for the standard deviation of the loan portfolio value in one year.

   (c) After generating some scenarios in MATLAB, compute a point estimate of mean and standard deviation of the loan portfolio value in one year. At a chosen confidence level, $1 - \alpha\%$, construct a confidence interval for the mean loan portfolio value in one year.

   (d) If the desired accuracy for the mean loan portfolio value is 5% at a confidence level of 99%, how many scenarios are needed for this level of accuracy?

(e) Create a measure for assessing the tail risk of the value of the loan portfolio in a year. Construct a point estimator for this measure, and compute an estimate for this tail risk measure.

(f) How well does a normal distribution fit the value of the loan portfolio in a year?

Chapter 5

Modeling Risk Evolving over Time

Risks are not static, and least so in the world of finance. In the real-world, the risk factors discussed in Chapter 1 are dynamic, evolving quantities mimicking the forces of change in finance. For instance, after a one-period return modeled by a lognormal distribution, what happens in the next period? After a bank sees a surge in the number of defaults in its loan portfolio in a quarter, what would it expect of the risk in the subsequent quarters? This dynamic evolution of risk is an essential feature to capture for risk management.

In the development of methodologies for risk management, we are now ready to put together our knowledge of models of risk from Chapter 1 to create stochastic processes. Stochastic processes are, simply put, time-dependent extensions of random variables used to model risk. In this chapter, we will develop these time-dependent models of risk and discuss their properties and related concepts. This will help in the modeling of specific risks discussed in the rest of the book, along with developing a framework for their management.

5.1   Stochastic Processes

In Chapter 1, we defined a random variable as a variable representing a risk that takes on certain values, but every time one makes an evaluation of the variable, it takes on values randomly. The *sample space*, Ω, was described as the collection of possibilities for the risk, which was the domain for the random variable. A random variable mapped the (possibly non-numeric) elements in Ω to a set of real values ($X : \Omega \to \mathbf{R}$). A risk evolving over time can be depicted by a sequence of random variables, $X_1, X_2, X_3, \ldots$, often describing the evolution of a risky, probabilistic system over discrete instants of time, $t_1 < t_2 < t_3 < \ldots$. This set of random variables put together in a sequence to describe evolution of a risky system constitutes a stochastic process.

In some cases, we would be interested in observing a risk factor evolve in the continuum of time, or at least there is no logical discrete instants of time that are obvious. In these cases, we will model the evolution of risk in continuous-time. Therefore, if we continuously observe a system that continuously changes, the stochastic process will need to be defined for all time instants in a bounded interval, such as $[0, 1]$, or in an unbounded interval, such as $[0, \infty)$. Consistently, the nomenclature we use for a stochastic process described for discrete instants of time is called a discrete-time stochastic process, while a process describing the evolution of risk in a continuum of time is termed a continuous-time stochastic process.

A stochastic process is not an arbitrary combination of random variables, rather they form a family of random variables, where each random variable in the family is related to others by a defined relationship. The nature of these relationships results in the variety of stochastic processes. Let $\mathbb{T}$ be the time under consideration, which could be $\{t_1, t_2, \ldots\}$, $[0, 1]$, or $[0, \infty)$, and assume that there is a common underlying probability space (Ω, A, P) for the family of random variables. Therefore, a stochastic process $X = \{X(t, \omega), t \in \mathbb{T}, \omega \in \Omega\}$ is a function $X : \mathbb{T} \times \Omega \to R$ of two variables.

When we describe a stochastic process as a two-dimensional entity, it becomes imperative for the modeler to understand what either of the two dimensions imply for the stochastic process. When we instantiate for a specific time, t, leaving the other dimension free to take any value, i.e., $X(t, .)$, it is a random variable for each t. If, however, we instantiate on a specific $\omega \in \Omega$ and let t be free to take any value, i.e., $X(., \omega) : T \to R$, it is a **realization**, **sample path** or **trajectory** of the stochastic process for each $\omega \in \Omega$.

When working with a stochastic process, conditional probabilities can also be determined based on events defined on the basis of the stochastic process taking certain specific values at a prior time or time period. For example,

$$P(X_3 \in A_3 | X_2 \in A_2, X_1 \in A_1) \quad \text{for discrete} - \text{time}, \tag{5.1}$$

$$P(X_t \in B_2 | X_s \in B_1 \quad \text{for} \ \ 0 \leq s < t) \quad \text{for continuous} - \text{time}. \tag{5.2}$$

Based on the above conditional probability relations between a family of random variables within a stochastic process, some specialized stochastic processes are defined. We begin looking at some such specific stochastic processes, beginning with those defined for the discrete time.

5.2 Discrete-Time Evolution of Risk

Many time-evolving risks naturally unravel themselves at discrete points of time. For instance, a government's, firm's or customer's creditworthiness can change from one month to the next, demand for a firm's products can change from one quarter to the next, inventories for finished goods or raw materials can change from day to day, number of fraudulent charges on credit accounts can increase from one month to the next. Even when the risk doesn't seem to, strictly speaking, unravel itself at discrete points of time, a modeler may choose to model them by a discrete-time risk model due to their ease

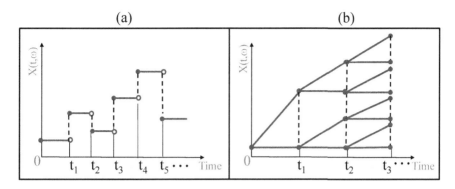

FIGURE 5.1: (a) A typical sample realization for a discrete-time stochastic process. (b) The binomial tree example of a discrete time stochastic process.

of use. For example, stock price evolution, interest rate changes are routinely modeled by binomial tree or trinomial tree models, which are discrete-time models.

In Figure 5.1, we display a typical sample path realization for a discrete-time stochastic process, as well as display the state space for a binomial tree. At each time point, $\{t_i\}$, the process takes the value of a node of the tree indicated for that time point. Therefore, at t_1 there are two possible outcomes for the stochastic process, while by t_3 the stochastic process can have eight (2^3) possible outcomes.

We next discuss and develop some technical concepts for a special kind of stochastic process, of which binomial tree is a special case, called a Markov chain.

5.2.1 Discrete-Time Markov Chains

Markov chains are named after a Russian mathematician, Andreyevich Markov, for his contributions to research in stochastic processes. Markov chains are stochastic processes that satisfy the Markov property. Therefore, in order to define Markov chains, we need to first explain what is meant by a stochastic process being Markovian. Moreover, Markov chains can be discrete-time, i.e., evolving over discrete points of time, or continuous-time, implying they could evolve in the continuum of time. However, in either case their outcomes are a discrete set of outcomes, no matter whether they are finite or infinite in number. The reader may recall that in Chapter 1, we had made a distinction between finite-valued discrete random variable (e.g., binomial distribution) versus infinite-valued random variable (e.g., Poisson distribution). In the discussion of this section, we will focus on discrete-time Markov chains.

Discrete-time Markov chains are useful for modeling risks that seem to 'forget' what happened in the distant past (or even recent past), and the

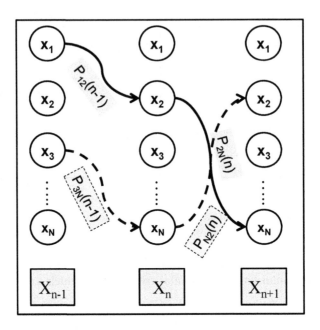

FIGURE 5.2: A pictorial depiction of states of a Markov chain, transitions following Markovian property, and transition probabilities.

values it takes at the immediate future time depends only on the state of the process at the present time. This is a key temporal property of stochastic processes, known as the **Markov property**. Therefore, only the present value of X_n need be known to determine the future value (and distribution) of X_{n+1}; the past values of $X_1, X_2, \cdots, X_{n-1}$ don't directly determine X_{n+1}. They only indirectly determine the value of X_{n+1} through X_n. More rigorously, this can be summarized in terms of conditional probabilities as follows,

$$P(X_{n+1} = x_j | X_n = x_i) = P(X_{n+1} = x_j | X_1 = x_{i_1}, X_2 = x_{i_2}, \cdots, X_n = x_i), \quad (5.3)$$

for all possible $x_i, x_j, x_{i_1}, x_{i_2}, \cdots, x_{i_{n-1}}$ in a given state space for X and all $n = 1, 2, 3, \cdots$. Therefore, what values the stochastic process took for $t = 1, 2, \ldots, n$, has no bearing on the conditional probability distribution of X_{n+1} once X_n is known. This is depicted pictorially in Figure 5.2. When a stochastic process satisfies this property, we call the process **Markovian**.

In other words, one can say that all the information needed to describe what happens at the next time point, $n+1$, is already accumulated in the value of X_n, therefore nothing more of the past is needed to be known. The examples of discrete-time stochastic processes listed earlier can all be reasonably modeled by Markov chains, albeit under certain assumptions. For instance, a government or a firm's changing creditworthiness, change in demand for a firm's products, inventories for finished goods or raw materials, or number of fraudulent charges on credit card accounts. Even if the assumptions made to utilize this dynamic model of risk could appear somewhat simplistic, Markov chains have many desirable properties that aid the study of these risks. We discuss some of the properties and develop some constructs that help in the study of Markov chains.

For a discrete-time Markov chain with a finite number of states, $\Omega = \{x_1, x_2, \ldots, x_N\}$, we can define an $N \times N$ **transition matrix**, $P(n) = [P_{ij}(n)]$, which summarizes the probability of transition from state i to state j. The components of the matrix are given by,

$$P_{ij}(n) = P(X_{n+1} = x_j | X_n = x_i), \tag{5.4}$$

for $i, j = 1, 2, \ldots, N$ and $n = 1, 2, 3, \ldots$. Properties of probabilities in general lead to the conditional probabilities of the transition matrix to satisfy,

1. $0 \leq P_{ij}(n) \leq 1$

2. $\sum_{j=1}^{N} P_{ij}(n) = 1$

for all i, j and n. The conditional probabilities, $P_{ij}(n)$, are called the transition probabilities of the Markov chain at time n, shown in Figure 5.2. Besides the conditional probability of the transitions, we would also like to identify the unconditional probability of the stochastic process being in any of its states at a given time, or the probability distribution of X_n for all n. This is summarized in an **(unconditional) probability vector** for X_n, as a row vector $p(n) = (p_1(n), p_2(n), \ldots, p_N(n))$. Each element, $p_i(n)$ of $p(n)$ defines the probability of random variable X_n reaching a state i at time n, i.e., $p_i(n) = P(X_n = x_i)$ for $i = 1, 2, \ldots, N$.

The unconditional probabilities are dependent on the conditional transition probabilities, which is a dependency we can utilize to compute the unconditional probability for all time, n. If we know the transition probability, $P(n)$, for all n, then we would obtain the unconditional probability by using the relation,

$$p(n + 1) = p(n)P(n), \tag{5.5}$$

where $p(n + 1)$ is the probability vector for X_{n+1}. Specifically, if the initial state is known for the Markov chain, i.e., $p(0)$ is known, we determine $p(1)$ by applying Eqn. (5.5) to obtain $p(1) = p(0)P(0)$. In Eqn. (5.5), by taking a step back in time and applying the equation to $n - 1$, we have $p(n) = p(n-1)P(n-1)$. On substituting this back in Eqn. (5.5), we obtain,

$$p(n + 1) = p(n - 1)P(n - 1)P(n). \tag{5.6}$$

On continuing this backward recursion, if we know the probability vector for $t = 0$ (the present time), $p(0)$, then we can obtain the unconditional probability vector for all the subsequent times using the transition probabilities by applying the following equation.

$$p(n + 1) = p(0)P(0)P(1)P(2)\ldots P(n - 1)P(n). \tag{5.7}$$

In its most general form, Eqn. (5.7) is quite demanding in terms of number of transition matrices that must be constructed to obtain the probability distribution for X_n at each time, n. One simplification often utilized is that the transition matrix does not depend on n, and is the same for all n. The modeler must assess if this assumption makes sense for their context. Even if it does not totally match the risk characteristics, it is a valuable simplification for the tractability it offers. The Eqn. (5.7) then simplifies to $p(n + 1) = p(0)P^{n+1}$, and we call such a Markov chain a **homogeneous Markov chain**. 'Homogeneous' refers to transition matrices being the same for all transitions.

For a homogeneous Markov chain, the probability vector for the Markov chain at time n, X_n, is $p(0)P^n$. If a sufficiently large amount of time elapses, a key question is what happens to the distribution of the Markov chain. If we can find a probability vector, $\bar{p} > 0$, such that $\lim_{n \to \infty} p(n) = \bar{p}$, it also satisfies $\bar{p} = \bar{p}P$, then such a probability vector is a **stationary probability vector** for the homogeneous Markov chain. Markov chains with unique stationary probability vector p often possess an important property called **ergodicity**. This property relates long-term averages of the Markov chain's realization to the spatial averaging with respect to the stationary distribution. We next define ergodicity formally.

Ergodic: For any bounded function $f : X \to R$, the time average of the values of $f(X_n)$ taken as a sequence of random variables $X_1, X_2, \ldots, X_n, \ldots$, generated by the Markov chain is given by,

$$\frac{1}{T} \sum_{n=1}^{T} f(X_n). \tag{5.8}$$

We say that a Markov chain is **ergodic** if for every initial X_0, the limits of the time average in Eqn. (5.8) exist and are equal to the average of f over X with respect to the stationary probability $\bar{p}$, that is, if

$$\lim_{T \to \infty} \frac{1}{T} \sum_{n=1}^{T} f(X_n) = \sum_{i=1}^{N} f(x_i)\bar{p}_i, \tag{5.9}$$

where the convergence is in distribution, i.e., the distributions of the right-hand and the left-hand side of Eqn. (5.9) become indistinguishable as T gets larger.

An alternate definition of ergodicity is given in terms of a Markov chain being *irreducible*, *aperiodic* and *positive recurrent*, but we leave readers to refer to a book on Markov chains to expand their knowledge on this topic [46, 47, 72]. In essence these properties impose on the Markov chain that it does not get stuck in some subset of the state space, doesn't start visiting states in set frequencies, and makes sure it visits all states. The relevance and significance of ergodicity for us is that when a Markov chain, for that matter any stochastic process, is ergodic, we can substitute spatial averages with a long-run time average for the stochastic process. This is valuable, since in the case of many time-evolving risks of interest, we do not have the luxury of observing more than one realization of the stochastic process. But we must routinely compute a variety of spatial risk measures to make risk management decisions. The property of ergodicity comes to the rescue.

We look at some useful examples of discrete-time Markov chains next.

5.2.2 Simple Random Walk

In 1973, a book rose to great fame and has since been published umpteen number of times. The book made the term 'Random Walk' more famous than it would ever have otherwise been. This was Burton G. Malkiel's *A Random Walk Down Wall Street* [58]. In his book, Malkiel defines a random walk as, 'a random walk is one in which future steps or directions cannot be predicted on the basis of past history.' The definition captures the essence of the Markovian property. The author goes on to say, 'when the term is applied to the stock market, it means that short-run changes in stock prices are unpredictable.' The ensuing discussion presents the difference in views on this statement among academics and practitioners. Our goal here, however, is to understand how the random walk model of risk is defined.

A simple random walk is a specific example of a discrete-time Markov chain, therefore it has a discrete set of outcomes and possesses the Markovian property. The state space of this process is all integers, which is the simple version of random walk. More general random walks can be constructed on similar principles. The simple random walk evolves with $S_0 = 0$ and $S_n = \sum_{i=0}^n X_i$, where X_i's are independent, identically distributed (i.i.d.) random variables with the following distribution,

$$
\begin{aligned}
X_i &= 1 \ \ w.p. \ p_1, \\
&= 0 \ \ w.p. \ p_2, \\
&= -1 \ \ w.p. \ (1 - p_1 - p_2),
\end{aligned}
\tag{5.10}
$$

where $0 \le p_1, p_2 \le 1$. If $p_1 = 1/2$ and $p_2 = 0$, the process is called a *standard symmetric random walk*. It is easy to check that S_n is a Markov chain. This is obtained from the fact that S_n is constructed using i.i.d. random variables, X_i's, therefore $P(S_n = j | S_{n-1} = i) = P(S_n = j | S_{n-1} = i, S_{n-2} = i_{n-2}, \ldots S_0 = 0)$. Once the transition probability is constructed, a stationary distribution can also be determined. This is provided as an exercise to the reader. A simple random walk is also ergodic.

In general, the independent identically distributed random variables used to construct a random walk don't have to be of the special kind with three possible outcomes, 1, 0, or −1. They can take a set of integer values, $P\{X_i = j\} = p_j$ for $j = 0, \pm 1, \ldots, \pm K$. The resulting process, S_n, is a general random walk. In some modeling contexts a more appropriate process may be a **bounded version** of a random walk. Let's consider a random walk that is bounded between $(-a, b)$, where $a, b > 0$. In a random walk with

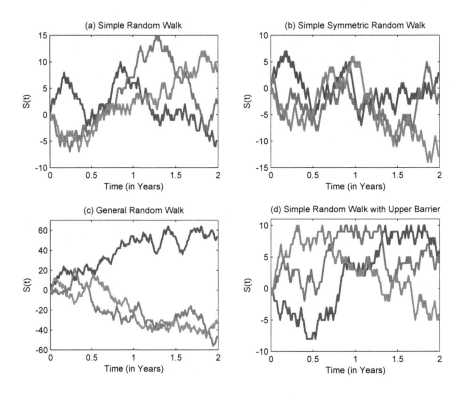

FIGURE 5.3: (a) Three realizations of a simple random walk. (b) Three realizations of simple symmetric random walk. (c) Three realizations of general random walk. (d) Three realizations of simple random walk with upper barrier set at 10.

barrier or bounds, once the process reaches either of its barriers, it is either 'reflected' back or gets 'absorbed' in the barrier. This is summarized in the special transition probabilities at the bound or barrier as follows,

$$P\{S_n = -a|S_{n-1} = -a\} = p_a,$$
$$P\{S_n = -a + 1|S_{n-1} = -a\} = 1 - p_a; \quad (5.11)$$
$$P\{S_n = b|S_{n-1} = b\} = p_b,$$
$$P\{S_n = b - 1|S_{n-1} = b\} = 1 - p_b. \quad (5.12)$$

Eqn. (5.11) indicates what happens when the random walk reaches the lower barrier, $-a$. The process either stays at $-a$ with probability p_a, or returns to the nearest higher outcome, $-a + 1$ with probability $1 - p_a$. Eqn. (5.12) depicts a similar behavior at the upper barrier, b. Two special cases of bounded random walk are when a barrier perfectly absorbs ($p_a = 1$ or $p_b = 1$) or perfectly reflects ($p_a = 0$ or $p_b = 0$).

In Figure 5.3, we display sample path realizations for several versions of random walks discussed in this section. These simulations are achieved by utilizing the random variate generation of two-outcome and n-outcome random variables discussed in Section 4.3.1 of Chapter 4. Comparing Figure 5.3(a) with 5.3(b), one can see how the simple random walk differs from simple symmetric random walk, where symmetric random walk sample paths are more jittery since the process must either go up or down at each time point. The process in Figure 5.3(c), which is a general random walk, has outcomes at each time point ranging from -4 to 4, therefore the levels the process reaches are much larger. We see the obvious truncations (bounds) to the process in Figure 5.3(d), when compared to Figure 5.3(a).

5.2.3 Geometric Random Walk

The random walk stochastic processes discussed in the previous section are all capable of becoming negative in their realizations, as is evident from the plots in Figure 5.3. For some risks it is necessary that the model does not let the risk factor become negative, since in reality the risk factor is inherently non-negative. For modeling the evolution of such a risk factor a variant of the random walk model is considered; this is the **geometric (or lognormal) random walk**.

In this process the changes are not added (or subtracted), but instead are multiplied. For two parameters r_u, r_d, where $r_u > 1$ and $0 < r_d < 1$, the stochastic process evolves as follows,

$$S_n = S_{n-1} * r_u \quad \text{with probability } p$$
$$= S_{n-1} * r_d \quad \text{with probability } (1 - p), \quad (5.13)$$

where the process starts at $S_0 = 1$ and $0 < p < 1$. Since the parameters, r_d and r_u are positive, S_n is always positive. Geometric random walk is a Markov chain, and a stationary distribution for the process can be determined. Many of the discrete-time models, such as binomial tree (Figure 5.1(b)) or trinomial tree models, used for market risk factors essentially utilize the geometric random walk.

5.3 Continuous-Time Evolution of Risk

Many risk factors evolve in a continuum of time, therefore to address certain risk management problems capturing this nature of risk using continuous-time models is a necessity. In other cases, a modeler chooses to model a risk using a continuous-time model for their flexibility in instantiating to any time granularity. For instance, a continuous-time model can be observed at any desired frequency, with no additional effort. In this section, we will discuss a range of continuous-time stochastic processes, beginning with the continuous-time counterpart of Markov chains. We will also develop the widely used continuous-time stochastic process of Brownian motion, extensively used in finance, and study some of its properties and extensions.

5.3.1 Continuous-Time Markov Chains

A continuous-time Markov chain is a Markov chain that evolves in the continuum of time. This modification from discrete-time Markov chain makes the technical description of the stochastic process a bit more involved. This higher level of technical content is well worth it, since these models of risk find widespread use. The Poisson process and the birth-death process are perhaps the most popular examples of continuous-time Markov chains, which we will study in this section. Most constructs from the discrete-time case carry over to the continuous-time Markov chains, albeit with some necessary modifications.

Let the continuous-time Markov chains be, $X(t)$, with its state space continuing to be a finite or infinite discrete set of outcomes, $\Omega = \{x_1, x_2, \ldots, x_N\}$ or $\Omega = \{x_1, x_2, \ldots, x_n, \ldots\}$, respectively. The Markov chain continues to have an N-dimensional (or infinite-dimensional) probability vector $p(t)$ for each $t \geq 0$, whose elements $p_n(t)$ define the probability, $P(X(t) = x_n)$.

The Markovian property applied to continuous-time Markov chain takes the form

$$
\begin{aligned}
P(X(t) = x_j \quad | \quad & X(s_1) = x_{i_1}, X(s_2) = x_{i_2}, \ldots, X(t_0) = x_i) \\
= \quad & P(X(t) = x_j | X(t_0) = x_i),
\end{aligned}
\tag{5.14}
$$

for all $0 \leq s_1 \leq s_2 \leq \ldots \leq s_n < t_0 \leq t_1$ and all $x_i, x_j, x_{i_1}, x_{i_2}, \ldots, x_{i_n}$ in Ω, where $n = 1, 2, 3, \ldots$. Therefore, now the transition matrix doesn't convey the move from one discrete time point to the immediate next one of the process.

Instead any two arbitrarily picked time points must be summarized in the transition matrix, and the Markovian property relates this pair of time points to any prior time point in the process's history. Therefore, the essence of the Markovian property is accurately translated. The transition matrix can now be defined for each pair of time points, t_0, t_1 $(0 \leq t_0 \leq t_1)$, component-wise by the following:

$$P^{i,j}(t_0; t_1) = P(X(t_1) = x_j | X(t_0) = x_i), \tag{5.15}$$

for all $i, j = 1, 2, 3, \ldots, N$. The transition probabilities are summarized in the transition matrix, $P(t_0; t_1)$, with its $ij-$th element given in Eqn. (5.15).

Clearly $P(t_0; t_0) = I$ (I is the identity matrix), i.e., instantaneously the Markov chain can't translate to a different location. Any transition requires a finite non-zero time to transpire. As in the discrete case, the **probability vectors** $p(t_0)$ and $p(t_1)$ are related by $p(t_1) = p(t_0)P(t_0; t_1)$. For any time points t_0, t_1, t_2, such that $t_0 \leq t_1 \leq t_2$, we have $p(t_2) = p(t_0)P(t_0; t_2)$, $p(t_1) = p(t_0)P(t_0; t_1)$, and $p(t_2) = p(t_1)P(t_1; t_2)$. Combining the latter two relationships, we obtain $p(t_2) = p(t_0)P(t_0; t_1)P(t_1; t_2)$. Therefore, there are two ways by which transition from t_0 to t_2 can be described for any $t_0 \leq t_1 \leq t_2$. Moreover, since this is true for any probability vector $p(t_0)$, we can conclude,

$$P(t_0; t_2) = P(t_0; t_1)P(t_1; t_2), \tag{5.16}$$

for all t_0, t_1, t_2, a non-decreasing sequence of times. This gives a useful guideline for how transition matrices for a longer time-span can be constructed from those of the shorter subset of durations we already know.

As in the discrete case, we will benefit from bringing in some simplicity to the definition of the transitions rule for the Markov chain, otherwise we must define the transition matrix for every choice and combination of times t_0, t_1. When the transition matrix $P(t_0; t_1)$ depends only on the time difference, $t_1 - t_0$, that is $P(t_0; t_1) = P(0; t_1 - t_0)$, for all $0 \leq t_0 \leq t_1$, we say that the continuous-time Markov chain is **homogeneous**. We write $P(t)$ for $P(0; t)$, since it is no longer important what the start time is, the only thing that matters is the duration of time for which the transition is being described. The relation in Eqn. (5.16) reduces to $P(s + t) = P(s)P(t) = P(t)P(s)$, for all $s, t \geq 0$.

Besides defining a transition matrix, homogeneity can also help summarize the transition rule into a rate of transition. We define an $N \times N$ **intensity matrix** $A = (a_{i,j})$ (or transition rates) for a homogeneous continuous-time Markov chain with components defined as follows,

$$a_{i,j} = \lim_{t \to 0} \frac{P_{i,j}(t)}{t} \quad i \neq j, \tag{5.17}$$

$$v_i = \lim_{t \to 0} \frac{1 - P_{i,i}(t)}{t} = -\sum_{j \neq i} a_{i,j}, \tag{5.18}$$

where v_i is the transition rate at which the chain exits state i. These transition rates, together with the initial probability vector $p(0)$, completely characterize a homogeneous continuous-time Markov chain. The **waiting time** for a homogeneous continuous-time Markov chain, that is the time between transition from a state x_i to any other state, is exponentially distributed with intensity parameter $\lambda_i = \sum_{j \neq i} a_{i,j}$.

We say a continuous-time Markov chain is **ergodic** if for each bounded function f,

$$\lim_{T \to \infty} \frac{1}{T} \int_0^T f(X(t))dt = \sum_{i=1}^N f(x_i)\bar{p}_i, \tag{5.19}$$

where $\bar{p}$ is the stationary probability vector, such that $\bar{p}P(t) = \bar{p}$ for all t, and $\bar{p} \simeq p(t)$ for all large enough $t(\geq 0)$. The time average now has an integral form, and the convergence is still taken in the distribution. The stationary probabilities may be computed by solving the system of equations, $\bar{p}A = 0$, where A is the intensity matrix for the Markov chain and $a_{i,i} = -v_i$, as defined in Eqn. (5.18).

We next consider several examples of stochastic processes that are continuous-time Markov chains. These examples are key ingredients for modeling risks whose impact appears episodically, however the random lengths of intervals between episodes is a key characteristic of the risk. We begin with the widely-used Poisson process, which as the reader would have guessed, is an extension of the Poisson distribution discussed in Section 1.2.2.5.

5.3.2   Poisson Process

The Poisson process is an example of a continuous-time Markov chain, and is often referred to as a counting process. It serves as a process for counting the number of times an event occurs up to time t. This suggests that this process has a discrete set of outcomes, and more precisely, the set of outcomes is the set of non-negative integers. The Poisson process is said to have an intensity $\lambda > 0$, which is the key term in the intensity matrix of this continuous-time Markov chain process. Additionally, we will define this stochastic process, $X = \{X(t), t \geq 0\}$, to have the following properties.

Property 1. $X(0) = 0$; this suggests that the counting by the Poisson process begins at zero.

Property 2. $X(t) - X(s)$, which is generally called the *increments of the stochastic process*, is a Poisson distributed random variable with parameter $\lambda(t - s)$, for all $0 \leq s < t$. This is the direct connection between Poisson distribution and Poisson process.

Property 3. The increments $X(t_2) - X(t_1)$ and $X(t_4) - X(t_3)$ are independent for all $0 \leq t_1 < t_2 \leq t_3 < t_4$. This property is crucial for the process to have the Markovian property.

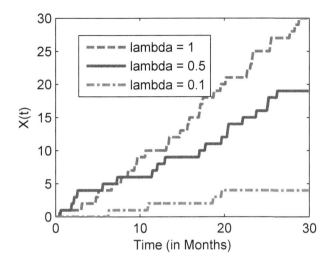

FIGURE 5.4: Three sample path realizations of a Poisson process with varied levels of λ.

The above three properties fully characterize the Poisson process, and can be used to simulate realizations of the Poisson process.

Simulation of a sample path realization of a Poisson process holds a challenge in that we don't know how much time to advance before the next event occurs, which the process counts and increases its level by one on the event's occurrence. This is especially an issue when the intensity parameter is time-dependent, λ_t, for a non-homogeneous Poisson process. If, however, the intensity parameter is a constant, we are able to utilize the relationship between Poisson process and exponential distribution with the same parameter, λ. This relationship states that the inter-arrival time between events in a Poisson process have the exponential distribution, with parameter λ. Taking advantage of this relationship, we can simulate paths of the Poisson process by the following steps.

Step 1. Set $X(0) = 0$. Set $t^* = 0$.

Step 2. Generate an exponential random variate (Y) with parameter λ. Set $X(t^* + Y) = X(t^*) + 1$, and $X(s) = X(t^*)$ for $t^* < s < t^* + Y$.

Step 3. Set $t^* = t^* + Y$.

Figure 5.4 displays some trajectories of a Poisson process with different choices of λ. For $\lambda = 1$ per month, the counting process mounts up rapidly, while for $\lambda = 0.1$ per month, only a few event realizations are observed, so the counting process rises gradually.

Poisson process is an example of a non-stationary stochastic process with independent increments. It is possible, however, to determine the spatial mean, variance and covariances of the Poisson process as follows,

$$\mu(t) \;=\; E[X(t)] = \lambda t; \quad \forall \; t > 0 \qquad (5.20)$$

$$\sigma^2(t) \;=\; E[(X(t) - \mu(t))^2] = \lambda t; \quad \forall \; t > 0 \qquad (5.21)$$

$$C(s,t) \;=\; E[(X(s) - \mu(s))(X(t) - \mu(t))]$$

$$=\; \lambda \min\{s,t\} \; \forall \; s,t > 0. \qquad (5.22)$$

We next study an extension of the Poisson process, which simultaneously counts events of two kinds. This is the birth-death process. The Poisson process is actually a special case of a birth-death process, and in this connection can be called either a pure birth process or a pure death process.

5.3.3 Birth-Death Process

The trajectories of a simple random walk process should be a reminder of another process, the birth-death process. The primary difference between a random walk and the birth-death process is, while a random walk changes its values at fixed points of time, a birth-death process changes values at stochastic points of time. The process is thought to represent the size of a population, and when the process increases by 1, a 'birth' event is said to have happened, and when it decreases by 1, a 'death' event has occurred. This suggests that the process takes integral values, but since it indicates the size of a population, it is restrained from becoming negative. When the process is at state i (the population size is i), the time until the next birth is exponentially distributed with rate λ_i and is independent of the time until the next death, which is also exponentially distributed with rate μ_i.

In the case of a birth-death process the intensity matrix can be constructed based on the above description as,

$$a_{i,i+1} \;=\; \lambda_i, \forall i \geq 0, \qquad (5.23)$$

$$a_{i,i-1} \;=\; \mu_i, \forall i \geq 1, \qquad (5.24)$$

$$a_{i,i} \;=\; -(\lambda_i + \mu_i), \forall i \geq 1. \qquad (5.25)$$

If all the terms in the intensity matrix are not state-dependent, i.e., $\lambda_i = \lambda$ and $\mu_i = \mu$ for all i, the birth-death process would be a homogeneous continuous-time Markov chain. A classic example of a birth-death process is the number of customers in an M/M/1, or more generally, in an M/M/s queuing system. These are systems studied in queueing theory, which has widespread application in real-world applications.

Simulation of a birth-death process is similar to that of a Poisson process. The major difference is that in a birth-death process, for each time that an event must be registered, and its impact on the process determined, two independent exponentially distributed random variates must be generated for the inter-arrival times between events. One of the two exponentially distributed random variates is used to indicate inter-arrival of a 'birth' event, and the other the inter-arrival of a 'death' event. The smaller of the two exponentially distributed random variates dictates whether the birth-death process will next experience an increment or a decrement.

5.3.4 Markov Process

In our progression of studying stochastic processes of greater complexity, we now arrive at studying stochastic processes that are defined for a continuous-time and continuous-space evolution of risk. A Gaussian process is an example of a **continuous-time, continuous-space** stochastic process. For a process that evolves both in time and spatial domain in continuum, as a continuous-time, continuous-space stochastic process does, the technical definitions for the rules of evolution of the stochastic process must be appropriately advanced. We develop some of these technical definitions before exploring specific processes at greater depth.

The counterpart of a Markov chain when the spatial dimension also becomes continuous is a Markov process. As described before, a continuous-time, continuous-space stochastic process qualifies to be called a **Markov Process** when it satisfies the Markov property, similar to Eqn. (5.14). The Markovian property equation must be customized for the continuous-space case as follows,

$$
\begin{aligned}
P(X(t) \in B_j \quad | \quad & X(s_1) = x_{i_1}, X(s_2) = x_{i_2}, \dots, X(t_0) = x_i) \\
= \quad & P(X(t) \in B_j | X(t_0) = x_i),
\end{aligned}
\tag{5.26}
$$

for all $0 \le s_1 \le s_2 \le \dots \le s_n < t_0 \le t_1$, all values realized by the process in the past $x_i, x_{i_1}, x_{i_2}, \dots, x_{i_n}$ in Ω, and any set of values it may realize at time t, $B_j \subset \Omega$.

For Markov processes, we can't define transition matrices or intensity matrices, as done for discrete-time or continuous-time Markov chains, since the spatial dimension is a continuum of values. We must instead define **transition densities**. We define the transition probabilities of a Markov process, as you would expect, as,

$$
P(s, x; t, B) = P(X(t) \in B | X(s) = x), \quad 0 \le s < t.
\tag{5.27}
$$

The transition probability, $P(s, x; t, B)$, is associated with a **transition probability density**, given by

$$
P(s, x; t, B) = \int_B p(s, x; t, y) dy.
\tag{5.28}
$$

This relation is similar to the relationship between a cumulative distribution function and probability density function of a continuous random variable. The only difference is that Eqn. (5.28) is for conditional distribution based on two instants of observation of a continuous-time stochastic process.

A Markov process with transition probability $p(s, x; t, y)$ is called a **diffusion process** if the following three limits exist for all $\epsilon > 0$, $s \geq 0$ and $x \in \mathbf{R}$.

Property 1. $\lim_{t \downarrow s} \frac{1}{t-s} \int_{|y-x| < \epsilon} p(s, x; t, y) dy = 0$,

Property 2. $\lim_{t \downarrow s} \frac{1}{t-s} \int_{|y-x| < \epsilon} (y - x) p(s, x; t, y) dy = a(s, x)$,

Property 3. $\lim_{t \downarrow s} \frac{1}{t-s} \int_{|y-x| < \epsilon} (y - x)^2 p(s, x; t, y) dy = b^2(s, x)$.

Property 1 prevents a diffusion process from experiencing instantaneous jumps. The quantity $a(s, x)$ is called the **drift** of the diffusion process and $b(s, x)$ its **diffusion coefficient** at time s and position x. Property 2 implies that

$$a(s, x) = \lim_{t \downarrow s} \frac{1}{t - s} E[X(t) - X(s) | X(s) = x]. \tag{5.29}$$

This makes $a(s, x)$ the instantaneous rate of change in the mean of the process, given that $X(s) = s$. Similarly, it follows from property 3 that the squared diffusion coefficient is given by,

$$b(s, x)^2 = \lim_{t \downarrow s} \frac{1}{t - s} E[(X(t) - X(s))^2 | X(s) = x]. \tag{5.30}$$

This denotes the instantaneous rate of change of the squared fluctuations of the process, given that $X(s) = x$. Therefore, properties 1-3 summarize the instantaneous evolutionary characteristics of the process at any time and location. This helps visualize how the realizations of a diffusion process might look; they are continuous, with no jumps, and have a tendency of drifting from a location depending on the present time and the location, but this tendency is appended with a variability summarized in the diffusion coefficient.

5.3.5 Gaussian Process

Stochastic process, $X(t)$, is a Gaussian process if the joint probability distribution of any finite observations of the process is Gaussian, or normally distributed. The reader is reminded that a Gaussian or normal random variable is a continuous random variable that takes values in the interval $(-\infty, \infty)$. Therefore, a Gaussian process can be indicated as, $X(t) \sim N(\mu(t), \sigma^2(t))$, for all $t \in \mathbf{T}$, where $\sigma(.)$ and $\mu(.)$ are some given functions of time.

If there is a constant μ and a function $c : \mathbf{R} \to \mathbf{R}^+$ such that the mean, variance and covariance of a stochastic process satisfy $\mu(t) = \mu$, $\sigma^2(t) = c(0)$, and $C(s, t) = c(t - s)$, for all $s, t \in \mathbf{T}$, then we call the process **wide-sense stationary**. This means that the process is stationary with respect to its first and second moments only. Therefore, a wide-sense stationary Gaussian process has a constant mean and variance, and a correlation structure that depends

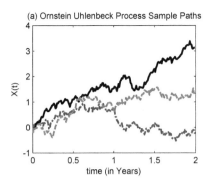
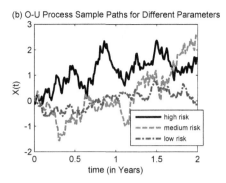

FIGURE 5.5: (a) Three sample path realizations of an Ornstein-Uhlenbeck process. (b) Three sample path realizations of Ornstein-Uhlenbeck process with different risk levels.

on the temporal distance between observations made of the stochastic process.

Ornstein-Uhlenbeck Process: A special example of a Gaussian, Markov process is the Ornstein-Uhlenbeck process, in short, O-U process. It is defined as, $X = \{X(t), t \geq 0\}$ with parameter $\gamma > 0$ and initial value $X_0 = N(0,1)$. It is a Gaussian process with mean and covariances given by,

$$\mu(t) = 0, \qquad C(s,t) = \frac{\sigma^2}{2\gamma}e^{-\gamma|t-s|}, \tag{5.31}$$

for all $s, t \geq 0$. This implies that besides being Gaussian and a Markov process (not demonstrated here), the O-U process is wide-sense stationary. In fact, an O-U process is a diffusion process and is strictly stationary. In Figure 5.5, we display some realization paths for the O-U process. We display several paths for one process in panel (a) of the figure, and a path with three different risk levels (σ) in panel (b). The sample paths with high volatility diverge rapidly, while low risk fluctuates about the initial value of zero. We will revisit the O-U process in later chapters, both as a model and in context of applications, since the O-U process finds widespread application in finance. It is utilized for modeling interest rates, currency exchange rates, commodity prices, and even certain trading strategies.

One important diffusion process is the standard Brownian motion, also known as the Wiener process, named after botanist Robert Brown and mathematician Norbert Wiener, respectively. The Brownian motion is perhaps the most popular stochastic process used in continuous-time finance and risk management. We will next discuss the Brownian motion in detail.

5.3.6   Brownian Motion

Standard Brownian motion or Wiener process is another example of a continuous-time Markov process that is also a diffusion. Some like to make a distinction between the Brownian motion, to denote the physical process denoted by the erratic motion of a grain of pollen on a water surface due to its being continually bombarded by water molecules. This was the botanist Robert Brown's observation. Wiener process denotes the mathematical representation of this process. However, we will interchangeably use either name for the process. Wiener process is a Gaussian process, hence the process has an infinite state space, $\Omega = (-\infty, \infty)$.

We will define the **standard Wiener process** or **Brownian motion**, $W = \{W(t), t \geq 0\}$ to be a Gaussian process with **independent increments** such that

$$W(0) = 0, \text{ with probability } 1, \tag{5.32}$$
$$E[W(t)] = \mu(t) = 0, \tag{5.33}$$
$$Var(W(t) - W(s)) = t - s, \tag{5.34}$$

for all $0 \leq s \leq t$. Therefore, the Wiener process is a Gaussian process that is wide-sense stationary. By this definition, increments of the Wiener process, $W(t) - W(s)$, are distributed by the normal distribution, $N(0, t-s)$ for all $0 \leq s < t$. Given $W(t)$ is a Gaussian process, the joint distribution of $(W(s), W(t))$ is Gaussian, for all $0 \leq s < t$. Therefore, $W(t) - W(s)$ is also Gaussian, with $E[W(t) - W(s)] = 0$ from Eqn. (5.33) and $Var(W(t) - W(s)) = t - s$ from Eqn. (5.34). By definition, the increments $W(t_2) - W(t_1)$ and $W(t_4) - W(t_3)$ are independent for all $0 \leq t_1 < t_2 \leq t_3 < t_4$.

In Figure 5.6, we show a few realizations of sample path of the Wiener process. The steps for the simulation of the Wiener process are as follows,

Step 1. Set $W(0) = 0$. Set the duration for which the Wiener process evolves, $[0, T]$.

Step 2. Set a time increment, Δt, for advancing time, and construct a time discretization $\{t_0, t_1, \ldots, t_N\}$, such that $t_{i+1} - t_i = \Delta t$, for every i, $t_0 = 0$, and $t_N = T$.

Step 3. Generate a normal random variate $Y \sim N(0, \sqrt{\Delta t})$. Set $W(t_{i+1}) = W(t_i) + Y$.

Step 4. Set $i = i + 1$ while $i < N$. Go to Step 3.

The above definition is sufficient for simulating paths of the Wiener process, and developing an understanding of the probabilistic properties of the process. However, in the next section we demonstrate a construction of the Wiener process from a simpler process we studied earlier, the simple symmetric random walk. This construction serves two purposes: first, it provides a better visualization of the Wiener process, and second, it relates a continuous-time process

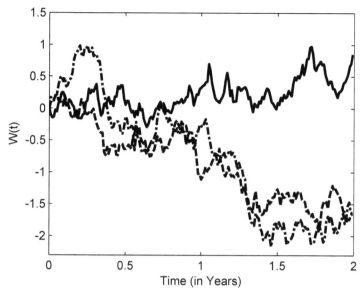

FIGURE 5.6: Three sample path realizations for the standard Brownian motion or the Wiener process.

as a limit of discrete-time processes. The latter objective sets the stage for the development in Chapter 6.

5.3.6.1 Approximating Brownian Motion by a Random Walk

Often visualization of Brownian motion path realizations are aided by constructing Brownian motion using limiting random walk processes. This is because we can approximate a standard Wiener process in distribution on any finite time interval by means of a sequence of **scaled** random walk processes. We start with a simple random walk, S_N, constructed in Section 5.2.2 using independent random variables, X_k, taking values $+1$ or -1 with **equal** probability. Utilizing the i.i.d. random variables, we make a minor modification to the process, S_N, to define

$$S_N(t_k^{(N)}) = (X_1 + X_2 + \cdots + X_k)\sqrt{\Delta t}, \tag{5.35}$$

where $0 < t_0^{(N)} < t_1^{(N)} < \ldots < t_N^{(N)} = 1$. This scaling modification is done since we are trying to approximate the Wiener process over the interval, $[0, 1]$, on N equidistant points taken in this interval of length, $\Delta t = \frac{1}{N}$, each. The values between these points are obtained by a linear interpolation as follows,

$$S_N(t) = S_N(t_k^{(N)}) + \frac{t - t_k^{(N)}}{t_{k+1}^{(N)} - t_k^{(N)}}(S_N(t_{k+1}^{(N)}) - S_N(t_k^{(N)})), \tag{5.36}$$

for $t_k^{(N)} \leq t \leq t_{k+1}^{(N)}$ and for $k = 1, 2, \ldots, N - 1$, where $S_N(0) = 0$. $S_N(t)$ constructed by the above method is a step-wise continuous symmetric random walk, with independent, equally probable steps of length $\pm\sqrt{\Delta t}$ in each subinterval of time. It can be shown by invoking the **central limit theorem** that $S_N(t)$ converges in distribution to the standard Wiener process as $N \to \infty$. A review of the central limit theorem was provided in Section 1.2.2.3. Implementing this approximation is provided as an exercise at the end of this chapter.

Before we proceed to study additional properties of the Wiener process, we take a short digression to define the different convergence rules for sequence of random variables. In this section, and earlier for the definition of ergodicity, we have utilized the definition of convergence in distribution, without formally defining it. In the next chapter, these definitions will be required again.

5.3.6.2 Convergence of Random Variables

Assume there is a sequence of random variables, $\{X_n\}$. These can also be a sequence of stochastic processes observed at a specific time point, t, i.e., $\{X_n(t)\}$. When we wish to conclude that the sequence of random variables converges to a random variable, X, or a stochastic process observed at the same time, t, i.e., $X(t)$, the convergence can be interpreted and applied in a few different ways. For instance, convergence can be in distribution, in probability, almost sure convergence, or convergence in mean or moment. We describe the meaning of each of these manners of convergence.

1. **Convergence in Distribution**: $\{X_n\}$ converges in distribution to X, if their cumulative distribution functions $F_n(x)$ converge to the cumulative distribution function $F(x)$ of the random variable, X, at all points of continuity of F.

2. **Convergence in Probability**: $\{X_n\}$ converges in probability to X, if for any $\epsilon > 0$, $P(|X_n - X| > \epsilon) \to 0$ as $n \to \infty$.

3. **Almost Sure Convergence**: $\{X_n\}$ converges almost surely (a.s.) to X, if for all ω outside a set of zero probability $X_n(\omega) \to X(\omega)$ as $n \to \infty$. This implies that the actual realizations of X_n fall arbitrarily close to those of X, for almost all ω.

4. **Convergence in Mean**: $\{X_n\}$ converges in mean to X, if $E[|X_n|] < \infty$ and $E[|X_n - X|] \to 0$ and $n \to \infty$. In general, convergence in r^{th} $(r \geq 1)$ moment may be defined, if $E[|X_n|^r] < \infty$ and $E[|X_n - X|^r] \to 0$ and $n \to \infty$.

The strongest criterion for convergence of the above is 'almost-sure convergence'. Therefore, if almost sure convergence is known for a sequence of random variables, this directly implies that the sequence converges in probability as well. Similarly, if we have convergence in probability for a sequence of random variables, this implies that the sequence also displays convergence in distribution. Finally, if we have convergence in r^{th} moment, for all values of r, it can be shown to be equivalent to convergence in probability. This comparison also points out that convergence in distribution is the weakest criterion for convergence of random variables.

5.3.6.3 Properties of the Wiener Process

Standard Brownian motion or Wiener process was defined as a Gaussian process with independent increments, $W(0) = 0$, and $W(t) - W(s)$ is distributed by normal distribution, $N(0, t - s)$ for $0 \leq s < t$. We now discuss some additional properties of the Wiener process.

1. $Var(W(t)) = t$;

    By the definition above, $Var(W(t) - W(0)) = Var(W(t)) = t - 0$. Therefore, variance of the Wiener process grows without bound as time passes, while the mean always stays at zero. So, typical sample paths must take larger and larger positive and negative values as time increases. This is evident in the trajectories of Figure 5.6 as they diverge away with time.

2. Using the Law of Large Numbers we get $\lim_{t \to \infty} \frac{W(t)}{t} = 0$ in the **mean square sense** or as convergence in 2nd moment. Therefore, in an asymptotic sense (in the sense of a limit) this gives an estimate for the growth rate of the Wiener process.

3. Sample paths of the Wiener process are **continuous, but not differentiable**. You can test this numerically. Theoretically it is because the increments of the Wiener process are independent and behave like $\sqrt{\Delta t}$, and not Δt. Therefore, the limit that defines the derivative, $\lim_{s \to 0} \frac{W(t+s) - W(t)}{s}$, is not finite.

4. $C(t, s) = Cov(W(t), W(s)) = \min(t, s)$. This can be shown by using the independence of increments property of Brownian motion.

5. Brownian motion possesses Markov property. This can be shown using the moment generating function.

6. Brownian motion is a martingale.

We need to elaborate on the last property, and in particular, define a martingale and its significance. This property tells us whether at any point of time it can be stated that the process is expected to go up or down relative to its current value, given its past history. In our discussion of Malkiel's book [58] *Random Walk down Wall Street*, we pondered if the stock market can in fact be predicted, or if nothing of great conviction can be said about its expected bias of movement, up or down, from where it is right now. In order to define a martingale formally, we will need to develop the technical construct of a 'filtration'.

We define $\mathcal{F}_t = \sigma(W(u), u \leq t)$ as the smallest σ-**field** (which is a synonym for a σ-algebra) that contains all sets of the form $\{a \leq W(u) \leq b\}$ for all $0 \leq u \leq t$, $a, b \in \mathbf{R}$. A σ-field is a collection of events for a random variable which is closed, i.e., also contains, with respect to unions and intersections of events for that random variable (this was formally defined in Section 1.2.1). Here our concern is the family of random variables of a stochastic process, therefore a filtration denotes the information available to an observer of W up to the time t. A **filtration F** is a family of increasing σ-fields (or σ-algebras), $\mathcal{F}_t$, for all time, t for the family of random variables of a stochastic process. Therefore, **F** specifies how the information is revealed in time regarding a stochastic process. As a mathematical construct, the property that a filtration is increasing corresponds to the fact that once the information is revealed, it is not forgotten.

To say that a stochastic process, $W(t)$, is a **martingale** implies that the stochastic process satisfies the following relation,

$$E[W(t) - W(s)|\mathcal{F}_s] = 0, \quad \forall t > s. \tag{5.37}$$

or

$$E[W(t)|\mathcal{F}_s] = W(s), \quad \forall t > s. \tag{5.38}$$

This implies that the stochastic process is non-anticipatory, that is, knowing the information of its past does not give you an inkling of its expected value in the future, more than knowing its current value. Not only can we show that the Wiener process, $W(t)$, is a martingale, we can also show that the stochastic process, $W(t)^2 - t$, is also a martingale. Therefore, $E[W(t)^2 - t|\mathcal{F}_s] = W(s)^2 - s$. As is the exponential function of the standard Brownian motion, $S(t) = e^{uW(t) - \frac{u^2}{2}t}$, for any real number u. $S(t)$ is related to the moment generating function of the Wiener process, and is called an exponential martingale.

Similarly to the concept of a martingale is the notion of a supermartingale and submartingale. Each of these deviations from the notion of a martingale is that the expectation of future values of a stochastic process is lower or higher than its current value, given the history of the process. A stochastic process, X_t, may be a supermartingale if $E[X_t|\mathcal{F}_s] \leq X_s$, and a submartingale if $E[X_t|\mathcal{F}_s] \geq X_s$. Therefore in the case of a supermartingale the stochastic process is expected to go downhill from its present value, while it is expected to go uphill in case of a submartingale. If we invoke the Jensen's inequality for a random variable, X, and any convex function $f(.)$, given by,

$$E[f(X)] \geq f(E[X]), \tag{5.39}$$

we can conclude that a convex function of a martingale is a submartingale, and similarly, a concave function of a martingale is a supermartingale.

Finally, we define the quadratic variation of a stochastic process. This definition supports the advancement we present in Chapter 6 for being able to define new processes based on the standard Wiener process. **Quadratic variation of Brownian motion** in a duration $[0, T]$ is defined on a discrete set of points of time, $0 = t_0^n < t_1^n < \ldots < t_n^n = T$, as follows,

$$QV(W) = \lim_{n \to \infty} \sum_{i=1}^{n} |W(t_i^n) - W(t_{i-1}^n)|^2, \qquad (5.40)$$

where the limit is taken over all partitions of $[0, T]$, such that $\delta_n = \max_i(t_i^n - t_{i-1}^n) \to 0$ as $n \to \infty$. The quadratic variation of Brownian motion on interval $[0, T]$ can be shown to be T. Quadratic variation of other stochastic processes can be similarly defined and one can study the properties of this variation. In the case of the Wiener process, its quadratic variation lends desirable properties to the advancement explored in Chapter 6.

5.3.7 Brownian Motion with Drift and Geometric Brownian Motion

Two useful new processes can be constructed using the Wiener process. One of them is the geometric Brownian motion, which we have already referred to in the context of discussion of a martingale. We define it more formally as a stochastic process obtained as a function of the Wiener process as, $Y_t = e^{W(t)}$. This modification of the Wiener process is similar in nature as the geometric random walk was from the simple random walk. The process is always positive, and is useful for modeling risks when it is the percentage change in the quantity being modeled that is independent and identically distributed, i.e., Y_t/Y_{t-1}, and not the absolute increments, just as in the geometric random walk. An example of this consideration is the rate of return of stocks or any other financial instrument, where the per period rate of return is assumed to be independent, identically distributed.

To work with this process, one needs to be reminded of the **moment generating function** of a normal distribution, noting that $W(t) \sim N(0, t)$.

$$E[e^{sW(t)}] = e^{ts^2/2}. \qquad (5.41)$$

With this knowledge, one can compute the mean of the process at any time, t, as $E[Y_t] = e^{t/2}$, and $Var(Y_t) = e^{2t} - e^t$. As stated earlier, a special version of geometric Brownian motion is a martingale. This is, $S(t) = e^{uW(t) - \frac{u^2}{2}t}$, for any real number u.

We define a new process, $Y_t = W(t) + \mu t$, and call it understandably, Brownian motion with drift. This new process is still a continuous Gaussian process with independent increments, with $Y(0) = 0$ with probability 1. Y_t, so defined, is normally distributed with mean μt and variance t. Therefore, Brownian motion with drift is a process that tends to drift off at a rate μ.

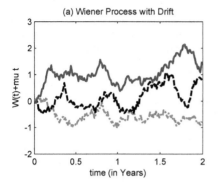 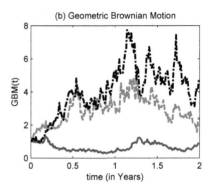

FIGURE 5.7: (a) Three sample path realizations for the standard Brownian motion or Wiener process with drift. (b) Three sample path realizations for geometric Brownian motion.

In Figure 5.7 we display trajectories for the two new processes introduced in this section. Panel (a) displays trajectories for Wiener process with drift. The sample paths look quite similar to those of the Wiener process in Figure 5.6, with the difference that they don't symmetrically diverge away; instead the sample paths have a tendency to increase in the positive direction. This is because the drift of the process is set at 10%. In panel (b), we display trajectories for geometric Brownian motion. These trajectories remain positive, as is expected. They sometimes increase in an exponentially rapid scale, while hovering at low levels in other cases. These correspond to whether the underlying Wiener process wandered off in the positive range or the negative one. The simulation of both these processes is a trivial modification of the simulation of the Wiener process. Once a Wiener process trajectory is obtained, we make the desired functional transformation to obtain sample paths of these new processes.

Both geometric Brownian motion and Brownian motion with drift can be shown to be solutions of certain Stochastic Differential Equations (SDEs). This will be developed in detail in Chapter 6.

5.3.8 Additional Concepts for Stochastic Processes

When stochastic processes are used to model the evolution of risks for any risk management objective, we often consider the stochastic process reaching some level, or exceeding some value to be an indicator of an important risk event. For instance, when the equity of a firm drops too low, it can be a sign of distress for the firm, or when the firm value decreases to a level close to the firm's debt level, this can trigger a significant credit rating drop. Several derivative instruments are defined so they generate a pay-off, or become worthless, when the underlying risk factor hits a barrier. Therefore, depending on the dynamic models of risk under consideration, such events are important to analyze, and their evaluation must be done either in analytical terms or by simulation.

An important concept from this point of view for a stochastic process is that of a **stopping time**. Stopping time is a stochastic time, τ_a, that records the following event for a stochastic process $X(t)$.

$$\tau_a = \min t|X(s) < a \text{ for } 0 \leq s < t; X(t) = a. \tag{5.42}$$

Therefore, it is the (stochastic) time when the stochastic process reaches the level B for the first time. Stopping time in its own turn is a random variable, with a state space, $\Omega_{tau_a} = [0, \infty]$. One can analytically derive the distribution of the stopping times based on one's understanding of the stochastic process defining the stopping time. Simulation is an available alternative for determining this distribution, especially when analytical results are not accessible.

Examples of stopping time are **exit times** or **hitting times**. Define a (random) hitting time by, T_a, the first time a Wiener process (or any other process) hits level a. One may want to develop an understanding of the **distribution of hitting times**, depending on the application, i.e., $P(T_a \leq t)$. We will begin with $P(W(t) \geq a)$.

$$
\begin{aligned}
P(W(t) \geq a) &= P(W(t) \geq a|T_a \leq t)P(T_a \leq t) \\
&+ P(W(t) \geq a|T_a > t)P(T_a > t). \tag{5.43}
\end{aligned}
$$

$$\text{note that } P(W(t) \geq a \mid T_a > t) = 0, \tag{5.44}$$

$$\text{and } P(W(t) \geq a \mid T_a \leq t) = 1/2, \tag{5.45}$$

since by symmetry, after hitting a, it is just as likely for the process to be above a or below a at time t (increments are normally distributed). Therefore,

$$P(T_a \leq t) = 2P(W(t) \geq a). \tag{5.46}$$

This final probability can be computed, since we know $W(t) \sim N(0, t)$. Therefore, in the case of standard Brownian motion the exact distribution of the hitting times can be obtained. In general, simulation analysis may be necessary to estimate these distributions.

In some applications, it is the **maximum** or the **minimum** of the Wiener process, or any other diffusion process, that is important, i.e., $Y_t = \max_{0 \leq s \leq t} W(s)$. In others, it is the **arithmetic or geometric average** of the process that becomes important, i.e., $Y_t = \frac{1}{t} \int_0^t W(s)ds$. Therefore, an understanding will need to be developed for distribution and properties of these new processes. Simulation analysis is versatile for being applied to assess the properties of these new processes.

5.4 Modeling Correlation

So far we have focused on a single stochastic process, focusing on a single risk and its evolution in time. In real-world applications, we encounter many risk factors simultaneously evolving with possible co-dependence and correlation. We need to define ways by which this simultaneous evolution of multiple risk factors can be modeled. In Section 1.2.1.2, we had developed the definition of correlation and covariance between multiple random variables. We need to extend this to stochastic processes.

5.4.1 Correlated Brownian Motion

Consider an N-dimensional independent Wiener process, $[W_t^i]_{i=1:N}$, with mean $[0]_N$ and standard deviation $[\sqrt{t}]_N$. If instead of independent Wiener processes, we need to model a correlated set of Wiener processes, we will need to make a transformation to introduce the correlation between original N independent Wiener processes. A similar transformation can also help transform N-dimensional correlated Wiener processes into independent ones.

We wish to create an N-dimensional correlated Wiener processes, $[B_t^i]_{i=1:N}$, with mean $[0]_N$ and standard deviation $[\sqrt{t}]_N$ with covariance for a pair of Brownian motions to be given by,

$$E[B_i(t)B_j(t)] = \rho_{ij}t \ \ for \ i = 1\ldots N; j = 1\ldots N. \tag{5.47}$$

The parameter ρ_{ij} can be identified as the correlation coefficient, since standard deviation of each Brownian motion $B_i(t), B_j(t)$ is $\sqrt{t}$. This also implies that $\rho_{ii} = 1$ for all $i = 1\ldots N$.

For transforming the N-dimensional independent Wiener process, $[W_i(t)]_{i=1:N}$, to N-dimensional correlated Wiener processes, $[B_i(t)]_{i=1:N}$, we must first construct a Cholesky factorization of the correlation matrix $[\rho_{ij}]_{i=1:N;j=1:N}$, given by,

$$[\rho_{ij}] = RR^T. \tag{5.48}$$

Once the Cholesky factors are available, the correlated Brownian motion is obtained by making the following transformation of the independent Brownian motion,

$$[B_i(t)]_{i=1:N} = R * [W_i(t)]_{i=1:N}. \tag{5.49}$$

It can be shown that each Brownian motion in $[B_i(t)]_{i=1:N}$ thus obtained is indeed a standard Brownian motion, but collectively they are no longer independent.

Let's consider a simple example of $N = 2$, where we seek two correlation Brownian motion processes, and wish to construct them from two independent

Brownian motion processes. Their desired correlation coefficient is ρ, therefore the correlation matrix is,

$$[\rho_{ij}] = \begin{pmatrix} 1 & \rho \\ \rho & 1 \end{pmatrix}. \tag{5.50}$$

The Cholesky factor of the above correlation matrix is,

$$R = \begin{pmatrix} 1 & 0 \\ \rho & \sqrt{1-\rho^2} \end{pmatrix}. \tag{5.51}$$

Therefore, the new correlated Brownian motion processes are obtained as $B_1(t) = W_1(t)$, $B_2(t) = \rho W_1(t) + \sqrt{1-\rho^2}W_2(t)$.

This derivation can be extended to non-constant or time-dependent correlation (or covariance) relationship, $\rho_{ij}(t)$, between the $i-j$ pair of correlated Wiener processes [81]. Rather than be exhaustive in recounting how correlation can be incorporated in all the stochastic processes discussed in this chapter, we have provided the Brownian motion case as a demonstration here. The reader should explore for their specific context how correlation can be introduced, say between random walks, Poisson processes, Markov chains, etc.

5.4.2 Copulas for Correlation

We now discuss an approach to model correlation between N random variables, known as copulas, and generating random variates from N-dimensional distribution. This is possible to do directly for a small set of N-dimensional distribution function, such as normal distribution or t-distribution. In general, especially also when the marginal distributions are not all the same, generating N-dimensional joint distributed random variates is not straightforward.

A copula is a distribution function that allows combining univariate distributions to create a joint distribution function with a particular, chosen dependence structure. Thus, using a copula to build multivariate distributions is a flexible and powerful technique. It decomposes the choice of dependence between random variables from the choice of each random variable's marginal distribution, with no restrictions placed on the marginal distributions.

The word 'copula' is a Latin noun, meaning 'something that connects,' 'a link or a bond.' Mathematically, a copula is itself a distribution function, defined on $[0,1]^N$, with uniform marginal distributions. Each of the marginal distributions produces a probability of the one-dimensional events. The copula function then takes these probabilities and maps them to a joint probability, enforcing a certain relationship on the probabilities. If we know the marginal distribution of the N random variables, $\{X_i; i = 1 : N\}$, copulas help in constructing the joint distribution of the N random variables with the desired correlation and dependence characteristics. Therefore, a copula is a function that describes the joint distribution of the N random variables in terms of the marginal distributions of the random variables. A copula is called a normal copula, a t-copula, etc. depending on the type of distribution used to construct the joint distribution. In our description, we will initially follow the normal copula function, followed by providing directions for how other copulas may be utilized.

We assume that the marginal distributions of the random variables are given by their cumulative distribution functions, $\{F_i(x_i); i = 1 : N\}$, which don't all have to be the same for all i. The desired joint distribution function is, $F(x_1, x_2, \ldots x_N)$. We define a **copula** C as a distribution function on $[0, 1]^N$ with a uniformly distributed marginal distribution. A copula, defined as such, is utilized to transform the marginal distributions, $\{F_i(x_i); i = 1 : N\}$, to the desired joint distribution, $F(x_1, x_2, \ldots x_N)$, as follows,

$$F(x_1, x_2, \ldots x_N) = C(F_1(x_1), F_2(x_2), \ldots, F_N(x_N)). \qquad (5.52)$$

A key question remains, given the marginal and desired joint distribution for random variables, $\{X_i; i = 1 : N\}$, how is the copula constructed. The answer lies in a re-hash of Eqn. (5.52), by defining the copula distribution as,

$$C(u_1, u_2, \ldots u_N) = F(F_1^{-1}(u_1), F_2^{-1}(u_2), \ldots, F_N^{-1}(u_N)), \qquad (5.53)$$

where F_i's are the N marginal distributions and F is the correlated joint distribution function. Even if this re-hash is not immediately more instructive, it gives us a guideline for steps for generating random variates by the N-dimensional joint distribution. Therefore, in order to simulate dependent (correlated) multivariate random variates using a copula, we need to specify and follow the following steps.

1. Identify the copula family by which to generate N-dimensional random variates. This includes any correlation or shape parameters for the distribution.

2. Generate random variates by the N-dimensional distribution. (This assumes one knows how to do this.)

3. Compute the component-wise (one-dimensional) cumulative distribution function for the N generated random variates, to obtain u_i's. This is the $N \rightarrow 1$ decomposing stage.

4. Invert the marginal distribution, F_i, to obtain the desired multi-dimensional dependent random covariate with the set marginal distributions, $X_i = F_i^{-1}(u_i)$.

In Figure 5.8, we display two marginal distributions in panels (a) and (b). Panels (c) and (d) of the figure provide the two-dimensional random variates generated for the pair of marginal distributions in panels (a) and (b), by utilizing Gaussian and t-copula, respectively. The effect of positive correlation of $\rho = 0.7$ is quite visible in both plots (c) and (d). In Figure 5.8(d), the

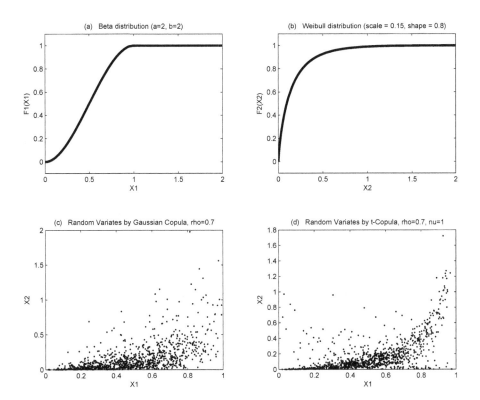

FIGURE 5.8: (a) Marginal CDF for first random variable, chosen to be beta distribution with parameters, a=2, b=2. (b) Marginal CDF for second random variable, chosen to be Weibull distribution with parameters, a=0.15, b=0.8. (c) Scatter plot of 1000 random variates generated by Gaussian copula with $\rho = 0.7$. (d) Scatter plot of 100 random variates generated using t-copula with $\rho = 0.7, \nu = 1$.

impact of tail characteristics due to t-Copula are distinct from those of the Gaussian copula of panel (c).

Copulas have become popular for modeling correlation and codependence between random variables, and are utilized for correlation modeling in credit risk and risk underlying insurance contracts. It is important to note that correlation is only one measure for codependence, other measures include tail dependence and rank correlation. Different types of copula functions may be used for these other kinds of dependency as desired. A wide range of copula functions exist, and a practitioner must choose which one to use. Frequently the choice is dictated by the usual criteria of familiarity, ease of use and analytical tractability. The most commonly used copulas are the Gaussian copula for linear correlation, the t-copula for dependence in the tail, and the Gumbel copula for extreme distributions.

5.5   MATLAB Tools for Modeling Risk Evolving over Time

MATLAB mathematical software has a vast array of functions and programming constructs to support writing code efficiently. Simulation of stochastic processes done in this chapter were performed in MATLAB, using a few lines of code. We list a few of functions used, and also list functions available for copula modeling. The reader is advised to look up the extensive help documentation available with MATLAB to see the details of these and other related functions. At the bottom of each function description in MATLAB help documentation, look for 'See Also' to explore other related functions. Resources such as MATLAB Primer [20] are also useful.

Conditioning: `if-elseif-else-end`, `if-else-end`

Loop: `for-end`

Multivariate distribution: `mvnrnd`, `mvtrnd`

Copula: `copulaparam`, `copulacdf`, `copulapdf`, `copularnd`

5.6   Summary

Many types of risks in the risk typology need to be considered and managed in a dynamic setting. To facilitate this requirement, in this chapter, we developed some time-dependent models of evolution of risk. These time-dependent models utilized the one-period model of risk developed in Chapter 1. Depending on the goals of a risk management problem and solution techniques the modeler plans to utilize, the models of risk can be constructed to evolve in discrete-time or continuous-time. We studied a set of examples from both categories, along with their properties. The examples considered in this chapter are well-known dynamic models of risk, and will be extensively utilized in the rest of the chapters in this book.

5.7 Questions and Exercises

Review Questions

1. What are stochastic processes? How and when can they be useful for serving risk management objectives?

2. How do discrete-time stochastic processes differ from their continuous-time counterparts? When and why would a modeler choose to model using discrete-time versus continuous-time stochastic process models?

3. What is the Markovian property of stochastic processes? What is the significance of this property?

4. When is a Markov chain considered homogeneous? When will the need for modeling risk using a non-homogeneous Markov chain arise? What are the related challenges?

5. What is the relationship between conditional and unconditional probabilities of a Markov chain? How can this relationship be used to derive the unconditional distribution of multiple transitions of the Markov chain?

6. What is the meaning of a stationary distribution for a stochastic process? How can it be determined for a discrete-time Markov chain?

7. What is ergodicity? Why is important for the study of dynamically evolving risk models?

8. What is a simple random walk? How can the process be generalized into a general random walk?

9. How is a random walk modified into a bounded random walk? Why is this modification useful? What properties can the boundaries of the bounded random walk possess?

10. Why are continuous-time Markov chains summarized by an intensity matrix? How do the diagonal entries of an intensity matrix differ from the off-diagonal entries?

11. What is meant when a process is said to have independent increments?

12. What is the Poisson process? When is a Poisson process called non-homogeneous? Give examples of risks appropriately modeled using homogenous and non-homogeneous Poisson processes.

13. What is a birth-death process? Give examples of risks appropriately modeled using a birth-death process.

14. What is a Gaussian process? Give an example of a Gaussian process which also satisfies the Markovian property.

15. What is a diffusion process? What are the properties of a diffusion?

16. What is meant by wide-sense stationary? Give an example of a wide-sense stationary process.

17. What is a standard Wiener process? Discuss the most important properties of the Wiener process.

18. What are the different manners in which convergence of random variables can be described?

19. What is a martingale? What is the implication of this property of a stochastic process? Give some examples of a martingale.

20. What is quadratic variation of a stochastic process? What is the quadratic variation of Brownian motion?

21. What is hitting time of a stochastic process? What modeling purpose may this concept serve in risk management?

22. What is a copula? How is it used for modeling correlation between risks?

Exercises

1. A frog hunting a fly moves between rock 1 and 2 (in a pond) according to a Markov chain with transition matrix,

$$P_1 = \begin{pmatrix} 0.8 & 0.2 \\ 0.2 & 0.8 \end{pmatrix} \tag{5.54}$$

starting on rock 1. The fly, unaware of the frog, starts on rock 2 and moves according to a Markov chain with transition matrix,

$$P_2 = \begin{pmatrix} 0.3 & 0.7 \\ 0.7 & 0.3 \end{pmatrix}. \tag{5.55}$$

The frog catches the fly and the hunt ends whenever they meet on the same rock.

    (a) What is the probability that the hunt will end in one transition? What about in two transitions?

    (b) Show that the progress of the hunt, except for knowing the rock on which it ends, can be described by a three-state Markov chain. Where the hunt ends represents what is known as an *absorbing state*, and the other two that the frog and the fly are on different rocks.

    (c) Obtain the one-step transition matrix of this new chain.

    (d) What is the probability that the hunt will continue for 10 transitions?

2. Set up a function in MATLAB to generate sample paths of a simple random walk. Generalize the code to create sample paths of a more general random walk, with i.i.d. X_i's distributed as,

$$
\begin{aligned}
X_i &= 2 \ \ w.p. \ p_1, \\
&= 1 \ \ w.p. \ p_2, \\
&= 0 \ \ w.p. \ p_3, \\
&= -1 \ \ w.p. \ p_4, \\
&= -2 \ \ w.p. \ (1 - p_1 - p_2 - p_3 - p_4), \quad\quad (5.56)
\end{aligned}
$$

with $0 < p_1, p_2, p_3, p_4 < 1$.

3. Create a MATLAB function to simulate a bounded simple random walk. Estimate the probability distribution and expected value of time it takes the random walk to hit the upper boundary, τ_b, for different choices of the level of upper boundary, b.

4. Construct the transition matrix and determine the stationary distribution for a simple random walk, S_n.

5. If S_n is a simple random walk, what kind of process is $|S_n|$?

6. Demonstrate that a geometric random walk is a Markov chain.

7. Simulate a Poisson process after choosing an intensity parameter, λ.

8. Simulate a birth-death process after choosing appropriate birth rate, λ, and death rate, μ, parameters. Estimate the mean and standard deviation of the process using your simulation.

9. Develop a simulation of the converging sequence of interpolated symmetric random walks, and demonstrate in distribution the limiting sequence matches the distribution of the Wiener process.

10. Set up a MATLAB function to simulate the standard Wiener process. Additionally, implement the transformations that produce sample paths for Brownian motion with drift, as well as geometric Brownian motion.

11. Demonstrate for the Wiener process, $W(t)$, that
$C(t, s) = Cov(W(t), W(s)) = \min(t, s)$.

12. Demonstrate that the Wiener process, $W(t)$, and function of Wiener process, $W(t)^2 - t$, are both martingales.

Chapter 6

Building and Solving Models of Risk

Time-dependent evolution of risk is modeled using stochastic processes. In Chapter 5, we introduced and discussed several models for the temporal evolution of risk. Utilizing these models, it is possible to create structures that allow constructing new stochastic processes. These processes can be constructed with chosen, desired properties. The structure is developed by extending ordinary calculus to the stochastic case, by developing principles of stochastic calculus. We begin with introducing the basic construct of ordinary differential equations for modeling a risk-free asset, and then introduce stochasticity to the model. The rest of the chapter is devoted to determining solutions to the stochastic models, by mostly focusing on simulation based analysis.

6.1 Deterministic Financial Modeling

Assets that increase in value by a risk-free rate can be modeled in the most simple terms, if the risk-free rate and the frequency of its application to determine the accrued interest is known. A modeler takes advantage of the fact that some quantities can be modeled as evolving in discrete-time, while if it suits the context, shift to modeling them in continuous-time. This is the motivation for building a continuous-time model for a risk-free asset in this section, although one can debate whether the asset value evolves truly in a discrete or a continuous manner.

One classic extension of the model of risk-free asset evolution would be the price of a share of the stock of a firm. It is well known that stock prices evolve stochastically. Stocks are traded and quoted in 'tick' values, and prices are known only for times at which a trade occurs. Therefore, does it make sense to model a stock price evolution by a continuous-time, continuous-state process? Continuous models have been successfully used for financial modeling, therefore we explore the flexibility of modeling asset price evolution in continuous and discrete-time. A continuous-time model also gives the flexibility of migrating to a discrete-time model of any frequency of observation with ease, as will be seen in later sections of this chapter.

Assume a bank account that guarantees a certain fixed interest rate, $r\%$, applied daily to the balance in the account. Say we deposited a certain amount, B_0, in that account today, and left it that way for 100 days. On the nth day the balance in the account, assuming no interim withdrawals, will be

$$B(n) = (1+r)^n B_0. \qquad (6.1)$$

If, however, the interest is applied every hour at a rate $r_2\%$, the amount available in the account on the nth day, again assuming no interim withdrawals, will be

$$B(m) = (1+r_2)^m B_0, \qquad (6.2)$$

where m is the conversion of n days into hours, including the hour of the nth day we check the balance. Figure 6.1 displays the value of a risk-free asset for a progression of frequency of accrual of interest rate. Continuing this at a more granular level gives us in the limit,

$$B(t) = B_0 e^{at}, \qquad (6.3)$$

where 'a' is the instantaneous interest rate. We call this interest rate the continuously compounded interest rate. Given that the interest rate is applied continuously, change in the value of the account for an infinitesimally short period 'dt' is $rB(t)$. Therefore, value of the account, $B(t)$, is a solution of the deterministic ordinary differential equation,

$$\frac{dB}{dt} = aB(t), \qquad (6.4)$$

along with the initial condition, $B(0) = B_0$.

Bank deposits that earn guaranteed interest might be one vehicle of investment, while many other short-term bonds, loans, certificates of deposits, etc. are utilized in organized economies. These different vehicles of investment would have different levels of returns. Obviously the higher the return, the better the investment, but along with greater returns comes greater **risk**. The link between risk and reward was discussed at great length in Chapters 1 and 2. One example of such high risk, possibly higher return, investment is investing in the shares of the stock of a company traded in a stock exchange. To model the return or price of a stock, the above model will need to be appropriately extended.

This section provided a brief introduction to modeling of a variable that requires a continuous but deterministic approach to modeling. For some of these asset value variables, introducing randomness in their parameters will remarkably improve the quality of the model. We begin with discussing such enhancements in the context of risky assets, specifically focusing on price of the stock of a publicly traded firm.

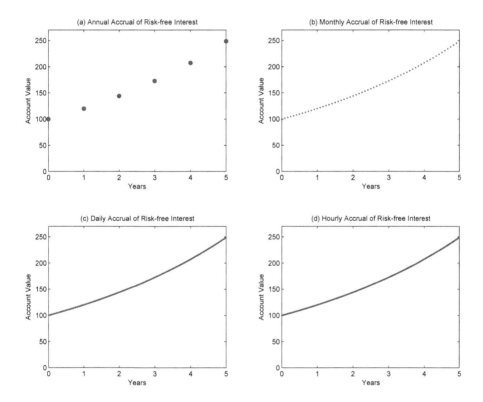

FIGURE 6.1: Applying different frequency of interest rate accrual for a risk-free investment. (a) Annual accrual applied for five years. (b) Monthly accrual applied for five years. (c) Daily accrual applied for five years. (d) Hourly accrual applied for five years.

6.2   Introducing Stochasticity in the Modeling

Let's begin with considering the model for stock price dynamics in which the rate of growth of the stock price at any point of time is proportional to the price of a share of the stock. In the previous section, we assumed this proportionality constant to be a deterministic interest rate, but the model will be more realistically applicable for stock price evolution if this proportionality constant has a random component. Since there is a deterministic component $r(t)$ that is known precisely, we introduce an additional random effects component to obtain, $a(t) = r(t) +$ "noise." Therefore, after a refinement of the model that incorporates the random fluctuations in the relevant quantities of the model, the model becomes,

$$\frac{dS}{dt} = (r(t) + \text{"noise"})S(t), \tag{6.5}$$

where $S(t)$ is the stock prices of a publicly traded corporation. We now need to identify an appropriate stochastic process that will represent the "noise" term in this model.

Let us denote the noise process by a quantity, V_t. If this quantity truly captures the noise-effects that happen at random, such as weather fluctuations, variations in demand, and variations in supply of goods, etc., then the following properties will be expected for the noise process:

1. For two time points, t_1, t_2, such that $t_1 < t_2$, we should have independence of V_{t_1} and V_{t_2}, since these are random fluctuations in the rate of return of the asset.

2. V_t is stationary, i.e., the distribution of the noise process remains the same for all time. Mathematically, distribution of the k-dimensional stochastic process $\{V_{t_1+t}, \ldots, V_{t_k+t}\}$ is independent of t.

3. Finally, being random fluctuations, the mean of the noise process at any time is zero, or $E[V_t] = 0$ for all t.

With these requirements imposed on what the noise term captures about the asset price evolution, we fall into some technical difficulties. There is no 'reasonable' stochastic process that can satisfy these requirements, especially requirements 1 and 2. The technical difficulty is that such a process, V_t, cannot have continuous paths. Moreover, if we require that the variance of the noise is a standard level, i.e., $Var(V_t) = 1$, then V_t cannot even be a measurable function.

In order to respond to these technical difficulties arising from our first attempt at introducing stochasticity in the asset value evolution model, we make an alteration by rewriting the equation in a form that will suggest a more appropriate stochastic process. The new form is one in which we will describe the price change in infinitesimal time interval 'dt', as we did for the risk-free asset in Eqn. (6.4). However, as an intermediate step, we approximate the change in asset price over a finite, positive time interval, Δt, and introduce the yet-to-be-finalized noise term in this approximation. The asset price change in this short time interval is approximated by creating an approximate of the derivative using the Euler method, as $\frac{\Delta S(t)}{\Delta t}$. As a second attempt of introducing stochasticity, the stochastic term is introduced as follows,

$$S_{k+1} - S_k = r(t_k)S_k\Delta t_k + S_k V_k\Delta t_k, \tag{6.6}$$

where $S_k = S(t_k)$, $V_k = V(t_k)$, and $\Delta t_k = t_{k+1} - t_k$, for $k \in \{0\dots N\}$ with $t_0 = 0$ and $t_N = T$.

We introduce a crucial modification to the last term, the noise term, by renaming it as $V_k\Delta t_k = \Delta W_k$. Now imposing the requirements 1, 2 and 3 listed above for the noise term implies that the W_t process should have stationary, independent increments with mean 0. We have a perfect candidate for this role, since as it turns out, the only process endowed with these properties that also has continuous paths is the **Wiener process**, or the **Brownian motion**. In Chapter 5, we had studied the Wiener process at length, including constructing it as a limit of random walks. With this modification, Eqn. (6.6) becomes,

$$S_{k+1} = S_k + r(t_k)S_k\Delta t_k + S_k\Delta W_k. \tag{6.7}$$

By applying Eqn. (6.7) to all values of j and adding from 0 to k gives,

$$S_{k+1} = S_0 + \sum_{j=0}^{k} r(t_j)S_j\Delta t_j + \sum_{j=0}^{k} S_j\Delta W_j. \tag{6.8}$$

We have succeeded in incorporating a noise term into the model, however, this is not the original model we wanted for the asset value evolution in continuous-time and continuous-space. It is only an approximate version of the original model constructed for a specific time discretization, $\{t_k\}$, for $k \in \{0\dots N\}$ with $t_0 = 0$ and $t_N = T$. In order to now retrieve the original, continuous counterpart, we need to find what happens if we take the Δt_k terms to be gradually smaller, or in the limit as $\Delta t_k \to 0$. Invoking the definition of the Riemann integral, we obtain,

$$S_t = S_0 + \int_0^t r(s)S(s)ds + \text{``} \int_0^t S(s)dW_s\text{''}, \tag{6.9}$$

where the last term is roughly defined, with a pending formal definition.

The second term in Eqn. (6.9) is an integral with respect to the Wiener process, and needs to be defined as an extension of the Riemann integral of Riemann calculus. This development will be the topic of the next section. Assuming for now we accept a meaningful existence of the second term in Eqn. (6.9), the model extension will prove useful to model stock dynamics, as well as other risk types, in later chapters.

6.3   Defining New Integrals

In Section 6.1, we developed a model for the evolution of a risk-free asset, $B(t)$. If in fact the asset experiences fluctuations in its rate of return, we would more likely expect the instantaneous interest rate a to be replaced by $a +$ "noise." This will transform the equation, $\frac{dB}{dt} = aB(t)$, into

$$\frac{dS}{dt} = (a + \text{"noise"})S(t). \tag{6.10}$$

We applied an argument in favor of desired properties of the noise term, transformed the structure of the equation, utilized the Wiener process to model the noise term and obtained,

$$S(t) = \int_0^t aS(s)ds + \int_0^t S(s)dW_s. \tag{6.11}$$

This model represents the dynamics of the value of a risky investment. In particular, this model may be used to model the stock price dynamics of a publicly traded company. It assumes a specific meaning to the term, '$\int_0^t S(s)dW_s$', which is the integral of the stock price path, $S(s)$, with respect to the Wiener process. We formally define this integral next.

6.3.1   Ito Integral

The construction of the new integral, leading on to a new calculus, is similar to how the Riemann integral is defined in real analysis in college-level calculus. Given a discretization of the time interval, $(0, T)$, the Riemann integral for a continuous function, $f(t)$, was defined as follows,

$$\sum_{j=0}^{N} f(t_j)\Delta t_j \to \int_0^T f(t)dt, \quad \text{as} \quad \Delta t_j \to 0, \forall j. \tag{6.12}$$

The Riemann integral is defined for sufficiently well-behaved functions. In particular, if $f(t)$ is a bounded function defined on a closed, bounded interval $[0, T]$ and $f(t)$ is continuous except at countably many points, then $f(t)$ is Riemann integrable.

In a similar manner, the integral can also be defined in terms of the Wiener process. Construction of this integral is necessary because in order to introduce randomness in the model considered in earlier sections, a derivative of the Wiener process cannot be used, since such a derivative does not exist. Instead we consider the integral of the Wiener process to do this job. The integral will be defined very similarly to the definition of the Riemann integral, with increments in time being replaced by increments in the Wiener process, as follows,

$$\sum_{j=0}^{N} f(t_j^*)[W_{t_{j+1}} - W_{t_j}] \rightarrow \int_0^T f(t)dW_t, \quad \text{as} \quad \Delta t_j \rightarrow 0, \forall j, \qquad (6.13)$$

where $t_j^* \in [t_j, t_{j+1}]$.

Unlike in the case of the Riemann integral, where t_j^* can be any point in the interval $[t_j, t_{j+1}]$, in the case of integral with respect to the Wiener process, it matters what value t_j^* is made to take in the interval. For instance, if $t_j^* = t_j$, i.e., the lower end of the interval, the integral is called an **Ito integral**, however if $t_j^* = \frac{t_j + t_{j+1}}{2}$, i.e., the middle point of the interval, the integral is a **Stratonovich integral**. The properties of the two integrals are quite different.

Physically realizable processes are often smooth with at least a small degree of autocorrelation. When the Wiener process is used as an idealization of a smooth real noise process, Stratonovich is the more appropriate definition to use. On the other hand, if the Wiener process is an idealization of a real noise process that is not necessarily autocorrelated, and, in fact, has a good reason not be so, then Ito definition is more appropriate.

In engineering and physical sciences, problems are studied using ordinary differential equations, which are obtained based on physical or phenomenological laws governing the system. Stochastic differential equations (SDEs), defined in terms of integral with respect to the Wiener process, are arrived at for improvement in models by including random fluctuations in the ordinary differential equations. The underlying systems being modeled are usually continuous both in time and space domains, hence the fluctuations are expected to be smooth. In contrast, for example in biological systems, the variables are discrete in either time or space, or both. In genetics or population dynamics, the population size is integer valued, successive generations may not overlap in time, breeding may occur in separate seasons, and environmental parameters may only change at discrete instants. In this case, an Ito description would be more appropriate.

From a mathematical viewpoint, both the Ito and Stratonovich definitions are correct. Of course, which one you use depends on such system specific extraneous reasons. Once the choice has been made regarding the integral definition to use, one needs to stay consistent in all the remaining development for the calculus one performs for the system. One can always use the other definition when it is considered advantageous, after appropriately modifying the stochastic differential equation.

The above limits in Eqn. (6.13), and therefore the integral, will be well-defined for a sufficiently well-behaved function, $f(t)$. To facilitate discussion of properties of the Ito integral with more care, we rigorously define the properties of function, $f(t)$, for which the Ito integral is well-defined. We define, $\Upsilon = \Upsilon(S,T)$ as a class of functions, $f(t,\omega) : [0,\infty) \times \Omega \rightarrow \mathbf{R}$, such that,

1. $f(t,\omega)$ is $\mathcal{B} \times \mathcal{F}$-measurable. Here $\mathcal{B}$ refers to the Borel σ-algebra on the

interval, $[0, \infty)$ and $\mathcal{F}$ is a σ-algebra defined on the state space, Ω. We employ usage of the word 'measurable' in a special mathematical sense. It refers to the structure-preserving property of the function, in a sense similar to the property of a continuous function mapping continuous intervals of the domain space to continuous intervals in the range space in the context of Riemann calculus.

2. $f(t, \omega)$ is $\mathcal{F}_t$-adapted, i.e., all events defined by function $f(t, \omega)$ up to time, t, belong in the σ-algebra, $\mathcal{F}_t$.

3. The function, $f(t, \omega)$, is square-integrable in time-domain, i.e., $E[\int_S^T f(t, \omega)^2 dt] < \infty$.

Such a class of functions, Υ, can be shown to have a well-defined Ito integral. It is worth noting in the above properties that it allows $f(t, \omega)$ to be a stochastic process in its own right. For greater depth and detail of definition of these concepts and constructs, refer to a book on real analysis [73, 74, 75] and stochastic calculus [68, 45].

6.3.2 Properties of the Ito Integral

There are many properties of the Ito integral that make working with them attractive. It is important for the risk modeler utilizing these models to be aware of these properties. We next discuss the major properties, and as before, the more inquisitive reader is advised to refer to the mathematical sources [68, 45] for more details. The Ito integral can be shown to satisfy the following properties for all functions, $f(t, \omega)$ and $g(t, \omega)$, for which the Ito integral is well-defined, where the first two properties are inherited from its Riemann-Stieltjes counterpart.

Property 1. When we define the Ito integral for a time range, $[S, T]$, we can compute the integral at once for the entire interval, or break it down in subintervals whose union is the original interval, $[S, T]$. Therefore, we have, $\int_S^T f(t) dW_t = \int_S^U f(t) dW_t + \int_U^T f(t) dW_t$ for $0 \le S < U < T$.

Property 2. Ito integral allows defining the integral of linear function of functions in terms of linear function of the Ito integral of the functions. We have, $\int_S^T (cf(t) + g(t)) dW_t = c \int_S^T f(t) dW_t + \int_S^T g(t) dW_t$.

Property 3. The remaining properties of the Ito integral are obtained from the fact that integral here is with respect to the Wiener process, and outcome of the integral is itself a random variable. From the property of increments of Wiener process, $E[\Delta W_t] = 0$, and definition of the Ito integral, we have that expectation of the Ito integral of any function is zero. Therefore, $E[\int_S^T f(t) dW_t] = 0$.

Property 4. The fact that all events defined by the function $f(t, \omega)$ up to time, t, belonged to the σ-algebra, $\mathcal{F}_t$, was required for the Ito integrability of the function, $f(t,\omega)$. If the Wiener process, W_t, is $\mathcal{F}_T$-adapted, the random variable defined by the Ito integral of $f(t,\omega)$ is $\mathcal{F}_T$-measurable. Therefore, $\int_S^T f(t)dW_t$ is $\mathcal{F}_T$ measurable.

Property 5. The concept of a martingale was defined in Section 5.3.6.3, where a process is a martingale if it is non-anticipatory about the future. From the independent increments property of the Wiener process and the definition of the Ito integral, we obtain that $M_t = \int_0^t f(t,\omega)dW_s$ is a martingale with respect to the σ-algebra, $\mathcal{F}_t$.

Property 6. The Ito Isometry property is derived from the quadratic variation of the Wiener process, discussed in Section 5.3.6.3. The quadratic variation of the Wiener process is defined as, $\sum_j (\Delta W_{t_j})^2$, over a time-discretization of the interval, $(0,t)$. It can be shown to converge to $\sum_j (\Delta W_{t_j})^2 \to t$ in mean-square sense, i.e., in $L^2(P)$. Utilizing this property, it can be shown that the expected value of squared Ito integral is the expectation of the Riemann integral of the squared function. Mathematically, $E[(\int_S^T f(t,\omega)dW_t)^2] = E[\int_S^T f^2(t,\omega)dt]$.

To develop a better feel for the Ito integral and how it is determined, let us find the Ito integral for some simple functions. We begin by considering the integral of the Wiener process itself. Therefore, we are seeking $\int_0^t W_s dW_s$. It can be shown that this integral is equal to $\frac{1}{2}W_t^2 - \frac{1}{2}t$. Note that by the definition of the Ito integral we have,

$$\int_0^t W_s dW_s = \lim_{\Delta t_j \to 0} \sum_j W_{t_j} \Delta W_{t_j}, \qquad (6.14)$$

where $\{\Delta t_j\}$ is a time-discretization for the interval, $[0,t)$. We observe that,

$$\Delta(W_j^2) = W_{j+1}^2 - W_j^2 = (W_{j+1} - W_j)^2 + 2W_j(W_{j+1} - W_j) \quad (6.15)$$
$$= (\Delta W_j)^2 + 2W_j \Delta W_j, \qquad (6.16)$$

where we have employed simplification of notation, $W_j = W_{t_j}$, for the time-discretization, $\{\Delta t_j\}$, of the interval, $[0,t)$. Since the standard Wiener process has $W_0 = 0$, we have,

$$W_t^2 = W_t^2 - W_0^2 = \sum_j \Delta(W_j^2) \qquad (6.17)$$

$$= \sum_j (\Delta W_j)^2 + 2\sum_j W_j \Delta W_j, \qquad (6.18)$$

where Eqn. (6.18) is obtained by applying Eqns. (6.15) and (6.16). We rearrange terms from Eqns. (6.17) and (6.18) to obtain the desirable expression to determine the integral we sought to find in Eqn. (6.14).

$$\sum_j W_j \Delta W_j = \frac{1}{2} W_t^2 - \frac{1}{2} \sum_j (\Delta W_j)^2. \tag{6.19}$$

As we let, $\Delta t_j \to 0$, required in the definition of Ito integral in Eqn. (6.14), we need to determine the impact on the second term on the right-hand side in Eqn. (6.19). We invoke the result that $\sum_j (\Delta W_j)^2 \to t$ in mean-square sense as $\Delta t_j \to 0$. We had earlier defined this as the quadratic variation of the Wiener process. Therefore, the value of the integral is established.

As another example, we compute the Ito integral of the simple function, $f(t) = t$. Therefore, we seek $\int_0^t s dW_s$. It can be shown that,

$$\int_0^t s dW_s = t W_t - \int_0^t W_s ds. \tag{6.20}$$

We first observe that the Ito integral of $f(t) = t$ is defined as, $\lim_{\Delta s_j \to 0} \sum_j s_j \Delta W_j$ and that we can express the terms in the summation as follows,

$$\sum_j \Delta(s_j W_j) = \sum_j s_j \Delta W_j + \sum_j W_{j+1} \Delta s_j. \tag{6.21}$$

After rearranging terms in Eqn. (6.21) and substituting in the definition of the Ito integral, we obtain the integral as we apply the limit of $\Delta s_j \to 0$. Following up on the above examples, it can be shown that $\int_0^t W_s^2 dW_s = \frac{1}{3} W_t^3 - \int_0^t W_s ds$.

Now that we have defined the new integral with respect to the Wiener process, and also seen it applied to compute the integral of some simple functions, we can start to utilize it to define more general models. The models will be written, with more generalization introduced later, as,

$$dX_t = \mu(t)X(t)dt + \sigma(t)X(t)dW_t, \tag{6.22}$$

which will refer to the more precise meaning of,

$$X_t = X_0 + \int_0^t \mu(s)X(s)ds + \int_0^t \sigma(s)X(s)dW_s. \tag{6.23}$$

However before we continue with building new models and solving them, we need to look at one important result of Ito calculus. This is the chain rule of Ito calculus.

6.3.3   Chain Rule of Ito Calculus - The Ito Formula

In Riemann calculus, when we define a function of a function, and wish to consider change in the composite function with respect to the underlying variable, we construct a chain rule. The chain rule for Riemann calculus for a function, $f(g(x))$, is constructed as,

$$\frac{df(g(x))}{dx} = \frac{df(g(x))}{dg}\frac{dg(x)}{dx}. \tag{6.24}$$

We wish to extend this to a process defined by the Ito integral. Consider an Ito process, X_t, defined by the solution of the model, $dX_t = a(t, X_t)dt + \sigma(t, X_t)dW_t$. Let us define a new process, Y_t, constructed as a function, $Y_t = f(t, X_t)$, of the original process, X_t. If the function, $f(t, x)$, is twice continuously differentiable, then the chain rule under Ito calculus gives,

$$
\begin{aligned}
dY(t) &= (\frac{\partial f}{\partial t} + a(t, X_t)\frac{\partial f}{\partial x} \\
&+ \frac{1}{2}\sigma^2(t, X_t)\frac{\partial^2 f}{\partial x^2})dt + \sigma(t, X_t)\frac{\partial f}{\partial x}dW_t.
\end{aligned}
\tag{6.25}
$$

This is popularly known as the **Ito Formula**. Although we don't provide a detailed derivation of the formula, we suggest a few important comments for why the chain rule works out as it does in Ito calculus. This can be seen by taking a Taylor expansion and looking at the leading terms. Roughly speaking, the $\frac{1}{2}\sigma^2(t, X_t)\frac{\partial^2 f}{\partial x^2}$ term in the Ito formula comes about since in the Taylor expansion the $dW_t.dW_t \simeq dt$, where as $dt.dW_t = dt.dt = dW_t dt = 0$.

6.4 Analytical Solutions

We have the ability to construct a variety of new models utilizing the newly defined integral with respect to the Wiener process. The keen reader will have noticed by now that when we define a process by a stochastic differential equation, as in Eqn. (6.22), we simply state that the concerned process must satisfy this equation. The actual process itself is left to be determined by solving the stated stochastic differential equation.

In the development since Section 6.2, an appropriately defined random component in the differential equation model provided us a well-defined mechanism of how risk can be modeled dynamically evolving in time. In Section 6.2, we motivated the model construction to describe the evolution of the value of a risky asset. In the notation developed since, the risky value of an asset, S_t, evolves as follows,

$$dS_t = \mu(t, S_t)dt + \sigma(t, S_t)dW_t, \tag{6.26}$$

where $\mu(t, S_t)$ is called the drift term and $\sigma(t, S_t)$ is called the diffusion term of the model. The reader may recall that the terms 'drift' and 'diffusion' were discussed in the context of a diffusion process in Section 5.3.4. Specific functions chosen for the drift and diffusion terms result in specific properties of the risky asset value.

A continuous-time version of the price dynamics equation lends a desirable property. It lets the modeler discretize the model to any chosen time frequency. This feature also offers an opportunity to computationally obtain a solution of the stochastic differential equation. We consider the Euler-method based discretization, which may be used to get a numerical solution of the system. If the current value of the asset is known, S_0, then the subsequent values can be obtained in an iterative procedure by the following equation,

$$S_{k+1} = S_k + \mu(t_k, S_k)\Delta t_k + \sigma(t_k, S_k)\Delta W_k, \qquad (6.27)$$

where $\Delta W_k \sim N(0, \Delta t_k)$ and $\{\Delta t_k\}$ is a time-discretization of the interval, $[0, T)$. Sample paths for the risky asset value, S_k, are constructed by generating the appropriate normally distributed random variates.

When we can exactly solve the stochastic differential equation, we call the solution an analytical solution. When, however, we utilize simulation to numerically realize the solution of the stochastic differential equation, the solution is a numerical solution. It is a challenge in general to analytically solve all stochastic differential equation models. The techniques and methods available constitute a vastly developed area of applied mathematics. We explore solutions to some of the simple equations in the next section. Since the emphasis in this book is to primarily discuss the numerical or simulation techniques for these models, we will not delve intp the general approaches for analytical solutions to stochastic differential equations. However, the reader is advised to refer to some good books for learning more about analytical methods for solving ordinary and stochastic differential equations, such as Tenenbaum and Pollard [85], Coddington and Landin [16], and Oksendal [68].

6.4.1 Solving the Model Exactly

We consider solving a simple version of the risky asset dynamics model exactly (or analytically). In practice, whenever it is possible to obtain the exact solution, the modeler should attempt to obtain it. The simplest stochastic model we have obtained thus far is,

$$dS_t = \mu S_t dt + \sigma S_t dW_t, \qquad (6.28)$$

where μ is the drift coefficient and σ is the volatility, which is usually a positive constant. Both coefficients can be generalized to vary with time, $\mu(t), \sigma(t)$. This is an Ito stochastic differential equation describing the evolution of a risky asset value over time.

In Section 6.1, we had demonstrated for the deterministic case that the solution for the related deterministic equation displays an exponential growth in the asset value over time (Figure 6.1). The deterministic case can motivate the solution exploration in the stochastic case. The equation we seek to solve is given in Eqn. (6.28), which is a much simplified version of Eqn. (6.26). We rewrite the Eqn. (6.28) in a favorable way that motivates the steps adopted thereafter to determine its solution.

$$\frac{dS_t}{S_t} = \mu dt + \sigma dW_t. \tag{6.29}$$

If we integrate both sides of Eqn. (6.29), we obtain,

$$\int_0^t \frac{dS_s}{S(s)} = \mu t + \sigma W_t. \tag{6.30}$$

Eqn. (6.30) shows that the integral of instantaneous return of the risky asset is $\mu t + \sigma W_t$. The structure of the left-hand side of Eqn. (6.30) motivates us to consider a guess function, $ln(S_t)$. From this motivation, we apply the chain rule of Ito calculus, in short the Ito formula, to the process $\ln(S_t)$. This gives us,

$$d(lnS_t) = \frac{1}{S_t}(\mu S_t dt + \sigma S_t dW_t) + \frac{-1}{2S_t^2}\sigma^2 S_t^2 dt. \tag{6.31}$$

Eqn. (6.31) can be simplified to obtain,

$$d(lnS_t) = \frac{dS_t}{S_t} - \frac{1}{2}\sigma^2 dt. \tag{6.32}$$

If we rearrange the terms in Eqn. (6.32), we obtain another form for the quantity $\frac{dS_t}{S_t}$, as follows,

$$\frac{dS_t}{S_t} = d(lnS_t) + \frac{1}{2}\sigma^2 dt. \tag{6.33}$$

Expressing the same quantity in two different ways (Eqns. (6.30) and (6.33)) gives us the opportunity to combine the two representations to obtain,

$$\mu dt + \sigma dW_t = d(lnS_t) + \frac{1}{2}\sigma^2 dt. \tag{6.34}$$

Rearranging the terms once again, and integrating both sides of the equation yields,

$$\ln(\frac{S_t}{S_0}) = (\mu - \frac{1}{2}\sigma^2)t + \sigma W_t \tag{6.35}$$

Finally, taking an exponential of both sides of Eqn. (6.35), and letting the initial asset price S_0 move to the right-hand side, gives the solution of the stochastic differential equation in Eqn. (6.28) as follows,

$$S_t = S_0 \exp((\mu - \frac{1}{2}\sigma^2)t + \sigma W_t). \tag{6.36}$$

The reader is reminded that in Section 5.3.7, we had introduced a stochastic process called geometric Brownian motion, which has a structure very similar to the process in Eqn. (6.36). We will also call the process in Eqn. (6.36) a **geometric Brownian motion**, with parameters μ and σ determining its time-evolution characteristics.

The trends of such a process can be summarized as follows. The process exponentially increases or decreases in time at the $(\mu - \sigma^2/2)$ rate, with dispersions determined by the size of the volatility parameter, σ. Additionally, we observe:

1. If $\mu > \frac{1}{2}\sigma^2$, then $S_t \to \infty$ as $t \to \infty$ almost surely.

2. If $\mu < \frac{1}{2}\sigma^2$ then $S_t \to 0$ as $t \to \infty$ almost surely.

3. If $\mu = \frac{1}{2}\sigma^2$, then S_t will fluctuate between arbitrarily large and small values as $t \to \infty$ almost surely.

4. Finally, if $\mu = 0$, the process S_t becomes a martingale, as discussed in Section 5.3.7.

As stated earlier, in general, analytical solutions are hard to come by for stochastic differential equations. They can be solved exactly only in special cases. Simulation techniques must be utilized to obtain the solution and understand their properties. This will be the emphasis of the rest of the chapter. The above simple risky asset dynamics model is not totally realistic! There are many improvements one can consider to bring greater realism to the model dynamics. With these improvements comes greater analytical intractability of the model, where a simulation-based solution will become more valuable.

Moreover, value of assets often does not evolve in isolation. They are affected by a basket of other risk factors and value of other risky assets. Such interactions may be beneficial, neutral or harmful to the value of each asset. These characteristics can be incorporated by considering a system of stochastic differential equations similar to the one in Eqn. (6.28), one for each asset (or risk factor) evolution. The system of equations together describe the co-evolution of all the assets hypothesized to be interrelated.

A stochastic model for the multi-asset context with the joint evolution of the asset price dynamics can be written as follows,

$$dS_t^i = (\mu_i S_t^i + \sum_{j=1}^{d} b^{i,j} S_t^j)dt + \sigma_i S_t^i dW_t^i, \quad \text{for} \quad i = 1 \ldots d, \qquad (6.37)$$

where we consider d assets, $b^{i,j}$ is the interaction drift coefficient that affects the rate of change of one risky asset based on the value of other risky assets. In Eqn. (6.37), we have retained the diffusion term to remain unaffected by other assets or risk factors in the basket. More generally, the diffusion term may be modeled as,

$$\sum_{j=1}^{d} \sigma^{i,j}(S_t^1, S_t^2, \ldots S_t^d) dW_t^j, \tag{6.38}$$

where each of the d Wiener processes, W_t^i, driving the system are allowed to interact with one another by the d-dimensional function, $\sigma(\cdot, \cdot, \ldots, \cdot)$. It is possible to write the combined model of Eqns. (6.37) and (6.38) in a single compact equation by utilizing matrix notation. The reader can judge from these general models that there is significant room for innovation in models by picking specific forms for the coefficient, $b^{i,j}$, and term, $\sigma^{i,j}(S_t^1, S_t^2, \ldots S_t^d)$. These generalizations put even greater burden on the ability to find analytical solutions. Therefore, numerical techniques must be developed and resorted to in order to obtain and analyze the solutions. We move on to this investigation next.

6.5 Solving Models Using Simulation

When analytical solutions of stochastic differential equations describing time-dependent evolution of risks are not readily available, we must resort to numerical techniques using simulation analysis. In this section, we develop these simulation techniques, as well as develop the terminology for assessing the accuracy of these numerical solutions. In order to develop the methodology for simulation-based solutions of stochastic differential equations, we will need to recall material covered in Chapter 4, as well as develop the basic principles of numerical solutions of deterministic differential equations.

The primary idea employed is to approximate the differential equation based representation of the model to a difference equation based model on a chosen time-discretization of the interval on which the solution is required. The solution can then be obtained by an iterative procedure, once the initial condition of the stochastic process is known. As stated earlier, the advantage of modeling risks by a continuous-time model, even though one is not able to solve these models analytically, is that an arbitrary choice of discretization can be picked to achieve the desired accuracy of the numerical solution. We begin with investigating the simplest of methods for simulation-based solutions of stochastic differential equations, which is the Euler method.

6.5.1 The Euler Method for Solving Differential Equations

We first apply the Euler scheme to solve differential equations in the deterministic case. Let $x = x(t; t_0, x_0)$ be the solution of an initial value problem

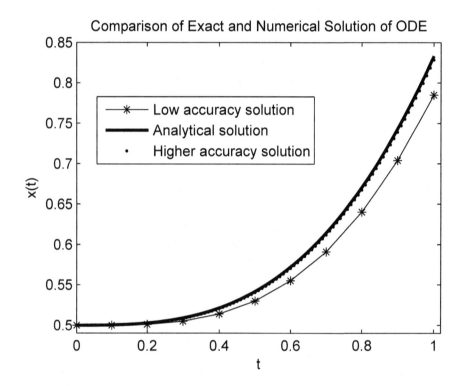

FIGURE 6.2: Comparison of exact and numerical solutions for an example ordinary differential equation.

(IVP)

$$\frac{dx}{dt} = a(t,x), \tag{6.39}$$

with $x(0) = x_0$. An initial value problem (IVP) consists of a deterministic differential equation based model, where the initial value of the function satisfying the differential equation is known. Suppose we want to solve the system given in Eqn. (6.39) on a time interval, $[0, T]$. We will begin with creating a discretization of the interval into subintervals, $0 = t_0 \le t_1 \le \ldots \le t_N = T$. We may keep the gap between these discretized time-points equidistant or not. If we choose to keep them equidistant, although this is not essential, we have for any, n, such that $0 \le n \le N$, the time-gap, $t_n - t_{n-1} = \Delta = \frac{T}{N}$. We will control the value of Δ to achieve the desired accuracy. In general, what is essential is that the maximum of all subinterval lengths gets smaller for improved accuracy.

In Eqn. (6.39), we first approximate the first derivative by a simple difference scheme. The difference scheme is, $\frac{dx}{dt} \simeq \frac{x(t_n) - x(t_{n-1})}{\Delta}$. This allows us to write the following approximation of the model in Eqn. (6.39),

$$y_{n+1} = y_n + a(t_n, y_n)\Delta \quad \text{for } n = 1 \ldots N, \tag{6.40}$$

where we are careful to indicate the approximate solution by a new symbol, indicating $y_n \simeq x(t_n)$. Using this approximate equation, the approximate solution y_n, for $n = 1 \ldots N$, can be obtained iteratively. Once we set $y_0 = x_0$, then y_1 can be obtained from Eqn. (6.40) by plugging in the value of y_0, and so on.

In Figure 6.2, we apply the Euler scheme to a problem for which we know the analytical solution. We also compute the simulation-based numerical solution to demonstrate how well the numerical solution approximates the analytical solution. In general, we don't have this luxury of comparing the analytical solution with the numerical one, since the reason we seek a numerical solution is that we do not know a way to obtain the analytical solution. In this simple case we get to make this comparison, and see that the numerical solution gets very close to the analytical solution for reasonably coarse discretization (here $N = 100$). Figure 6.2 displays the solution of the following problem and its discretized approximation.

$$\frac{dx}{dt} = t^2, \tag{6.41}$$

$$y_{n+1} = y_n + t_n^2\Delta \quad \text{for } n = 1 \ldots N, \tag{6.42}$$

where the initial condition is taken as, $x(0) = y_0 = \frac{1}{2}$.

We now consider a general stochastic differential equation model for a process, X_t, given as,

$$dX_t = a(X_t)dt + b(X_t)dW_t, \quad t \in [t_0, T], \tag{6.43}$$

Comparison of trajectory from analytical solution and simulation based solution of an SDE

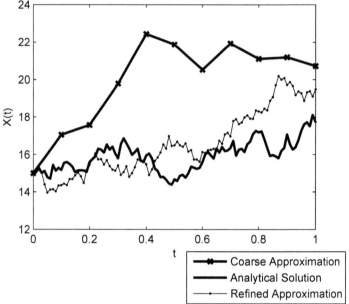

FIGURE 6.3: Comparison of exact and numerical solution for an example stochastic differential equation.

with the initial value of the process given as, $X_0 = x_0$.

An Euler approximation of the original model in Eqn. (6.43) can be obtained and simulated to obtain an approximate solution. Once again, we construct the discretization, $t_0, t_1, t_2, t_3, \ldots, t_{N-1}, t_N = T$, of equidistant points with a time-step of length, $\Delta = (T - t_0)/N$. The Euler approximation in this case is a continuous-time stochastic process $Y = \{Y(t), t_0 \leq t \leq T\}$ satisfying the iterative scheme,

$$Y_{n+1} = Y_n + a(Y_n)\Delta + b(Y_n)\Delta W_n. \qquad (6.44)$$

The main difference in the stochastic case is that here ΔW_n needs to be generated as a random variate with the appropriate distribution. Given the properties of the Wiener process, these Wiener increments are independent Gaussian random variables with mean, $E[\Delta W_n] = 0$, and variance, $E[(\Delta W_n)^2] = \Delta$.

In Figure 6.3, we apply the Euler scheme to a problem for which we know the analytical solution. We also compute the simulation-based solution to demonstrate how the numerical solution compares with the analytical one. Figure 6.3 displays the solution of the following problem we have studied before, as well as its discretized approximation.

$$
\begin{aligned}
dX_t &= \mu X_t dt + \sigma X_t dW_t, & (6.45) \\
Y_{n+1} &= Y_n + \mu Y_n \Delta + \sigma Y_n \Delta W_n, \quad \text{for } n = 1 \ldots N, & (6.46)
\end{aligned}
$$

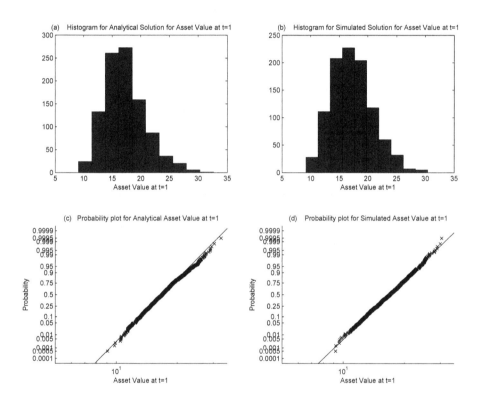

FIGURE 6.4: Comparison of distributional properties of the exact solution (in left panel) and numerical solution (in the right panel) for the example stochastic differential equation.

where the initial value of the asset is taken as, $X(0) = Y_0 = \$15$, the drift is taken as $\mu = 14\%$ and volatility is $\sigma = 20\%$. As seen in Figure 6.3, the refined approximation, which corresponds to $N = 100$, performs much better than the coarse approximation, and is already picking up the characteristics of the analytical solution.

In the simulation approach, not only is the $\{Y_n\}$ sequence obtained from applying the Euler scheme in Eqn. (6.44) an approximation of the actual solution, X_t, of the original model in Eqn. (6.43), it is also only observed at discrete time-points of the time discretization, $t_0, \ldots, t_{N-1}, t_N = T$. To make the $Y(t)$ process a continuous-time stochastic process, the value at the intermediate points must be determined by interpolation, for example, as follows,

$$Y_t = Y_{n_t}, \quad n_t = \min\{0, 1, 2, \ldots N; t_n < t\}. \tag{6.47}$$

The interpolation done in Eqn. (6.47) is a piece-wise constant interpolation. Instead of a piece-wise constant, linear interpolation can also be used, as is done for sample paths in Figure 6.3. In either case, whether we do a linear or a piece-wise constant interpolation, the fine structure of sample paths of the general diffusion processes that they inherit from the Wiener process will not be captured. The finer the grid-size, the finer will be the structure inherited. The question of quality of approximation in the approximate solution still remains.

In Figure 6.4, we plot the histogram of the approximate and analytical solution of the example problem in Eqns. (6.45) and (6.46) at the terminal time point, $t = 1$. Since we know the properties of the analytical solution, namely that it is the geometric Brownian motion, we also plot a lognormal probability plot for both the analytical and simulation based solution in Figure 6.4. In both cases, the lognormal distribution is supported. The difference in mean and standard deviation of asset value at $t = 1$ by the two solutions is of order 10^{-2}.

We next turn our attention to creating precise measures for accuracy of solutions, and how they respond to improved accuracy as time-discretization is refined.

6.5.2   Evaluating Simulation Solutions

When attempting to solve a differential equation using numerical techniques, two important questions must be addressed. For any specific method of discrete approximation of the continuous problem and for a specific selected time-discretization, how 'far' will the approximate solution end up being from the actual solution. Second, for a specific method of discrete approximation of the continuous problem, as the time-discretization is refined, how rapidly does the approximate solution become 'closer' to the actual problem. The first inquiry is termed **error analysis**, while the latter is called **rate of convergence** of an approximate method. We begin with defining the latter.

6.5.2.1 Convergence Properties of Solutions

We have so far looked at only one method of approximating the continuous problem, the Euler scheme. The Euler method was utilized to define an approximate solution in the deterministic and stochastic examples in the previous section. We can measure the goodness of the approximate solution in two ways. There is the **local discretization error**, which is the error made in a single time step, or a single iteration of the numerical procedure to obtain y_n in Eqn. (6.40). This is obtained assuming that the exact solution at t_n, $x(t_n)$, is known, and matches y_n, then the difference between $x(t_{n+1})$ and y_{n+1} is observed. This error is usually not going to be zero.

Once several such steps of the iteration are made to obtain a sequence of y_n values, the (local discretization) errors accumulate. **Global discretization error** is all the errors accumulated up to a time-point, t. Therefore, global discretization error can be seen as the sum of propagated truncation error and local discretization error at any point of the iterative procedure. The size of this error, and how rapidly it can go down as discretization is refined, provides us the definition of convergence rate.

Order of convergence: A method converges with order γ if there exists a constant K $(< \infty)$ such that the $|x(t_{n+1}) - y_{n+1}| = |e_{n+1}|$ can be bounded from above by $K\Delta^{\gamma}$, for all $\Delta \in (0, \delta_0)$, for some $1 > \delta_0 > 0$.

By this definition of order of convergence, the Euler method can be shown to have an order, $\gamma = 1.0$. However, it should be noted that the above definition of order of convergence is stated assuming that there are no round-off errors. In practice, due to round-off errors, there is a minimum time-discretization level, Δ_m, below which we cannot hope to improve accuracy by taking finer time-discretization levels. Therefore, in order to get more accurate approximate solutions, we need to consider methods with higher orders of convergence. Different discretization schemes may be explored and adopted, each with a level of accuracy and computational burden.

6.5.2.2 Error Analysis - Absolute Error Criterion

In Section 6.5.2.1, a definition of order of convergence was developed with a focus on deterministic problems. When instead we apply simulation to solve a stochastic differential equation, the solution is not a deterministic function. The solution is a realization of a stochastic process. From the point of view of accuracy of the solution, a question arises regarding the sense in which the approximate solution converges to the actual one. Is it path-wise? Or is it in distribution?

In fact, quality of a discrete-time approximate solution can only be judged on the basis of the main goal of simulation. There are two basic tasks connected with the simulation of solution of a stochastic differential equation: **1.** when a good **path-wise approximation** is required, and **2.** when an approximation of **expectation of a functional** of an Ito process is required, such as, probability distribution or its moments. Depending on the requirement, an appropriate error criterion needs to be used. We first define the criterion for path-wise convergence of approximate solution.

Absolute Error Criterion is defined as:

$$\epsilon = E[|X_T - Y(T)|] \tag{6.48}$$

which gives a measure of path-wise closeness at the end of the time interval $[0, T]$. In case of the deterministic initial value problem (IVP), where the function $b(.) = 0$ in Eqn. (6.43), the absolute error criterion coincides with the usual deterministic error criterion for the absolute global discretization error.

Method to estimate the absolute error: Generate N sample paths for the Wiener process in $[0, T]$ and compute the discrete solution $Y(t)$ for each sample path of the Wiener process. Using the same sample paths, compute the exact solution (provided this is known). Denote the quantities obtained from the kth simulation by $Y(T, k)$ and the corresponding exact solution by $X_{T,k}$. Then the absolute error can be estimated as:

$$\hat{\epsilon} = \frac{1}{N} \sum_{k=1}^{N} |X_{T,k} - Y(T, k)|. \tag{6.49}$$

Confidence interval for the Absolute Error Estimate: The estimated absolute error, $\hat{\epsilon}$ is a random variable. It will be asymptotically (or in the limit) Gaussian, if $|X_{T,k} - Y(T, k)|$ is believed to have the same distribution for all k. The estimated absolute error, $\hat{\epsilon}$, will converge in distribution to the non-random expectation ϵ as $N \to \infty$. These facts can be used to create a confidence interval estimate of the true absolute error, ϵ.

This requires getting an estimate of the standard deviation, σ_ϵ of the estimated absolute error, $\hat{\epsilon}$. This is accomplished by performing M batches of N simulations each. Now, $X_{T,k,j}$ will be the kth exact solution in the jth batch and $Y(T, k, j)$ will be the kth discrete approximate solution in the jth batch. Compute the absolute error for the jth batch as follows,

$$\hat{\epsilon}_j = \frac{1}{N} \sum_{k=1}^{N} |X_{T,k,j} - Y(T, k, j)|. \tag{6.50}$$

Then compute the overall absolute error across batches,

$$\hat{\epsilon} = \frac{1}{M} \sum_{j=1}^{M} \hat{\epsilon}_j = \frac{1}{MN} \sum_{j=1}^{M} \sum_{k=1}^{N} |X_{T,k,j} - Y(T, k, j)|. \tag{6.51}$$

Once the batch mean and grand mean across batches is computed, the estimated variance of the absolute error estimates can be computed as follows,

$$\hat{\sigma}_\epsilon^2 = \frac{1}{M-1} \sum_{j=1}^{M} (\hat{\epsilon}_j - \hat{\epsilon})^2. \tag{6.52}$$

This implies that,

$$t = \frac{\hat{\epsilon} - \epsilon}{\hat{\sigma}_\epsilon / \sqrt{M}} \tag{6.53}$$

will have a t-distribution with $M - 1$ degrees of freedom, since if we knew the variance of absolute error precisely, the ratio would be approximately normally distributed. We select a confidence level of $100(1 - \alpha)\%$ for getting a confidence interval for ϵ, then the interval will be obtained as $(\hat{\epsilon} - \Delta\epsilon, \hat{\epsilon} + \Delta\epsilon)$ where $\Delta\epsilon = t_{1-\alpha/2,M-1} \frac{\hat{\sigma}_\epsilon}{\sqrt{M}}$ and $P(-t_{1-\alpha/2,M-1} \le t \le t_{1-\alpha/2,M-1}) = 1 - \alpha$.

Convergence by Absolute Error Criterion

A discrete-time approximation Y with maximum time step size δ **converges strongly** to X at time T, if $\lim_{\delta \to 0} E[|X_T - Y(T)|] = 0$. In order to compare schemes for the quality of solutions they give, we develop the measure for their **order of strong convergence**. We say a discrete time approximation Y **converges strongly with order** $\gamma > 0$ at time T, if there exists a positive constant C, which does not depend on δ and a $\delta_0 > 0$, such that,

$$E[|X_T - Y(T)|] \le C\delta^\gamma, \tag{6.54}$$

for each $\delta \in (0, \delta_0)$. This is a direct extension of the definition in the deterministic case of order of convergence, and reduces to it when the diffusion term is zero. The Euler scheme studied earlier was the simplest useful scheme, but in general is not particularly efficient. The Euler scheme has an order of strong convergence, $\gamma = 0.5$.

Let's look at a new scheme that has a higher order of strong convergence, the **Milstein scheme**, which when applied to the model in Eqn. (6.26) gives,

$$Y_{n+1} = Y_n + a(Y_n)\Delta_n + b(Y_n)\Delta W_n + \frac{1}{2}b(Y_n)b'(Y_n)(\Delta W_n^2 - \Delta_n). \tag{6.55}$$

This is obtained by a Taylor expansion of $X(t)$ about $X(t_0)$, and truncating after the first three terms. The strong order of convergence for Milstein scheme is 1.0.

6.5.2.3 Error Analysis - Mean Error Criterion

At other times we may be interested not in the paths of an Ito process, but instead might want to capture some distributional information about them, such as specific moments, probability of events, etc. Under such circumstances, the requirements are less stringent than for the path-wise case. Here we are seeking a good approximation in moments of the solution, and will need to define order of weak convergence. Let's begin with estimating the error in approximate solution of a stochastic differential equation when the criterion is being able to match the mean of the Ito process. Therefore, we are interested in computing the $E[X_T]$ using the discrete-time approximation, $Y(T)$, and its mean $E[Y(T)]$.

Mean Error Criterion: We define the error by,

$$\mu = E[Y(T)] - E[X_T]. \tag{6.56}$$

Note that the mean error can take both negative and positive values. The estimated values of **mean error** will be obtained by running the simulation N times and calculating the mean of these outcomes of the $Y(T)$, as follows.

$$\hat{\mu} = \frac{1}{N} \sum_{k=1}^{N} Y(T, k) - E[X_T]. \tag{6.57}$$

As in the case of estimating absolute error, if we want to construct a confidence interval for the mean error estimate, we will generate M batches with N simulations each and estimate the above mean error for each batch.

$$\hat{\mu}_j = \frac{1}{N} \sum_{k=1}^{N} Y(T, k, j) - E[X_T]. \tag{6.58}$$

Using this set of estimates, we compute the overall average across batches,

$$\hat{\mu} = \frac{1}{M} \sum_{j=1}^{M} \hat{\mu}_j = \frac{1}{MN} \sum_{j=1}^{M} \sum_{k=1}^{N} Y(T, k, j) - E[X_T]. \tag{6.59}$$

Combining the batch mean and overall mean, we compute the standard deviation of mean error estimate as follows:

$$\hat{\sigma}_\mu^2 = \frac{1}{M-1} \sum_{j=1}^{M} (\hat{\mu}_j - \hat{\mu})^2. \tag{6.60}$$

This implies that,

$$t = \frac{\hat{\mu} - \mu}{\hat{\sigma}_\mu / \sqrt{M}} \tag{6.61}$$

follows t-distribution with $M - 1$ degrees of freedom. As before, if we knew the theoretical variance of absolute error, the ratio in Eqn. (6.61) would be approximately normally distributed. Now if we select a confidence level of

$100(1 - \alpha)\%$ for constructing a confidence interval for μ, then the confidence interval will be obtained as,

$$(\hat{\mu} - \Delta\mu, \hat{\mu} + \Delta\mu), \tag{6.62}$$

where $\Delta\mu = t_{1-\alpha/2, M-1} \frac{\hat{\sigma}_\mu}{\sqrt{M}}$ and $P(-t_{1-\alpha/2, M-1} \leq t \leq t_{1-\alpha/2, M-1}) = 1 - \alpha$.

Convergence by Mean Error Criterion

In general, there could be a general functional to be estimated for the Ito process, say, $E[g(X_T)]$. Then we would do a similar computation using the simulated value $Y(T)$ as we would do for the mean, $E[X_T]$. We will say that a general discrete-time approximation $Y(T)$ with maximum step-size, δ, **converges weakly** to X at time T as $\delta \to 0$ with respect to a class $\mathcal{C}$ of test functions $g : R \to R$, if we have

$$\lim_{\delta \to 0} |E[g(X_T)] - E[g(Y(T))]| = 0, \tag{6.63}$$

for all $g \in \mathcal{C}$. So, if $\mathcal{C}$ contains all polynomials, then all moments of $Y(T)$ converge to the moment of the true process.

We say a discrete time approximation Y **converges weakly with order** $\beta > 0$ to X at time T as $\delta \to 0$, if for each polynomial $g(x)$ there exists a positive constant C, which does not depend on δ, and a finite $\delta_0 > 0$, such that,

$$\mu(\delta) = |E[g(X_T)] - E[g(Y(T))]| \leq C\delta^\beta, \tag{6.64}$$

for each $\delta \in (0, \delta_0)$. The Euler scheme has an order of weak convergence of 1.0. We will next look at another scheme with a higher order of weak convergence.

The important point to note is that weak and strong convergence criteria lead to development of different discrete time approximations, which are only efficient with respect to one of the two criteria. This fact makes it **important to clarify the aim of a simulation before choosing the approximation scheme**. The question to ask is whether a good path-wise approximation of the Ito process is required or if the approximation of some functional of the Ito process is the real objective.

Weak Higher Order Methods

The Euler scheme has a weak order of convergence of 1.0. We consider a discretization scheme with a higher order truncation of Ito-Taylor expansion. The resulting scheme is as follows,

$$Y_{n+1} = Y_n + a(Y_n)\Delta_n + b(Y_n)\Delta W_n + \frac{1}{2}b(Y_n)b'(Y_n)(\Delta W_n^2 - \Delta_n)$$

$$+ a'(Y_n)b(Y_n)\Delta Z_n + (\frac{1}{2}a(Y_n)a'(Y_n) + \frac{1}{2}a''(Y_n)b^2(Y_n))\Delta_n^2$$

$$+ (a(Y_n)b'(Y_n) + \frac{1}{2}b''(Y_n)b^2(Y_n))(\Delta W_n\Delta_n - \Delta Z_n), \tag{6.65}$$

where ΔZ_n is normally distributed with mean $E[\Delta Z_n] = 0$, $Var(\Delta Z_n) = \frac{1}{3}\Delta_n^3$ and $cov(\Delta Z_n, \Delta W_n) = \frac{1}{2}\Delta_n^2$. This method has an order of weak convergence of 2.0.

6.5.3 Higher Order Methods

When solutions with better quality of approximation are required, new methods for discretization must be developed. In Section 6.5.2, we developed two different error criteria by which solutions may be assessed depending on the intended use of the approximate solution. We also presented a higher order scheme for an improvement over the Euler scheme by both the error criteria. Developing higher order methods in general is not always strict science (or math). Some of it is just more akin to art. The reader may have heard of the concept of 'art of simulation'! Here we will look at some types of higher order methods.

6.5.3.1 Trapezoidal Method

In the case of deterministic differential equations, the trapezoidal method is an improvement over the Euler scheme. It belongs to the class of **implicit** schemes, since iteration at time t requires knowing the next value, y_{n+1}. Therefore, in an actual implementation, each iteration step requires solving a non-linear equation. A trapezoidal scheme applied to the problem in Eqn. (6.39) is given as,

$$y_{n+1} = y_n + \frac{1}{2}(a(t_n, y_n) + a(t_{n+1}, y_{n+1}))\Delta \quad \text{for } n = 1 \ldots N, \qquad (6.66)$$

where Δ is the equidistant time step-size. The (deterministic) order of convergence for the trapezoidal scheme can be shown to be 2, for small enough step-size. The trade-off for the higher accuracy, however, is the higher level of computations required, due to the need to solve a non-linear equation in each iteration. Also, in general this non-linear system may not be solved algebraically (analytically).

To avoid this computational burden of solving a non-linear equation, we consider a modification. We will first use the Euler scheme to approximate the y_{n+1} on the right-hand side, then use this approximate $\bar{y}_{n+1}$, to obtain the next iterate value, y_{n+1}. This is the **modified trapezoidal method** given by,

$$\bar{y}_{n+1} = y_n + a(t_n, y_n)\Delta, \qquad (6.67)$$

$$y_{n+1} = y_n + \frac{1}{2}(a(t_n, y_n) + a(t_{n+1}, \bar{y}_{n+1}))\Delta \quad \text{for } n = 1 \ldots N. \quad (6.68)$$

This method is also known as improved Euler or Heun method. It is a simple example of a **predictor-corrector method**, the first equation being the predictor, and the second one the corrector. The order of convergence for modified trapezoidal is still 2.

A further higher order accuracy can be obtained by using information from previous discretization intervals. An example of this is the 3-step **Adams-Bashford method**, applied to the problem in Eqn. (6.39), given by,

$$y_{n+1} = y_n + \frac{1}{12}(23a(t_n, y_n) - 16a(t_{n-1}, y_{n-1}) + 5a(t_{n-2}, y_{n-2}))\Delta$$

$$\text{for } n = 1 \ldots N. \quad (6.69)$$

This method has a 3rd order global discretization error. A one-step method may be used to generate values to get the multi-step procedure started.

Other higher order schemes can be obtained by truncating Taylor expansions of $x(t)$. These are sometimes not very practical because they involve higher order derivatives of $a(t, x)$, which may be complicated and hard to compute. The Taylor expansion applied to $x(t)$ of problem in Eqn. (6.39) would look like,

$$x(t_{n+1}) = x(t_n) + a(t_n, x(t_n))\Delta + \frac{1}{2!}(\frac{\partial a}{\partial t} + a\frac{\partial a}{\partial x})\Delta^2 + O(\Delta^3). \quad (6.70)$$

Truncating this before Δ^3 term gives a 2nd order truncated Taylor expansion method, with an order of convergence of 2. Truncating after the third term, after it was explicitly written, would give the 3rd order Taylor expansion method with an order of convergence of 3, and so on. However, note again that the later terms will have multiple derivatives of the $a(t, x)$ function, which may not be easy to compute in all cases.

Runge-Kutta methods avoid the use of derivatives to provide higher order accuracy. The classical 4th order Runge-Kutta method is an important explicit method. This method applied to the problem in Eqn. (6.39) is given as,

$$y_{n+1} = y_n + \frac{1}{6}\{k_n^{(1)} + 2k_n^{(2)} + 2k_n^{(3)} + k_n^{(4)}\}\Delta, \quad (6.71)$$

$$\text{where } k_n^{(1)} = a(t_n, y_n), \quad (6.72)$$

$$k_n^{(2)} = a(t_n + \frac{1}{2}\Delta, y_n + \frac{1}{2}k_n^{(1)}\Delta), \quad (6.73)$$

$$k_n^{(3)} = a(t_n + \frac{1}{2}\Delta, y_n + \frac{1}{2}k_n^{(2)}\Delta), \quad (6.74)$$

$$k_n^{(4)} = a(t_{n+1}, y_n + k_n^{(3)}\Delta). \quad (6.75)$$

Another important issue about using simulation methods for solving deterministic systems is **instability**. Instability occurs when the errors end up oscillating with increasing amplitude. Convergence of a simulation approach is guaranteed when the method may be demonstrated to be stable and consistent, approximating the true equation to a degree. Therefore, convergence is result of a combination of consistency and stability of an approximation scheme.

In this section, we have given an overview of a variety of improved methods for deterministic differential equations, along with principles on which these improved methods get constructed. For additional improved schemes for stochastic models, the reader is advised to refer to Kloeden and Platen [48].

6.6   Estimating Parameters

In this chapter, we have constructed a framework for developing new models for dynamic evolution of risks utilizing the standard Wiener process. Similar constructions can be done using other processes, those discussed in Chapter 5 and beyond. These will likely involve developing new definition of integrals, as we did with the Ito integral in this chapter, and the related calculus. We will explore utilizing the Poisson process for this purpose in later chapters of the book; however, developing new integrals beyond the Ito integral is beyond the scope of this book.

Once a model is constructed, described by a stochastic differential equation, it can be solved analytically or numerically. The reader will have noticed that each model gets stated in terms of some crucial parameters. Unless some meaningful values of these parameters can be identified, the models cannot be used. In this section, we present some methods for calibrating the models developed in this chapter and in Chapter 5.

6.6.1   Geometric Brownian Motion

The simplest stochastic model we obtained in earlier discussions of this chapter was,

$$dS_t = \mu S_t dt + \sigma S_t dW_t, \tag{6.76}$$

which was used to describe price dynamics, S_t, for a risky asset, along the initial value of the asset know to be, S_0. Analytical solution was obtained for this model as, $S_t = S_0 \exp((\mu - \sigma^2/2)t + \sigma W_t)$. However, the two parameters in the model, namely μ and σ, remain to be estimated in order to use the model.

Estimation of these parameters requires data, say M historical observations of the risky asset price, $\{S_{t_1}, S_{t_2}, \ldots, S_{t_M}\}$. It is noted that for these observations, given they are all realized by the model in Eqn. (6.76), we have

$$S_{t_n} = S_{t_{n-1}} \exp((\mu - \frac{\sigma^2}{2})(t_n - t_{n-1}) + \sigma(W_{t_n} - W_{t_{n-1}})), \tag{6.77}$$

for all values of $n \in \{1, 2, \ldots, M\}$. Rearranging the terms in Eqn. (6.77) gives,

$$\ln(\frac{S_{t_n}}{S_{t_{n-1}}}) = (\mu - \frac{\sigma^2}{2})(t_n - t_{n-1}) + \sigma(W_{t_n} - W_{t_{n-1}}), \tag{6.78}$$

where $\ln(\frac{S_{t_n}}{S_{t_{n-1}}}) = r_{t_n}$ is called the log-return of the asset. If t_n's are equi-spaced, $(\mu - \sigma^2/2)(t_n - t_{n-1}) + \sigma(W_{t_n} - W_{t_{n-1}})$ are independent, identically distributed (i.i.d.) by the normal distribution. The estimation of the parameters can be accomplished by first estimating the variance of the i.i.d. normal observations, followed by their mean. We obtain the following estimates,

$$\hat{\sigma}^2 = \frac{1}{(t_n - t_{n-1})} variance(\ln(\frac{S_{t_n}}{S_{t_{n-1}}})), \tag{6.79}$$

$$\hat{\mu} = \frac{1}{(t_n - t_{n-1})} mean(\ln(\frac{S_{t_n}}{S_{t_{n-1}}})) + \frac{\hat{\sigma}^2}{2}. \tag{6.80}$$

These estimates utilize the fact that the distribution of log-returns are known to be normal in this geometric Brownian motion model. Since the estimates are obtained utilizing the first and the second moments of the normal distribution, we can call this approach the method of moments. We discuss this method in more detail later. However, there is one other method these estimates can be obtained by; we discuss this important method next.

6.6.2 Method of Maximum Likelihood

We describe the maximum likelihood method for calibrating models in general terms, which utilizes the knowledge of distributional properties of the variable being modeled. Assume that the model describes the risky asset dynamics, S_t, and we can describe the probability distribution of a function of these risky asset price dynamics. For example, the log-returns described in the previous section, i.e., $r_t = \ln \frac{S_t}{S_{t-\Delta t}}$. Consider a sample of observations for log-return of the risky asset are available, $\{r_{t_0}, r_{t_1}, ..., r_{t_N}\}$, where $t_0 = 0$ and $t_N = T$ are two time end-points of the sample. Desirably the observations are equi-spaced, i.e., Δt apart, which will make the notation simpler.

If we know the distributional properties of r_{t_k}, even if they are conditional on precisely knowing the value of $r_{t_{k-1}}$, we can make this useful in this parameter estimation method. Let's say the conditional distribution for r_{t_k}, given $r_{t_{k-1}}$, is denoted by, $f(r_k|r_{k-1};\theta)$, for $k = 1 \dots N$ and with the set of parameters to be estimated in the model being, $\theta = [a; b; c; d]$. For instance, if the conditional distribution of r_{t_k}, given $r_{t_{k-1}}$, is normally distributed, we would have,

$$f(r_k|r_{k-1};\theta) = \frac{1}{\sqrt{2\pi}\sigma(r_{k-1}, \theta, \Delta t)} e^{-\frac{[r_k - \mu(r_{k-1}, \theta, \Delta t)]^2}{2\sigma^2(r_{k-1}, \theta, \Delta t)}}, \tag{6.81}$$

for $k = 1 \dots N$, and $\mu(r_{k-1}, \theta, \Delta t)$ and $\sigma(r_{k-1}, \theta, \Delta t)$, the mean and standard deviation expressed in terms of the parameters, $\theta = [a; b; c; d]$, and a given value of r_{k-1}. In particular, in the geometric Brownian motion case, we can describe $\sigma(r_{k-1}, \theta, \Delta t) = c\sqrt{\Delta t}$, and $\mu(r_{k-1}, \theta, \Delta t) = a\Delta t$.

We construct the likelihood function utilizing the conditional distribution of all the N observations as follows,

$$L(\theta) = \prod_{k=1}^{N} f(r_k | r_{k-1}; \theta) f(r_0).$$ (6.82)

The maximum likelihood estimate of parameters attempts to find those values of the parameters, θ, for which the likelihood that the observations resulted from the purported conditional distribution is the highest (maximized). Noting that the maximizers of a function also maximize the logarithm of that function, we define the log-likelihood function. Taking the logarithm of the likelihood function gives us the log-likelihood function as,

$$\ln L(\theta) = \sum_{k=1}^{T} \ln f(r_k | r_{k-1}; \theta) + \ln f(r_0).$$ (6.83)

The advantage of a log-likelihood function over the likelihood function is that the former has less cumbersome summations of conditional density, while the latter has a product of conditional density. In particular, for the normal distribution case, we have

$$\ln L(\theta) = \sum_{k=1}^{T} \frac{-1}{2} \ln[2\pi\sigma(r_{k-1}, \theta, \Delta t)]$$
$$- \frac{[r_k - \mu(r_{k-1}, \theta, \Delta t)]^2}{2\sigma(r_{k-1}, \theta, \Delta t)} + \ln f(r_0).$$ (6.84)

In order to maximize the log-likelihood function, which is the same as maximizing the likelihood function, we take the first derivative of the log-likelihood function with respect to all the four parameters, $\theta = [a; b; c; d]$, and equate each of them to zero to obtain,

$$\frac{\partial \ln L(\theta)}{\partial a} = 0,$$ (6.85)

$$\frac{\partial \ln L(\theta)}{\partial b} = 0,$$ (6.86)

$$\frac{\partial \ln L(\theta)}{\partial c} = 0, \quad \text{and finally,}$$ (6.87)

$$\frac{\partial \ln L(\theta)}{\partial d} = 0.$$ (6.88)

For the specific geometric Brownian motion case, these equations become,

$$\frac{\partial \ln L(\theta)}{\partial a} = \sum_{k=1}^{N} \frac{2\Delta t[r_k - a\Delta t]}{2c^2 \Delta t} = 0,$$ (6.89)

and

$$\frac{\partial \ln L(\theta)}{\partial c} = \sum_{k=1}^{N} \frac{-1}{c} - \frac{[r_k - a\Delta t]^2}{c^3 \sqrt{\Delta t}} = 0.$$ (6.90)

Solving Eqns. (6.89) and (6.90) yields, $\hat{a}\Delta t = \frac{1}{N}\sum_{k=1}^{N} r_k$, and $\hat{c}^2\Delta t = \frac{1}{N}\sum_{k=1}^{N}(r_k - a\Delta t)^2$. These estimates are not too different from those obtained in Eqns. (6.79) and (6.80) utilizing the method of moments, and are in fact identical if the variance is defined as a biased sample variance, $\frac{\sum_{i=1}^{N}(x_i - \bar{x})^2}{N}$. Therefore, in the case of normal distribution, the estimates obtained from method of moments and maximum likelihood method are the same.

6.6.3 Method of Quasi-Maximum Likelihood

The method of maximum likelihood is applicable only when the exact density of the risk factor is known. When this is the case, the method produces the most efficient way to determine the parameters that drive the risk. However, in many cases, the exact density for the risk factor, as suggested by the model, may not be determined or stated in the closed-form. In such cases, one can benefit from making an approximation to the density, by picking a density which is tractable, and yet in some way not too far improved from the true density of the risk factor.

This approximated method of maximum likelihood is generally called the quasi-maximum likelihood method or the pseudo-maximum likelihood method. Once the true density is approximated by an approximate density, the actual procedure for determining the quasi-maximum likelihood estimates follows the steps of the maximum likelihood method. Suppose the true density, $f(r_k|r_{k-1}; \theta)$ is approximated by the density,

$$g(r_k|r_{k-1}; \theta), \tag{6.91}$$

with $k = 1 \ldots T$ and $\theta = [a; b; c; d]$. Using the approximate density, the quasi-maximum likelihood function is constructed in the same manner as for the likelihood function, as follows,

$$L^Q(\theta) = \prod_{k=1}^{T} g(r_k|r_{k-1}; \theta)g(r_0). \tag{6.92}$$

Taking the logarithm to simplify the product into a summation yields,

$$\ln L^Q(\theta) = \sum_{k=1}^{T} \ln g(r_k|r_{k-1}; \theta) + \ln g(r_0). \tag{6.93}$$

We then maximize the log-quasi likelihood function to determine the parameters that best describe the data coming from this approximate Likelihood function. The reader may be rightly unconvinced about this method generating good estimates. The benefit of the method lies in making a good approximation of the density, and also in conducting an analysis of the consistency of the estimated parameters. Detailed analysis of consistency of the estimated parameters is beyond the scope of this book. The reader is referred to books on econometric analysis [33] and financial econometric analysis [13].

6.6.4 Method of Moments

In some cases, it may be difficult to extract the exact, or even approximate, density for the process, S_t. Therefore, applying maximum likelihood or quasi-maximum likelihood methods will not be an option for estimating the model parameters. One may still be able to determine some key moments of the process, S_t, in terms of the model parameters, $\theta = [a; b; c; d]$, which may lead to constructing a system of equations. The solution of the system of equations would yield parameter estimates of the model, $\hat{\theta} = [\hat{a}; \hat{b}; \hat{c}; \hat{d}]$.

Consider the first four moments for the process, S_t, are known functions of the parameters, $\theta = [a; b; c; d]$. We will construct the following four equations:

$$E[S_t] = f_1(\theta), \tag{6.94}$$

$$E[S_t^2] = f_2(\theta), \tag{6.95}$$

$$E[S_t^3] = f_3(\theta), \quad \text{and finally,} \tag{6.96}$$

$$E[S_t^4] = f_4(\theta), \tag{6.97}$$

and solve them simultaneously to obtain the parameter estimates. Sometimes the moments known are not pure moments, but functions of the process S_t or ΔS_t, etc. The method of moments can accommodate these variations, as long as the right-hand side of the system of equations can be written in terms of just the parameters, and perhaps uniform time increment, Δt. This will allow solving the system of equations for the model parameters. An example of application of method of moments is considered next.

6.6.4.1 Ornstein-Uhlenbeck Process

In Section 5.3.5, we introduced a specific Gaussian process, some of whose properties were presented in that section. We had stated that the O-U process, besides being an example of a Gaussian process, is also a Markov process and is strictly stationary. In fact, the O-U process is a diffusion process and can be shown to satisfy the following stochastic differential equation,

$$dX_t = -\gamma X_t dt + \sigma dW_t. \tag{6.98}$$

This stochastic differential equation can be solved analytically by guessing an integrating factor, $e^{-\gamma t}$. Considering $d(e^{-\gamma t} X_t)$, the solution can be obtained as, $X_t = X_0 e^{-\gamma t} + e^{-\gamma t} \int_0^t \sigma e^{\gamma s} dW_s$. For the solution of the SDE, thus obtained, it can be shown that the mean of the process is given by,

$$E[X_t] = X_0 e^{-\gamma t}, \tag{6.99}$$

which comes straight from the properties of the Ito integral (Property 3) discussed in Section 6.3.2. While the variance is given by,

$$Var(X_t) = \frac{\sigma^2}{2\gamma} e^{-2\gamma t}(e^{2\gamma t} - 1), \tag{6.100}$$

obtained from the Ito isometry property of the Ito integral (Property 5) discussed in Section 6.3.2. These two moments in Eqns. (6.99) and (6.100) can be used to estimate the parameters, γ and σ. Once an estimate of γ is obtained from Eqn. (6.99), it can be applied in Eqn. (6.100) to obtain an estimate for σ.

A variant of the simple O-U process is the mean-reverting O-U process, given to satisfy the following stochastic differential equation,

$$dX_t = (\mu - X_t)dt + \sigma dW_t. \tag{6.101}$$

A more general version of a mean-reverting O-U process is given by the stochastic differential equation,

$$dX_t = \gamma(\mu - X_t)dt + \sigma dW_t. \tag{6.102}$$

For both models in Eqns. (6.101) and (6.102), we can find the explicit solution. It is more convenient, however, to work with the moments of the process. It can be shown that the mean, variance and covariance of the mean-reverting O-U process are as follows,

$$E[X_t] = X_0 e^{\gamma t} + \mu(1 - e^{-\gamma t}), \tag{6.103}$$

$$Var(X_t) = \frac{\sigma^2}{2\gamma} e^{-2\gamma t}(e^{2\gamma t} - 1), \tag{6.104}$$

$$Cov(X_s, X_t) = \frac{\sigma^2}{2\gamma}(e^{-\gamma(t-s)} - e^{-\gamma(t+s)}), \tag{6.105}$$

for $s < t$. These moments can be utilized by the method of moments to generate estimates of the parameters, μ, γ, and σ.

6.7 MATLAB Tools for Building and Solving Models of Risk

MATLAB mathematical software has a vast array of functions for working with differential equations, where the emphasis is on deterministic differential equations. The reader will benefit from browsing these functions and their descriptions to improve their general grasp of differential equations. We list a few of these functions here. As always, the reader is advised to look up the extensive help documentation available with MATLAB to see the details of these and other related functions. At the bottom of each function description in MATLAB help documentation, look for 'See Also' to explore other related functions. Resources such as MATLAB Primer [20] are also useful.

Deterministic differential equations: ode23, ode45, ode113, ode15s, ode23s, ode23t, ode23tb, pdepe

Calibrating models: `polyfit, roots, fzero, fsolve, normlike, explike, wbllike, lognlike`

6.8   Summary

In this chapter, we developed the mathematical framework to define new dynamic models of risk. These dynamic models of risk were constructed based on a new integral, the Ito integral, and calculus developed for this new integral. Ito processes described in terms of the Ito integral offer versatility of creating various continuous-time dynamic models of risk with desired properties. After constructing and defining the Ito integral and Ito processes, we discussed analytical solution of Ito stochastic differential equations. In absence of analytical solutions, numerical solutions can be sought for the stochastic differential equations by means of simulation. We developed techniques for solving stochastic differential equations using simulation, and evaluated the performance of the simulated solutions. To complete the general discussion of developing models for risk management, we concluded this chapter with procedures for estimating model parameters. With this chapter we end our general discussion of models to support risk management framework. For the remaining chapters in the book, we will look at risk management problems of specific types of risk.

6.9   Questions and Exercises

Review Questions

1. How are ordinary differential equations used for modeling dynamic changes in a system? How are these models modified to incorporate randomness arising in system behavior?

2. How is the Ito integral with respect to the Wiener process defined? How does it differ from the Stratonovich integral?

3. In modeling using integrals with respect to the Wiener process, when is it advised to use the Ito definition versus the Stratonovich definition?

4. What are the most important properties of the Ito integral? Discuss each property for its implication.

5. What is the chain rule of Ito calculus? Why does the unexpected term, $\frac{1}{2}\sigma^2(t, X_t)\frac{\partial^2 f}{\partial x^2}$, appear in the chain rule?

6. Discuss the properties of the following process for different values of parameters, μ, σ.

$$S_t = S_0 \exp((\mu - \frac{1}{2}\sigma^2)t + \sigma W_t). \tag{6.106}$$

7. How is the Euler scheme applied to a deterministic ordinary differential equation and to a stochastic differential equation?

8. What is local discretization error, and how does it compare with global discretization error?

9. What is the definition of order of convergence of a numerical method for solving an ordinary differential equation?

10. What is absolute error criterion for solution of a stochastic differential equation? When should one use this criterion to judge the accuracy of a method?

11. How does absolute error criterion compare with the mean error criterion? When should one use the mean error criterion to judge the accuracy of a method?

12. Discuss some of the higher order methods for solving stochastic differential equations, and when is it appropriate to use them.

13. Discuss some of the higher order methods for solving ordinary differential equations. What are the orders of convergence of these methods?

14. What is the method of maximum likelihood for estimating parameters of a model? What is the principle behind the method?

15. What is the quasi-maximum likelihood method for estimating parameters of a model? When does this method work well?

16. Discuss how the method of moments is applied to estimate parameters of a model.

Exercises

1. Show that the Ito integral of the square of the Wiener process is as follows, $\int_0^t W_s^2 dW_s = \frac{1}{3}W_t^3 - \int_0^t W_s ds$.

2. Solve the following model of a stock price evolution in MATLAB using the Euler scheme in the interval, $[0, 1]$. Apply a higher order method to solve the problem in order to obtain a better path-wise accurate solution.

$$dX_t = \gamma(\mu - X_t)dt + \sigma\sqrt{X_t}dW_t, \tag{6.107}$$

where $\mu = 30$, $\sigma = 1.5$ and $\gamma = 6.0$; the initial value of the stock is $X_0 = \$32$.

3. Identify three time points in the interval, $[0, 1]$, at which you must evaluate the accuracy of the solution obtained in Problem 2. For different increasing refinement of time-discretization, Δt, compute point estimates of the absolute error and confidence interval of absolute error. Comment on how the absolute error changes with:

   (a) the time points selected in the interval, $[0, 1]$,

   (b) improved refinement of the time-discretization, and

   (c) the two different solution methods utilized.

4. Solve the following model for return evolution of a risky asset in MATLAB using the Euler scheme in the interval, $[0, 1]$. Apply a higher order method to solve the problem in order to obtain a better approximation of some functional, such as $E[\frac{1}{R_t}]$, of the Ito process.

$$dR_t = \mu e^{-R_t} dt + \sigma R_t^2 dW_t, \tag{6.108}$$

   where $\mu = 0.2$, $\sigma = 0.1$; The initial value of return is $R_0 = 1.042$.

5. Identify three time points in the interval, $[0, 1]$, at which you must evaluate the accuracy of the solution in Problem 4. For different increasing refinement of time-discretization, Δt, compute point estimates of the mean error and confidence interval of mean error. Comment on how the mean error changes with:

   (a) the time points selected in the interval, $[0, 1]$,

   (b) improved refinement of the time-discretization, and

   (c) the two different solution methods utilized.

6. Download the daily closing price of your favorite stock for a duration of a year. In MATLAB, write a function to calibrate the following model using these data.

$$S_t = S_0 \exp((\mu - \frac{1}{2}\sigma^2)t + \sigma W_t). \tag{6.109}$$

   The model may or may not be a good representation for your favorite stock, however our goal here is to practice estimation of parameters.

7. Utilizing the data from the previous question, calibrate the following model for evolution of the stock price.

$$dX_t = \gamma(\mu - X_t)dt + \sigma X_t dW_t. \tag{6.110}$$

Part III

Risk Management

Chapter 7

Managing Equity Market Risk

In the chapters thus far in the book, we have discussed the general notion of risk and developed frameworks for the management of risks. We extensively developed rigorous constructs for the modeling of risks of different types, and developed methods for solving and analyzing these models. In this segment of the book, we turn our attention to specific types of risks. We will develop models for specific types of risks, with the objective of serving specific goals of risk management for these risk types. Along the way, we will take advantage of the contexts to further advance the development of risk management frameworks for these risk types. We begin in this chapter with equity risk.

Stock of an enterprise as a way to finance the ventures of the enterprise was an exciting financial innovation, and has been around for centuries. The British and the Dutch East India Companies successfully embraced this financing model to fund their risky voyages in the sixteenth and seventeenth centuries. The modern corporation is where the financing model got more comprehensively tested and has enjoyed much success. Shares of the stock of a firm offer the possibility to general investors of bite-sized ownership of the firm. Well-developed stock markets in the developed economies, and rapidly developing ones in emerging markets, allow investors to acquire or relieve themselves of the ownership stakes in these corporations.

Shares of a firm offer the exciting ownership proposition, but are designed to have a residual claim on the firm's assets. Therefore, the risks underlying the firm's projects and ventures are primarily borne by the equity holders. One can only imagine the extremely wide range of risks a diverse range of firms and their projects expose the investors to. Nevertheless, the enthusiasm for investing in stocks of firms ranks high among asset classes, both by individual and institutional investors. The justification for this high enthusiasm lies in the potential of high returns for the shareholders, should the firms' projects turn out to be successful. From the comfort of the trading desk (or online brokerage account) investors can partake in ownership stake in firms, provided they have comprehended the risk-return prospect offered by the firm and its projects.

Analysis of equity valuation of a firm can be pursued on two threads, and combinations thereof. The first thread comprises a fundamental approach, hence '*fundamental analysis*,' where an investor evaluates the firm's financial statements, current projects, and prospects of future project to assess the equity valuation of the firm. The second thread of analysis takes a stand by a version of efficient markets hypothesis, in the claim that markets are promptly incorporating available information for the firm into the firm's stock price. *Technical analysis* focuses on the trajectory of evolving stock prices of the firm to project what the future trends, levels and risk of the stock price may be.

In this chapter, we begin analysis of equity risk from the perspective of the mitigate objective of risk management. In the process, we will develop a rigorous framework for risk mitigation. The rest of the chapter will be devoted to developing methodologies for valuation and utilization of instruments designed for transferring equity risk.

7.1   Mitigating Equity Risk

Investors expose themselves to equity risk in an attempt to take advantage of the potential return they may receive from this investment. The precise approach for how the investor may choose to determine the investment strategy would depend on a variety of factors. The approach could be completely subjective, based on the intuition the investor has developed for the markets, sectors, industries and specific firms. Alternatively, the approach could be objectively driven, strictly by model-driven quantitative analysis set to respond to stated goals of the investment strategy. More practically, the approach tends to be a mix of model-driven quantitative analysis, complemented by expert judgement and intuition of how the investment strategy may fare.

Unless the investment strategy is constructed entirely intuitively, the investor must explicitly state the goals of the investment strategy in order to construct it. As discussed in Chapter 2, the risk preference versus aversion of the investor guides the construction of risk measures. Risk measures may be gainfully used to construct and state the investment objectives.

The other dimension by which an investment strategy is defined, which may in fact require renaming the activity as a trading strategy, is the time dimension. This is the frequency with which actions are expected to be taken to respond to the risk-return profile of the investment assets to achieve the investor's desired objectives. The frequency can be low enough to be stated in months or years, to high enough so as to be restricted only by the speed of light traveling on communication networks for implementing trades. At the lowest frequency, the investment strategy may be constructed in a static analysis; however at the high frequencies, a dynamic analysis is inevitable. We begin evaluating the process in the static setting.

7.1.1   Portfolio Diversification

Let W_0 be the total resources or funds available for investment for a period of time, $[0, T]$. The investor must allocate the wealth among N stocks whose

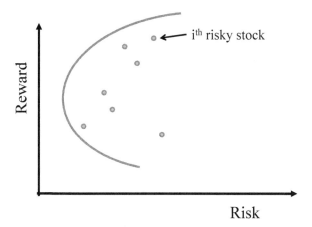

FIGURE 7.1: Plot of risk-reward trade-off of individual stocks. The combination of the individual stock helps mitigate the risk in the frontier.

T-period return is stochastic, depicting the risk underlying the investment. A model-driven quantitative approach will require describing the T-period return of the N stocks by an appropriately chosen model. Lets say the T-period return of the ith stock is r_i. If the investor chooses to invest w_i fraction of the initial wealth, W_0, in the ith stock, the overall T-period return from the investment, r_π, can be written in terms of w_i and r_i as follows.

$$r_\pi = \sum_{i=1}^{N} w_i r_i, \tag{7.1}$$

where the wealth at time, T, is given as, $W_T = (1 + r_\pi)W_0$.

The goal of mitigation or diversification of risk is to expose oneself to many more risks, so that the joint impact of the risks is an improvement over each individual risk. In order to achieve this goal, a precise measure of risk, as well as reward, must be identified. In Figure 7.1, individual stocks are placed in the graph for the risk-reward pairing they offer, along with the risk-reward curve achievable by combining the stocks in order to mitigate the individual risks.

7.1.1.1 Classical Mean-Variance Reward-Risk Measures

The classical portfolio theory, credited to Markowitz, Sharpe, Lintner, and Mossin, chooses the mean of a portfolio return, $E[r_\pi]$, for the reward measure of risk, and variance of the return, $Var(r_\pi)$, as the risk measure. With this selection of risk and reward measures, if the mean, variance, and covariance information is known for the individual stock returns and all pairs of stock returns, respectively, the portfolio measures can be conveniently created. These are given by applying Eqn. (2.12) in the more general form, as follows,

$$E[r_\pi] = \mu_\pi = \sum_{i=1}^{N} w_i E[r_i], \tag{7.2}$$

$$Var(r_\pi) = \sigma_\pi^2 = \mathbf{w}^T Cov \ \mathbf{w}, \tag{7.3}$$

where $\mathbf{w} = [w_1; w_2; \ldots, w_N]$ and Cov is the covariance matrix of stock returns. Each entry of the covariance matrix must be constructed as, $Cov(i,j) = Cov(r_i, r_j)$.

In the mean and variance choice of reward-risk measures, it is possible to easily demonstrate the benefit of risk mitigation. In Chapter 2, this was displayed graphically (Figure 2.5). We now illustrate it mathematically, where it suffices to show that going from one stock to two stocks helps reduce risk without necessarily lowering the reward, or can improve reward without taking on greater risk. The more general N stock case can be similarly illustrated by considering portfolios of $N-1$ stocks and N^{th} stock.

Consider two stocks with T-period return of r_1 and r_2, respectively. The investment weight for the first stock is w, while that of the second one is $(1-w)$. The correlation between the two stock returns is given by ρ. For any $w \in [0,1]$, by applying Eqns. (7.2) and (7.3) to this case, we have the portfolio mean return and variance of return given by,

$$E[r_\pi] = \mu_\pi = wE[r_1] + (1-w)E[r_2], \tag{7.4}$$

$$\sigma_\pi^2 = w^2\sigma_1^2 + (1-w)^2\sigma_2^2 + 2\rho w(1-w)\sigma_1\sigma_2, \tag{7.5}$$

where $\sigma_1^2 = Var(r_1)$, $\sigma_2^2 = Var(r_2)$. Without loss of generality, assume $E[r_1] \leq E[r_2]$ and $\sigma_1^2 \leq \sigma_2^2$. The other case of $E[r_1] \leq E[r_2]$ and $\sigma_1^2 \geq \sigma_2^2$ can be similarly illustrated, even though with such risk-reward profile, r_1 appears to be a rather undesirable stock.

For $0 < w < 1$, we have $E[r_1] < E[r_\pi] < E[r_2]$, since

$$E[r_\pi] = wE[r_1] + (1-w)E[r_2]$$
$$> wE[r_1] + (1-w)E[r_1] = E[r_1], \text{ and} \tag{7.6}$$
$$E[r_\pi] = wE[r_1] + (1-w)E[r_2]$$
$$< wE[r_2] + (1-w)E[r_2] = E[r_2]. \tag{7.7}$$

Therefore, the reward from the portfolio exceeds that of one of the two stocks. We now show that the risk of the portfolio is lower than the riskier stock.

$$\sigma_\pi^2 = w^2\sigma_1^2 + (1-w)^2\sigma_2^2 + 2\rho w(1-w)\sigma_1\sigma_2$$
$$\leq w^2\sigma_2^2 + (1-w)^2\sigma_2^2 + 2\rho w(1-w)\sigma_2\sigma_2$$
$$= 2(1-\rho)(w^2 - w)\sigma_2^2 + \sigma_2^2 < \sigma_2^2, \tag{7.8}$$

since $1 - \rho > 0$ and $w^2 - w < 0$, as $0 < w < 1$.

It is also worth noting how the portfolio reward and risk depend on the correlation between the two stocks. As seen in Eqn. (7.5), the portfolio reward, i.e., mean return, does not depend on the correlation between the stocks. Variance of portfolio return, however, depends on the correlation, and for the correlation range of $-1 \leq \rho \leq 1$, we can determine the lower and upper bound of portfolio risk. For any portfolio weight, w, if correlation between stocks is perfect, i.e., $\rho = 1$, we have,

$$
\begin{aligned}
\sigma_\pi^2 &= w^2\sigma_1^2 + (1-w)^2\sigma_2^2 + 2w(1-w)\sigma_1\sigma_2 \\
&= (w\sigma_1 + (1-w)\sigma_2)^2.
\end{aligned}
\tag{7.9}
$$

Therefore, the standard deviation of the portfolio return is a linear combination of the standard deviation of return of the individual stocks. For any investment weight, w, this is the upper bound on the portfolio risk. Similarly, if the two stocks have a perfect negative correlation, i.e., $\rho = -1$, we obtain the lower bound on portfolio risk for any investment weight.

$$
\begin{aligned}
\sigma_\pi^2 &= w^2\sigma_1^2 + (1-w)^2\sigma_2^2 - 2w(1-w)\sigma_1\sigma_2 \\
&= (w\sigma_1 - (1-w)\sigma_2)^2.
\end{aligned}
\tag{7.10}
$$

For a specific choice of investment weight, in presence of perfect negative correlation, it is possible to make the standard deviation (or variance) of portfolio return to be zero. This weight choice is obtained by equating Eqn. (7.10) to zero, as $w^* = \frac{\sigma_2}{\sigma_1 + \sigma_2}$. In reality, a perfect positive or negative correlation between stocks is hard to find. Therefore, these bounds remain theoretical bounds, that one can attempt to approximate, but not exactly achieve.

In Figure 7.2, we depict the feasible region in a mean-standard deviation plot of reward-risk as it depends on the investment weight and correlation between the two stocks. Perfect correlation, negative and positive, defines the borders of the region.

7.1.1.2 Dynamic Investment Strategy

From the static, single-period case discussed thus far, we move to the consideration of dynamic investment strategies. If the total resources available for investment initially are W_0, the investor must allocate the wealth among N stocks. Now instead of making one decision for investments for a T-period horizon, the intention is to make a sequence of decisions. At certain time points, $\mathcal{T}$, in the T-period of time, the investor must determine the investment weights for the N stocks. The time points can be discrete, $0 \leq t_0 < t_1 < \dots < t_M = T$, or a continuum, $[0, T]$. The dynamic investment decisions are, $\{w_i(t) | t \in \mathcal{T}\}$.

As in the static case, a model-driven quantitative approach will require describing the price or return evolution of the N stocks by an appropriately chosen model. Let's say the initial price per share of the N stocks is given by, $S_i(0)$. We will utilize the modeling approach developed in Chapter 6 for the price evolution of the N stocks. Each of the N stock's price evolves by the

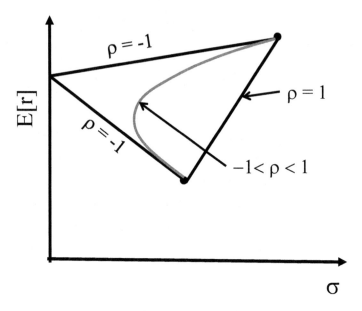

FIGURE 7.2: Plot of mean and standard deviation of two stock returns. The correlation of $\rho = 1$ and -1 define the right and left extents of the region, respectively.

following stochastic differential equation model,

$$dS_i(t) = \mu_i S_i(t)dt + \sigma_i S_i(t)dB_i(t), \tag{7.11}$$

where the N Wiener processes or standard Brownian motion processes, $B_i(t)$, are correlated. We employ a different notation for the Wiener process in this section, in order to reserve 'W' for the wealth process. The correlation between the N Wiener processes can be described by,

$$E[dB_i(t)dB_j(t)] = \rho_{ij}dt \ \ for \ i = 1 \dots N; j = 1 \dots N. \tag{7.12}$$

The parameter ρ_{ij} is the correlation coefficient between increments of Wiener processes $B_i(t)$ and $B_j(t)$, with $\rho_{ii} = 1$ for all $i = 1 \dots N$.

If the investment weight for the ith stock at a time, t, is maintained at, $w_i(t)$, then we can derive that the wealth, $W(t)$, will accumulate or evolve by the following equation:

$$dW(t) = \sum_{i=1}^{N} w_i(t)W(t)(\mu_i dt + \sigma_i dB_i(t). \tag{7.13}$$

If $w_i(t)$'s are $\mathcal{F}_t$-adapted, then $W(t)$ is also $\mathcal{F}_t$-adapted, where $\mathcal{F}_t$ is the filtration of the N Wiener processes, $\{B_i(t); 0 \leq i \leq N\}$. Moreover, if the fund intends to maintain a withdrawal rate of $C(t)$ as part of this investment strategy, for either immediate consumption or for running expenses, the wealth evolves by the following model,

$$dW(t) = -C(t)dt + \sum_{i=1}^{N} w_i(t)W(t)(\mu_i dt + \sigma_i dB_i(t)). \qquad (7.14)$$

Once the wealth evolution based on an investment strategy is defined, as in Eqns. (7.13) or (7.14), various investment strategies can be evaluated and compared for selection of the 'best' one. Simulation of the wealth evolution model in Eqns. (7.13) or (7.14) can be an alternative for this task, utilizing the discretization based simulation schemes developed in Chapter 6. Applying the simplest Euler scheme to Eqns. (7.13) and (7.14), respectively, with discretization, $0 = t_0 < t_1 < \ldots < t_M = T$, yields,

$$W(t_{k+1}) = W(t_k) + \sum_{i=1}^{N} w_i(t_k)W(t_k)(\mu_i \Delta t_k + \sigma_i \Delta B_i(t_k)), \ (7.15)$$

$$W(t_{k+1}) = W(t_k) - C(t_k)\Delta t_k + \sum_{i=1}^{N} w_i(t_k)W(t_k)(\mu_i \Delta t_k + \sigma_i \Delta B_i(t_k)), (7.16)$$

for $k \geq 0$, where $W(t_0) = W_0$, $w_i(t_k)$ are the investment weight decisions made for i^{th} stock at time, t_k, and as usual, $\Delta t_k = t_{k+1} - t_k$ and $\Delta B_i(t_k) = B_i(t_{k+1}) - B_i(t_k)$. As done for the static mean-variance risk-reward framework, assessment of an investment strategy must also be done against the goals set for the investment strategy. Once the investment strategies' performance metrics are set, one can seek to do the best one can by them. We explore this next.

7.1.2 Portfolio Optimization

The goal of risk mitigation is to reduce risk. However, in the process of risk reduction, risk optimization can seek the best risk-reward profile to adopt in an investment strategy. For constructing the best risk-reward trade-off, one must first clearly define the criteria for 'best', followed by developing a methodology for seeking the best option. Moreover, the approach should be able to accommodate the mix of subjective and objective views of the investor for the markets, sectors, industries, and specific firms, while allowing definition of risk measures suitable for the investor's risk management objectives. We advance the development of this section thus far to construct portfolio optimization frameworks for optimal risk-reward trade-off, both in the static and the dynamic case.

7.1.2.1 Optimum Risk-Return Trade-Off

The classical mean-variance framework, developed in Section 7.1.1.1, views risk of a portfolio as the variance of the portfolio return, defined in Eqn. (7.3).

The reward is captured by the expected or mean portfolio return for the period, given by Eqn. (7.2). Expressed as a function of portfolio weights, $\mathbf{w} = [w_1; w_2; \ldots, w_N]$, variance of portfolio return is an N-dimensional quadratic function, while expected portfolio return is an N-dimensional linear function, provided the parameters in the two equations, μ_i's and $Cov(i,j)$'s, are identified.

An optimum risk-return trade-off can be constructed by selecting portfolio weights that minimize portfolio risk, while not letting the expected portfolio return fall below a certain chosen threshold, r_{th}. This optimum portfolio weight selection problem can be summarized in the following quadratic programming optimization problem.

$$Obj \quad : \quad \min Var(r_\pi) = \mathbf{w}^T Cov \ \mathbf{w}, \tag{7.17}$$

$$S.t. \quad : \quad E[r_\pi] = \sum_{i=1}^{N} w_i E[r_i] \geq r_{th},$$

$$\sum_{i=1}^{N} w_i \leq 1. \tag{7.18}$$

The final constraint in Eqn. (7.18) should be added to retain feasibility regarding the funds invested in all stocks to not exceed the initial funds, W_0, available.

The above risk-reward trade-off problem is the simplest statement for the problem, with portfolio weights, $\mathbf{w} = [w_1; w_2; \ldots, w_N]$, left free to take any value, provided they satisfy the constraint in Eqn. (7.18). Therefore, the portfolio weights can be negative, implying that the investor is allowed to borrow or *short-sell* any stock to any capacity. Unlimited level of short-selling of stocks may neither be permitted nor desirable, therefore lower bounds, l_i, will be introduced for each portfolio weight, w_i, in the portfolio optimization problem. When no short-selling is permitted, lower bound will be set at zero, i.e., $l_i = 0$, for each i.

Solution of a quadratic programming problem given in Eqns. (7.17)-(7.18) can be obtained numerically using any standard optimization software, including the optimization toolbox in MATLAB, with suggestions for specific functions provided in Section 7.4. Figure 7.3 shows the minimum risk of the optimum portfolio, σ_{th}^*, corresponding to the choice of expected portfolio return threshold, r_{th}. A range of expected return thresholds chosen generates the '*efficient frontier*' of optimum risk-return trade-off portfolios.

Variance is a central measure of risk, and as discussed in Section 2.2.1, due to squaring of deviations from the mean, it gets heavily influenced by extreme observations. This can be a concern if there are outliers present in the historical data used to estimate variance of equity returns for portfolio construction. Alternative central measures of risk, such as mean-absolute deviation, can be considered to alleviate this concern, if the investor's risk management objectives must be defined in terms of a central measure of risk. The advantage of

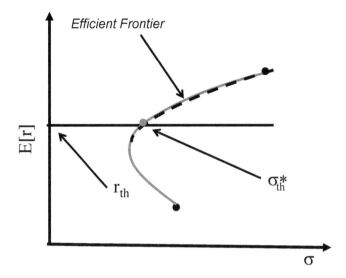

FIGURE 7.3: Plot of mean and standard deviation space spanned by return on portfolio of stocks. For a choice of expected portfolio return threshold, the optimum risk-return trade-off is made on the left most feasible points. The dashed curve is the efficient risk-return trade-off points, or the efficient frontier.

using mean-absolute deviation as the objective in Eqn. (7.17) is also that the portfolio optimization problem can be formulated as a linear programming problem.

When portfolio return is anticipated to not have a symmetric distribution, the characteristics of the upper tail of the portfolio return distribution can be quite different from that of the lower tail. Variance or other central measures of risk tend not to differentiate between positive and negative deviations from the mean (or other central tendencies chosen, such as the median). Therefore, these measures of risk would penalize both positive and negative deviations from the central tendency in the portfolio construction.

Empirical evidence for equity returns shows asymmetry in their distributions, therefore an un-diversified or 'lumpy' portfolio will retain the asymmetry in its return distribution. In such cases, tail measures of risk may be considered more appropriate to capture the adverse scenarios. Risk-reward trade-off can be considered by combining semi-variance or semi-absolute deviation versus mean of portfolio return. Mean versus Value-at-Risk (VaR) and mean versus Conditional VaR (CVaR) portfolio optimization have also been considered [71, 1, 2]. In the cases when the normal distribution provides a good representation of the portfolio return distribution, a direct relationship between mean-variance efficient frontier and the mean-VaR or the mean-CVaR efficient frontiers can be established.

7.1.2.2   Simulation Analysis for Portfolio Decisions

When mean-variance is used as the definition of reward and risk, respectively, expected return and variance of portfolio return can be computed explicitly using the formulas given in Eqns. (7.2) and (7.3), provided individual stock mean return and pair-wise covariance of equity returns were known. Moreover, as stated earlier, if the normal distribution provides a good representation of the portfolio return distribution, the central or tail measure of risk can be explicitly expressed in terms of mean and variance of portfolio return. In all other cases, simulation can prove to be useful to assess the risk and reward of a portfolio.

In the general distribution case, where return of each stock is governed by a chosen distribution that best fits the historical observations for the stock, generating random variates by the chosen distribution provides scenarios for future equity returns for each stock. Procedures for generating random variates by different distributions were described in Chapter 4. For a given choice of portfolio weights, $\mathbf{w} = [w_1; w_2; \ldots, w_N]$, the individual stock return scenarios can be combined to create portfolio return scenarios, r_π. Estimates of the reward-risk measures for the portfolio can be constructed based on these scenarios, with sufficiently large sample of scenarios providing desired accuracy. Confidence intervals created at a certain confidence level for the risk-reward measure estimates provides an indication of the accuracy.

An additional issue with mean-variance portfolio optimization, or any other parametric set-up for the risk-reward trade-off, is how reliable are the parameters used to state the problem. The optimum risk-reward trade-off obtained as the outcome of portfolio optimization is a function of these parameters, and a reliable implementation of the portfolio optimization results must ascertain that the decisions don't significantly change due to a small perturbation of the parameters. As the parameters are typically estimated based on historical return data, they are in fact prone to be imprecise.

Simulation analysis can help assess the parametric impreciseness for their impact on the portfolio risk-reward characteristics. Historical data available can be used to assess degree of parametric preciseness, through confidence intervals on accuracy of estimates for the parameters, as well as guided by the subjective view of how well the investor believes the past is representative of future stock returns. Parametric scenarios thus generated can feed into stock return random variates generation to eventually create portfolio return scenarios for given portfolio weights, $\mathbf{w} = [w_1; w_2; \ldots, w_N]$. As before, portfolio return scenarios can be used to create risk-reward assessments of the portfolio. Figure 7.4 shows this tiered assessment of portfolio risk-reward characteristics.

Simulation analysis can provide insights in case of parametric uncertainty and in attempts to incorporate investor-specific views about the parameters in portfolio construction and analysis. However, more rigorous approaches exist that are developed for specifically addressing both these cases. Techniques in robust portfolio optimization develop rigorous optimization frameworks that

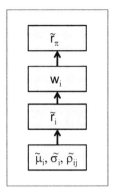

FIGURE 7.4: Simulation analysis of risk-reward of a portfolio based on equity returns scenarios and parametric scenarios.

integrally incorporate parametric uncertainty in the process of determining optimal portfolio weights. Construction of 'robust' portfolios using such frameworks has been an active and rapidly growing area, with an excellent summary provided in Fabozzi et al. [23].

On the other hand, investors often have independent, proprietary views for future stock returns. Black-Litterman portfolio framework [12] was created to incorporate these views, or 'priors', as an extension to the traditional Markowitz's mean-variance portfolio optimization framework. Such extensions follow the Bayesian learning view of formulating the portfolio optimization problem, that combines the long-term, short-term and investor-specific views into the portfolio construction process [92].

Finally, in Section 7.1.1.2, we presented the models for developing dynamic investment strategies. The aspects of portfolio analysis and optimization we have investigated so far in the single-period static case can be extended to the dynamic case. The simulation methodologies developed in Chapter 6 for models of risk evolving over time are applied to evaluate dynamic investment strategies. The risk-reward of the strategy must also be assessed in a dynamic manner, or if the context of the investment justifies, at the planning horizon. Utility function based performance measures, discussed in Chapter 2, can be constructed to evaluate the investment strategies. For instance,

$$U(W) = \int_0^T E[u(W(t))], \qquad (7.19)$$

where in the notation of Section 7.1.1.2, $W(t)$ is obtained as a result of investment decisions, $w_i(t)$, for each stock. The optimal investment strategy will be obtained that maximizes the utility, $U(W)$.

In this section we have examined the ways by which simulation analysis aids portfolio decisions for the purpose of risk mitigation. In Chapter 12, simulation-based optimization, or simply simulation optimization, will be developed that can be applied for performing single-period and dynamic portfolio optimization.

7.2   Transferring Equity Risk

Along with a high level of interest in equity investment comes the high level of need to transfer equity risk to carve out the desired risk-reward characteristics for the investment. Equity options, and generally equity derivatives, constitute key risk transfer instruments for equity market risk. Moreover, as residual claim on a firm's assets and profits from the firm's projects and ventures, the risks underlying shares of the stock of the firm are high. As discussed earlier, the enthusiasm for investing in stocks of firms ranks high among asset classes nevertheless, both by individual and institutional investors. Investors' high enthusiasm arises from the potential of high returns from equity investment. In this regard, beyond simply investing in the shares of the firm's stock, equity options allow a higher degree of speculation on a firm's future prospects. We will discuss these roles of equity options in detail in this section, along with making equity options a context for developing a theoretical framework for derivative pricing and hedging strategies.

The simplest of equity options are plain-vanilla European call and put options. A European call option gives the buyer of the option the right (and seller the obligation) to buy (sell) the underlying stock, S_t, at a set price, K, at a set time, T, in the future. The time T is called the maturity of the option and K is the exercise or strike price. Similarly, a European put option gives the buyer the right to sell the underlying stock at a set price, K, at a set time, T, in the future. The right to buy or sell the underlying stock is obtained by paying a premium, which is the price of the option. We will utilize these simple options as the context for developing a derivative pricing framework in Section 7.2.1. Pricing derivatives is crucial to developing risk transfer or hedging strategies.

The 'plain-vanilla' adjective is applied to distinguish these simplest options from more exotic ones that allow different kinds of transactions with respect to the underlying stocks. Once we have developed the analysis of pricing of plain-vanilla European options, we will extend the analysis in the coming sections to the exotic variety of options. Plain-vanilla American options are counterparts of corresponding European options, which give the buyer of the option the flexibility to exercise the option, i.e., buy or sell the underlying stock, at any time prior to maturity of the option, T. This order of developing analysis of types of options is inconsistent with their popularity and trading volume. In fact, the exchange traded equity options on single stocks are of American variety, hence command a dominant trading volume among a variety of options. After exploring a variety of options in this section, with the motivation of obtaining more accurate equity option prices, we will consider a few modeling enhancements for the underlying equity prices.

7.2.1 Option Pricing - Black-Scholes-Merton Approach

Let S_t be the price of a share of the stock of a firm; we need to describe in a model how the stock price may evolve with time. We are interested in being able to represent the price evolution of the stock in a time period relevant for the equity option we wish to price, hence up to the maturity of the option, T. At maturity of the option, given that it is a plain-vanilla European option, the pay-off in the case of a call option can be written as, $\max(S_T - K, 0)$, while in the case of a put option can be written as, $\max(K - S_T, 0)$. At maturity of the option, a rational investor will exercise the option only if purchase (or sale) price, i.e., the strike price K, is attractive relative to the market price of the stock at the time of maturity, i.e., S_T.

Based on whether the option is in the attractive range, that is if one had the flexibility of exercising it immediately, if it is worth exercising, the option is said to be '*in-the-money*'. In the case of a European call option, this coincides with $S_t > K$ scenario, while for the corresponding put option, option is in-the-money if $S_t < K$. In the unattractive zone, which is $S_t < K$ for call option and $S_t > K$ for put option, the option is said to be '*out-of-the-money*'. At the transition point, the option is said to be '*t-the-money*', i.e., $S_t = K$. In any scenario, option premium is the price for buying the option, denoted by $c(t, S_t)$ in the case of a call option, and $p(t, S_t)$ in the case of a put option. We have assumed that option premium must change with time, as well as depend on the price of the underlying stock.

Price of an option, or option premium, should depend on other factors also, such as the strike of the option, time to maturity of the option, etc. As we develop the option pricing framework, we will learn that the option premium does depend on these and some other parameters. From the moneyness of an option, which is its current status of either being in-the-money, at-the-money, or out-of-the-money, it may be concluded that the option premium should at least be as much as the moneyness of the option at any time, in other words, $c(t, S_t) \geq \max(S_t - K, 0)$, for all t in the case of a call option. In the case of an American option, this is very evident. For European options, being in-the-money at any time provides a stronger prospect of remaining in-the-money at the same or greater level at maturity. Hence, a commensurate premium. This implies that even some out-of-the-money options will command a non-zero premium when there is still some time to maturity for the prospect that at maturity they may turn in-the-money.

If an investor purchases a call option at t_0 for $c(t_0, S_{t_0})$ or a put option for $p(t_0, S_{t_0})$ with a strike price, K, and maturity, T, the pay-off of the option at maturity is shown in Figure 7.5. The option doesn't yield positive profit as soon as the option turns in-the-money at maturity when one accounts for the option premium paid. The option profit curve, also drawn for both

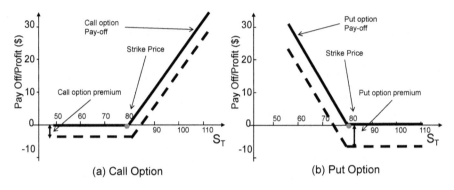

FIGURE 7.5: (a) Display of pay-off and profit curve for a plain-vanilla European call option with strike price, K=\$80. (b) Display of pay-off and profit curve for a plain-vanilla European put option with strike price, K=\$80.

call and put options in Figure 7.5, is $\max(S_T - K, 0) - c(t_0, S_{t_0})$ for the call and $\max(K - S_T, 0) - p(t_0, S_{t_0})$ for the put option, if time-value of money is ignored for the option premium. It is also instructive to see the pay-off and profit curve from the perspective of the seller or writer of the options, as shown in Figure 7.6. This is also called a short position in the call or the put option. From the figure, it is evident why a writer of an option will be motivated to take a short position in the option; it provides the opportunity of a positive profit with no further cash flow in all cases when the option matures out-of-the-money, at-the-money and slightly in-the-money.

Figures 7.5 and 7.6 provide the value of the European options at maturity, $c(T, S_T) = (S_T - K)_+$ in the case of a call option, and $p(T, S_T) = (K - S_T)_+$ in the case of a put option. We have utilized an alternative notation of $\max(x, 0) = x_+$ here, which we will use interchangeably in the rest of this section. The central goal here, however, is to develop a formal framework to determine the option premium for any time, $0 < t < T$. One may be tempted to price an option by applying the net present value concept, i.e., price of the option should be the discounted expected cash flow at maturity, given by,

$$c(t, S_t) = E[e^{-\alpha(T-t)}(S_T - K)_+]. \tag{7.20}$$

One of the problems with this approach is which discount rate, α, to use for the present value computation. For instance, if everyone applied their respective opportunity costs of capital for discount rate, there would be no agreement on trading price of the option. Therefore, for pricing derivatives we need to develop an alternative to the net present value framework.

We will develop the alternative asset valuation framework in an idealized economy, which will serve as a theoretically sound setting for discovering the fair value of an equity option. In this idealized economy, continuous trading of underlying stock and the equity option will be possible. There will be no

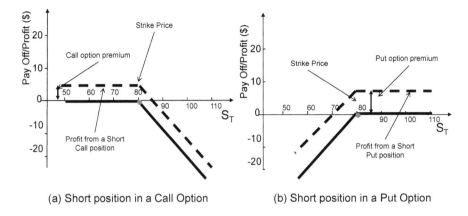

(a) Short position in a Call Option (b) Short position in a Put Option

FIGURE 7.6: (a) Display of pay-off and profit curve for a short position in a plain-vanilla European call option with strike price, K=\$80. (b) Display of pay-off and profit curve for a short position in a plain-vanilla European put option with strike price, K=\$80.

transaction cost associated with the trades. Moreover, the economy will allow risk-free short-term borrowing and lending at the same rate.

In this idealized economy, the price of the stock, S_t, will be modeled to evolve by the following stochastic differential equation model.

$$dS_t = \mu(t, S_t)dt + \sigma(t, S_t)dW_t, \tag{7.21}$$

where W_t is the standard Wiener process and $\mathcal{F}_t$ is its filtration. We have extensively studied the definition and simulation-based solution methods for such models in Chapter 6. For the special case of $\mu(t, S_t) = \mu S_t$ and $\sigma(t, S_t) = \sigma S_t$, we also derived the exact solution of the model in Section 6.4. The solution obtained was as follows,

$$S_t = S_0 \exp((\mu - \frac{1}{2}\sigma^2)t + \sigma W_t), \tag{7.22}$$

where S_0 is the initial or current price of the stock. The value of a risk-free short-term bond, B_t, was also modeled in Chapter 6 to evolve by the following deterministic differential equation model.

$$\frac{dB_t}{dt} = rB_t, \tag{7.23}$$

where 'r' is the continuously compounded interest rate. The solution of this deterministic differential equation was obtained as,

$$B_t = B_0 e^{rt}, \tag{7.24}$$

where B_0 is the initial value of the risk-free bond.

Since the equity option is a derivative defined on the underlying stock, price risk for the equity option is dictated by the price risk of the underlying stock. Therefore, it is possible to create a portfolio of investment in the stock and the stock option that is instantaneously risk-free. If this portfolio can be dynamically adjusted, it can be continuously maintained to be risk-free, where stock option position perfectly offsets the risk of the stock price. Let's construct such a portfolio, $\Pi(t)$, as follows.

$$\Pi(t, S_t, c(t, S_t)) = w_1 c(t, S_t) + w_2 S_t, \qquad (7.25)$$

where we have arbitrarily set the weights of equity option and the stock at w_1 and w_2, respectively. If we consider constructing the portfolio from the option writer's perspective, we would set $w_1 = -1$, or from the option buyer's perspective, we would set $w_1 = +1$. Let us proceed from the former perspective, with the knowledge that from the option buyer's perspective the opposite positions will be required.

In order to determine the weight on the stock in the portfolio, so that the portfolio becomes instantaneously free of risk, we need to utilize the Ito formula developed in Section 6.3.3 and apply it to the option price, $c(t, S_t)$. By the application of the Ito formula, we obtain that the option price must evolve by the following equation.

$$dc(t, S_t) = (\frac{\partial c}{\partial t} + \mu(t, S_t)\frac{\partial c}{\partial x} + \frac{1}{2}\sigma^2(t, S_t)\frac{\partial^2 c}{\partial x^2})dt + \sigma(t, S_t)\frac{\partial c}{\partial x}dW_t. \quad (7.26)$$

We have utilized the notation '$\frac{\partial c}{\partial x}$' simply to indicate that the partial derivative of function, $c(t, S_t)$, is with respect to the spatial (second) coordinate, x, even though the spatial coordinate in this case is the stock price, S_t. If we want the portfolio, $\Pi(t)$, to evolve risk-free, the diffusion terms in Eqns. (7.21) and (7.26) should match, and cancel each other. This implies that the portfolio weights for the portfolio, $\Pi(t, S_t, f(t, S_t))$, must be selected as follows.

$$c(t, S_t) \quad : \quad -1 \qquad (7.27)$$

$$S_t \quad : \quad \frac{\partial c}{\partial x}. \qquad (7.28)$$

With this choice of portfolio weights assigned to the equity and equity option in the portfolio, the portfolio becomes,

$$\Pi(t, S_t, c(t, S_t)) = -c(t, S_t) + \frac{\partial c}{\partial x}S_t. \qquad (7.29)$$

We again apply the Ito's formula to examine how the portfolio, $\Pi(t, S_t, f(t, S_t))$, evolves in time. We obtain,

$$dΠ(t, S_t, c(t, S_t)) = \frac{\partial c}{\partial x}dS_t - dc(t, S_t) \qquad (7.30)$$

$$= -(\frac{\partial c}{\partial t} + \frac{1}{2}\sigma^2(t, S_t)\frac{\partial^2 c}{\partial x^2})dt. \qquad (7.31)$$

By using two risky instruments, albeit one being a derivative of the other, we have successfully created a portfolio that evolves risk-free. As we stated earlier, in this economy a risk-free bond is already available to investors. A second risk-free instrument, in the form of the portfolio just constructed, creates an opportunity for investors. If either of the two risk-free investments, namely the risk-free bond and risk-free portfolio constructed, provides better return, investors can borrow as much as possible of the lower return risk-free instrument and invest in the higher return risk-free instrument, thus creating unbounded amounts of profit. Clearly such 'free-lunch' can't be available forever, since either of the two counterparties will wise up and change prices to eliminate the 'free-lunch'. In other words, when everyone attempts to profit from this free-lunch opportunity, prices will move in directions that eliminate the availability of two risk-free interest rates. Therefore, in equilibrium such profits will not be available, which gives us a relation for the constructed portfolio to satisfy. We take a pause to define such 'free-lunches', which is called an *arbitrage*.

Arbitrage:. Any trading strategy that allows the possibility of gaining positive cash flows now or in the future, with no net liabilities now or in the future, is an arbitrage. This clearly sounds like 'free-lunch'. Arbitrage profits are possible when two or more securities are mispriced with regards to fundamentals that relate the prices of the securities. In the above discussion, the fundamental relationship between equity price, equity option price and risk-free interest rate offers an opportunity for arbitrage profits due to mispricing. If two different investments or trading strategies produce the exact same payoff, they must also cost the same, failing which arbitrage can be constructed. On the basis of this principle, we will soon consider a relationship between vanilla European call option price and put option price with the same strike and maturity, called the put-call parity.

Mathematically, the notion of *arbitrage* can be summarized as a trading strategy yielding a portfolio, $\Psi(t)$, where the initial value of the portfolio is $\Psi(0) = 0$. However, at some time T in future, the strategy yields a portfolio value $\Psi(T)$, such that, $P(\Psi(T) \geq 0) = 1$ and $P(\Psi(T) > 0) > 0$. Therefore, this portfolio has no liability now or in the future, but offers a positive probability of positive gain in the future.

Arbitrage-free price of the option, as concluded from the above discussion, should imply that the risk-free portfolio, $\Pi(t, S_t, c(t, S_t))$, constructed by appropriately choosing positions in the stock and equity option, should evolve at the risk-free rate. In the idealized economy with risk-free borrowing and lending available for the same rate, r, in order to eliminate arbitrage opportunities, the risk-free portfolio should evolve at the economy's risk-free rate, r. Therefore, in addition to Eqn. (7.31), an alternative formula the portfolio, $\Pi(t, S_t, c(t, S_t))$, can be described to evolve by is,

$$d\Pi(t, S_t, c(t, S_t)) = r\Pi(t, S_t, c(t, S_t))dt. \tag{7.32}$$

We combine the two definitions for evolution of the portfolio, $\Pi(t, S_t, c(t, S_t))$, in Eqns. (7.31) and (7.32) to obtain,

$$-(\frac{\partial c}{\partial t} + \frac{1}{2}\sigma^2(t, S_t)\frac{\partial^2 c}{\partial x^2})dt = r(-c(t, S_t) + \frac{\partial c}{\partial x}S_t)dt, \quad (7.33)$$

or

$$(\frac{\partial c}{\partial t} + \frac{1}{2}\sigma^2(t, S_t)\frac{\partial^2 c}{\partial x^2} + \frac{\partial c}{\partial x}rS_t - rc(t, S_t))dt = 0. \quad (7.34)$$

Now Eqn. (7.34) should be true for any arbitrary choice of small time increment, dt. This will be the case if in fact the option price, $c(t, S_t)$, satisfies the following partial differential equation.

$$\frac{\partial c}{\partial t} + \frac{1}{2}\sigma^2(t, S_t)\frac{\partial^2 c}{\partial x^2} + rS_t\frac{\partial c}{\partial x} = rc(t, S_t). \quad (7.35)$$

This is the **Black-Scholes Partial Differential Equation (PDE)** for option pricing. For solving this partial differential equation, we will also need to specify an end condition, which describes what is known about the option price at maturity of the option. The end condition in the case of a vanilla European call option will be give by,

$$c(T, S_T) = (S_T - K)_+. \quad (7.36)$$

We have developed the derivation of pricing of options using European call option as the reference equity derivative, however we could have equally well taken European put option, $p(t, S_t)$, as the demonstrative example. In this case, the portfolio would require a short position in the put option and a position $\frac{\partial p}{\partial x}$ in the stock. The rest of the derivation would proceed exactly as done for the European call option, with European put option price, $p(t, S_t)$, satisfying the exact same partial differential equation, with one crucial difference. The end condition we would use for the European put option would be,

$$p(T, S_T) = (K - S_T)_+. \quad (7.37)$$

There is however an easier way available to find the European put option price once one has obtained the price of a call option of the same strike price and maturity. This is done by utilizing the following **put-call parity**.

$$c(t, S_t) + Ke^{-r(T-t)} = p(t, S_t) + S_t. \quad (7.38)$$

As stated earlier, the principle behind existence of the put-call parity is elimination of arbitrage. We will examine this further in Section 7.3, in the context of hedging (and arbitrage) strategies.

7.2.1.1   Solving Black-Scholes Partial Differential Equation

For solving the Black-Scholes partial differential equation, with the chosen end condition, we must invoke the **Feynman-Kac theorem**. For a process satisfying,

$$dS_t = rS_t dt + \sigma(t, S_t)dW_t, \quad (7.39)$$

a measurable function $h(y)$ and fixed time $T > 0$, we define $g(t, x) = E[e^{-r(T-t)}h(S_T)|S_t = x]$, then $g(t, x)$ satisfies

$$\frac{\partial g}{\partial t} + \frac{1}{2}\sigma^2(t, x)\frac{\partial^2 g}{\partial x^2} + \frac{\partial g}{\partial x}rx = rg(t, x), \tag{7.40}$$

with terminal condition,

$$g(T, x) = h(x), \forall x. \tag{7.41}$$

Applying the **Feynman-Kac theorem** gives us the Black-Scholes solution for call option price to be,

$$c(t, S_t) = E[e^{-r(T-t)}h(S_T)|S_t], \tag{7.42}$$

where in the case of European call option with strike, K, $h(S_T) = (S_T - K)_+$. The stock prices, S_t, satisfies the following Ito stochastic differential equation,

$$dS_t = rS_t dt + \sigma(t, S_t)dW_t. \tag{7.43}$$

One should observe two important points, first Eqn. (7.42) holds stark resemblance with Eqn. (7.96), however the discount rate used in Eqn. (7.42) is the risk-free rate. Second, the model to evolve the price of the underlying stock in Eqn. (7.43) for computing the option price has changed from the original model in Eqn (7.21). We will come back to discuss both these points later in this section.

We have done the entire derivation for option pricing in quite a general case, where the Feynman-Kac theorem is applied to a general cases of drift, $\mu(t, S_t)$, and diffusion, $\sigma(t, S_t)$, terms appearing in the underlying equity price evolution model. If specific simple cases are picked for drift and, more importantly, diffusion, the price can be more explicitly determined. We consider the case of a simpler diffusion term, $\sigma(t, S_t) = \sigma S_t$, in order to derive the well-known *Black-Scholes option price formula*. The European call option price is given by,

$$c(t, S_t) = E[e^{-r(T-t)}(S_T - K)_+|S_t], \tag{7.44}$$

with stock price evolving by the following equation,

$$dS_t = rS_t dt + \sigma S_t dW_t, \tag{7.45}$$

or,

$$S_t = S_0 \exp((r - \frac{1}{2}\sigma^2)t + \sigma W_t). \tag{7.46}$$

Once the exact solution of Eqn. (7.46) is substituted in Eqn. (7.44), after many steps of derivation utilizing the log-normal distribution of stock price at T, we obtain the call option price to be,

$$c(t, S_t) = S_t \Phi(d_1) - Ke^{-r(T-t)}\Phi(d_2), \tag{7.47}$$

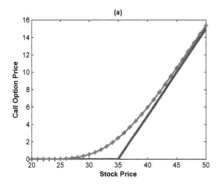
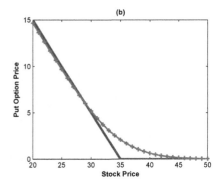

FIGURE 7.7: (a) Display of pay-off and price curve for a plain-vanilla European call option with strike price, K=\$35, $\sigma = 23\%$, $T - t = 1/2$ year, and short-term interest rate of $r = 2\%$. (b) Display of pay-off and price curve for a plain-vanilla European put option with the same set of parameters as the call option.

where $\Phi(x)$ is the cumulative distribution function of the standard normal distribution $(N(0,1))$, and

$$d_1 = \frac{\ln(\frac{S_t}{K}) + (r + \frac{\sigma^2}{2})(T - t)}{\sigma\sqrt{T - t}}, \tag{7.48}$$

and

$$d_2 = \frac{\ln(\frac{S_t}{K}) + (r - \frac{\sigma^2}{2})(T - t)}{\sigma\sqrt{T - t}}, \tag{7.49}$$

$$= d_1 - \sigma\sqrt{T - t}. \tag{7.50}$$

Applying the put-call parity given in Eqn. (7.38) provides the price of the corresponding put option as,

$$p(t, S_t) = Ke^{-r(T-t)}\Phi(-d_2) - S_t\Phi(-d_1), \tag{7.51}$$

where d_1 and d_2 are as defined in Eqns. (7.65) and (7.66), respectively.

For the interested reader, we formally state the Feynmann-Kac theorem in a somewhat more general form. The version below is for a single spatial dimension, however the theorem can be extended to multiple spatial dimensions, which we will utilize in some cases later in the chapter. For the development of the proof of the theorem, the reader should refer to Oksendal [68] or Shreve [81].

Feynmann-Kac theorem: Given a partial differential equation,

$$\frac{\partial g}{\partial t} + \frac{1}{2}\sigma^2(t, x)\frac{\partial^2 g}{\partial x^2} + r(t)x\frac{\partial g}{\partial x} = rg(t, x), \tag{7.52}$$

where $r(t)$ is time-dependent interest rate and the terminal condition is given by,

$$g(T, x) = h(x), \forall x, \tag{7.53}$$

for a measurable function $h(x)$ and fixed time $T > 0$, a function $g(t, x)$, defined as follows is a solution.

$$g(t, x) = E[e^{-\int_t^T r(s)ds} h(S_T) | S_t = x], \tag{7.54}$$

where S_t satisfies the following stochastic differential equation,

$$dS_t = r(t) S_t dt + \sigma(t, S_t) dW_t, \tag{7.55}$$

with S_0 given to be a constant.

For the proof of this result, the reader may refer to Oksendal [68].

We have developed the option pricing using the partial differential equations framework. An alternative pricing framework for an option, which utilizes probabilistic constructs instead of partial differential equations, can also be developed. This is developed by constructing an equivalent martingale measure (EMM) by the application of the Girsanov theorem, and obtains the option price in a similar form as in Eqn. (7.42) under the EMM. For more details, please refer to Shreve [81].

7.2.1.2 Estimating Option Price by Simulation

The option pricing formula given in Eqn. (7.42) is useful, but to determine the actual numeric value of the price, some additional work is required. In the case of a simple choice of diffusion coefficient, i.e., $\sigma(t, S_t) = \sigma S_t$, we were able to obtain the exact pricing formula, as given in Eqn. (7.64). For other more general diffusion coefficients, $\sigma(t, S_t)$, one would have to either analytically resolve Eqn. (7.42) or apply simulation to obtain estimates of the option price.

The algorithm for simulation-based option price estimation will have the following structure in the case $\sigma(t, S_t) = \sigma S_t$.

For i=1:n
    Generate $Z_i \sim N(0, 1)$;
    $S_i(T) = S_0 \exp((r - \frac{\sigma^2}{2})T + \sigma\sqrt{T} Z_i)$;
    $C_i = e^{-rT} \max(S_i(T) - K, 0)$;
End
$\widehat{C}(n) = \frac{\sum_{i=1}^n C_i}{n}$;

This may in itself not be so useful, since in this case the exact formula for the option price is known analytically. This simulation, however, provides the opportunity to test the accuracy of the simulation relative to the theoretical price. In the above algorithm, $\widehat{C}(n)$ is an **estimator** for the option price. As we know, for any estimation task, constructing a confidence interval around the estimate is essential to highlight the accuracy of the estimate. Confidence interval for a $(1 - \alpha)100\%$ confidence level will be obtained by adding the following steps to the above algorithm.

$\hat{\sigma}_C =$ standard deviation of C_i's;
$\alpha = 0.05$;
$z_{\alpha/2} =$ Inverse of Normal Distribution$(\alpha/2, N(0,1))$;
$\widehat{C}_l(n) = \widehat{C}(n) + z_{\alpha/2}\frac{\hat{\sigma}_C}{\sqrt{n}}$;
$\widehat{C}_u(n) = \widehat{C}(n) - z_{\alpha/2}\frac{\hat{\sigma}_C}{\sqrt{n}}$;

The $(1-\alpha)100\%$ confidence interval for the option price, $(\widehat{C}(n) \pm z_{\alpha/2}\frac{\hat{\sigma}_C}{\sqrt{n}})$, can be compared with the analytical price of the option obtained from Eqn. (7.64). In the more general case of the diffusion term, $\sigma(t, S_t)$, if the analytical solution of terminal stock price, S_T, as a solution of the stochastic differential equation in Eqn. (7.21) is not known, the above algorithm has to be adapted as follows.

$\Delta t = \frac{T}{M}$;
For i=1:n
    For $t = 0 : \Delta t : T$;
        Generate $Z_{it} \sim N(0,1)$;
        $S_i(t + \Delta t) = rS_i(t)\Delta t + \sigma(t, S_i(t))\sqrt{\Delta t}Z_{it}$;
    End
        $C_i = e^{-rT}\max(S_i(T) - K, 0)$;
End
$\widehat{C}(n) = \frac{\sum_{i=1}^{n} C_i}{n}$;

In the above, we have implemented an Euler scheme to develop numerical solution of the stock price evolution. More general schemes discussed in Chapter 6 can be applied as needed. The extension to compute confidence interval remains as presented above.

7.2.1.3 Making Model Simpler - Binomial Tree Approach

In this section, we explore simpler representation of the stock evolution, which is in fact consistent with the continuous-time stock evolution models we have considered thus far. We develop the option pricing framework under the binomial tree stock evolution model to reinforce the risk-neutral pricing approach developed in continuous-time setting of Section 7.2.1.1.

The binomial tree model of stock price evolution is a discrete-time stochastic process with discrete, specifically two ('bi'-nomial), outcomes in each time

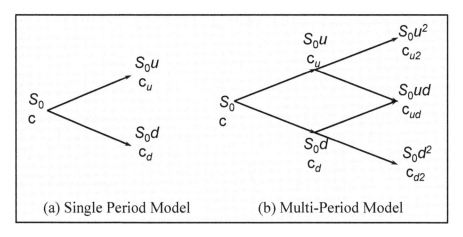

(a) Single Period Model   (b) Multi-Period Model

FIGURE 7.8: (a) Single period binomial tree model for stock price evolution. (b) Multi-period binomial tree model for stock price evolution.

step. Consider a time-step, Δt, where at the start of this time-step the stock price is, S_t, while at the end of the time-step, stock prices either go up with a factor of 'u' or go down with a factor, 'd'. We depict the first outcome by, S_u, which has the value, uS_t, while the second outcome is depicted by, S_d, with a value, dS_t. The probability of the 'up'-outcome can be taken to be p, while that of the 'down'-outcome would be, $(1 - p)$. Figure 7.8 displays the one time-step of the binomial tree in the left panel.

If we trace the evolution to another time-step, the realizations of the stock can be identified to be, $\{u^2 S_t, udS_t, d^2 S_t\}$, in the three distinct outcomes. The fact that two of the outcomes merge after two time-steps is why this tree would be called a *recombining* binomial tree. The right panel of Figure 7.8 displays the two-step binomial tree. Similarly, a multiple-step binomial tree can be constructed. An appropriate number of time-steps may be needed in the binomial tree depending on the size of the time-step, Δt, chosen, in order to span the time period, $[0, T]$, for the purpose of option pricing.

There are three parameters in the binomial tree stock evolution model, namely u, d and p, that must be estimated to calibrate the model before putting it to any use. Our intended use here is to price a stock option. In order to develop a binomial tree model that is consistent with the Black-Scholes-Merton stock evolution model discussed in Section 7.2.1.1, we can calibrate the parameters of the binomial tree to match the moments of the geometric Brownian motion model for stock evolution. We take the probability of the up outcome, p, to be half, in order to match the symmetry of the normal distribution of Wiener process increments that drive the geometric Brownian motion. Calibration by matching the first two moments yields the following two equations for u and d, which can be simultaneously solved to obtain estimates of u and d consistent with the geometric Brownian motion.

$$\frac{1}{2}(u+d) = e^{\mu \Delta t}; \tag{7.56}$$

$$\frac{1}{2}(u - \frac{1}{2}(u+d))^2 + \frac{1}{2}(d - \frac{1}{2}(u+d))^2 = e^{2\mu \Delta t}(e^{\sigma^2 \Delta t} - 1); \tag{7.57}$$

Alternatively, if the estimates of μ and σ are not readily available, the two equations (Eqns. (7.57) and (7.56)) can be modified to match the moments obtained from observations of the stock price at Δt interval. If stock price observations at Δt interval are provided as, $\{S_{t_0}, S_{t_1}, \ldots, S_{t_N}\}$, with $t_{i+1} - t_i = \Delta t$ for all i, the right-hand side of Eqn. (7.56) would be substituted by the sample mean of $\{\frac{S_{t_{i+1}}}{S_{t_i}}\}$. The right-hand side of Eqn. (7.57) will be substituted by sample variance of $\{\frac{S_{t_{i+1}}}{S_{t_i}}\}$.

Pricing an equity option in the binomial tree framework will employ similar arbitrage-free principle for derivative pricing as developed at the start of Section 7.2.1. We will construct a portfolio of the stock and risk-free short-term bond that matches the pay-off of the European call option at maturity. We assume that every dollar invested in the risk-free short-term bond yields \$R in a time-step of Δt. Assume that the maturity of the call option is in Δt time. Therefore, given a strike price K of the call option, the option will be worth, $(uS_t - K)_+$, at maturity in the up scenario. We label this outcome for the call option by, c_u. Similarly, the down outcome at maturity of the option will be, $c_d = (dS_t - K)_+$.

The replicating portfolio, Π, is constructed at time, t, by buying x dollars worth of stock and b dollars worth of risk-free asset. Therefore, the portfolio is worth, $\Pi_t = x + b$. The exact values of x and b will be determined so that the portfolio matches the up and down scenario of the option at maturity. The two equations we should simultaneously solve to obtain x and b are as follows.

$$\Pi_u = ux + Rb = c_u; \tag{7.58}$$
$$\Pi_d = dx + Rb = c_d. \tag{7.59}$$

Solving Eqns. (7.58) and (7.59) simultaneously yields the following values of x and b, respectively.

$$x = \frac{c_u - c_d}{u - d}; \tag{7.60}$$

$$b = \frac{1}{R}(\frac{uc_d - dc_u}{u - d}). \tag{7.61}$$

We have constructed a portfolio that matches the pay-off of the European call option in all scenarios of the option at maturity, Δt. If the portfolio value at t doesn't match the value of the call option at t, an arbitrage strategy can be constructed that takes advantage of this mismatch in prices. In order for the arbitrage to be eliminated, prices will need to modify so that we obtain the following.

$$\Pi_t = x + b \;\; = \;\; c_t, \tag{7.62}$$

where c_t is the price of the European call option at time, t. Substituting the values of x and b from Eqns. (7.60) and (7.61), after some rearrangement, yields that the European call option price at time, t, is as follows.

$$c_t = \frac{1}{R}\left(\frac{R-d}{u-d}c_u + \frac{u-R}{u-d}c_d\right). \tag{7.63}$$

One should note that this formula matches in principle the expected discounted pay-off of the option at maturity formula given in Eqn. (7.42), where expectation is taken with respect to probabilities, $q = \frac{R-d}{u-d}$ and $(1-q)$. These are the risk-neutral probabilities. Moreover, we have derived pricing in the binomial tree framework for a single time-step, Δt. In reality, one would want to include several time-steps of the binomial tree to span the time to maturity, T, of the option.

This extension would be achieved by applying the above derivation to each node of the binomial tree, starting with nodes at $T - \Delta t$. Specifically, in the right panel of Figure 7.8, we would first apply the above derivation to obtain the values of c_u and c_d, in terms of $\{c_{u^2}, c_{cd}, c_{d^2}\}$. The value of option, c, is obtained in terms of $\{c_u, c_d\}$ by applying the derivation once again. In multiple-time steps of the binomial tree, the price at the base node is obtained by recursively applying the derivation from $T - \Delta t$ backward.

The method developed for pricing of vanilla European call options can be extended to other European-style options by changing the terminal pay-offs, and conducting the backward recursive steps exactly as described above.

7.2.2 Implied Volatility and Calibration for Risk-Neutral Pricing

To say that volatility, σ, stays constant, as suggested in the Black-Scholes option price formula in Eqn. (7.64), would be a bit of an unrealistic stretch of imagination. On the other hand, of all the factors that determine the price of an option, volatility may very well be the most crucial factor, both for the valuation and for risk management of an option. In light of a time-varying volatility, historic volatility estimates may not be the most valuable predictor of future volatility. Therefore, we lay a note of caution here. While the calibration we have presented in the previous sections has been based on historic prices of the stock, for the purpose of option pricing, information for better predictive estimates of volatility would be needed.

The notion of implied volatility is very useful to assess suitability of historic volatility estimates in predicting future volatility. Using historic stock data alone to estimate future volatility would be quite risky if higher or lower volatility is expected in the market for the future. *Implied volatility* is the volatility **implied** by the market price of options traded in a liquid options market. This volatility offers crucial information about the market's expectation of future volatility.

Implied volatility would be computed, however, by first assuming a model by which the market is purportedly transforming volatility to the option price. This transformation could be explicit, as the analytical Black-Scholes option price formula, or indirect, in which case implied volatility must be iteratively inferred. In any case, implied volatility is not explicitly observed, it must be inferred from the observed market price of options. For instance, if we assume the market determines option price by the Black-Scholes option pricing formula, implied volatility will be obtained by solving the following for σ_{impv}.

$$c_{mkt}(t, S_t) \;=\; S_t \Phi(d_1) - K e^{-r(T-t)} \Phi(d_2), \tag{7.64}$$

$$\text{where} \quad d_1 \;=\; \frac{\ln(\frac{S_t}{K}) + (r + \frac{\sigma_{impv}^2}{2})(T-t)}{\sigma_{impv}\sqrt{T-t}}, \tag{7.65}$$

$$\text{and} \quad d_2 \;=\; d_1 - \sigma_{impv}\sqrt{T-t}. \tag{7.66}$$

In Figure 7.9, we plot the range of implied volatility corresponding to a range of strike prices, or moneyness, of the option defined on a stock with a historical volatility of $\sigma = 23\%$. We observe that the market anticipates a higher volatility for both upward and downward movement of the stock, with the downward move seemingly more volatile than the upward move.

The second insight obtained from a plot of implied volatility, as in Figure 7.9, is whether and to what extent does the lognormal distribution of the underlying stock evolution model capture the true tail risk of the stock price. The fact that the options deep in-the-money and out-of-the-money have much more enhanced implied volatility than the near at-the-money options suggests that the lognormal distribution is failing to capture the tail risk in the future stock price. More so for the lower tail than the upper tail of the stock price distribution. The implied volatility curve, often called a 'volatility smile' or 'volatility smirk' based on the shape of the curve, is indicative of this model inaccuracy.

Volatility smirks, a skew to the left, are indicative of investors' worry about sudden large downward movement of the stock, or more adversely, a market crash. Appropriate adjustments should be sought to make up for this model inaccuracy, some of which will be discussed in Section 7.2.6. Moreover, calibration of models for risk-neutral pricing of options should account for the importance of volatility estimates. With this in view, for risk-neutral pricing of options, the binomial tree model is often calibrated by modeling the up and down factor in a single-step of evolution as, $u = e^{\sigma\sqrt{\Delta t}}$ and $d = \frac{1}{u}$, respectively. This singles out the volatility parameter, allowing the modeler to utilize appropriate volatility estimates, instead of relying on historical stock price based estimates alone.

Volatility smiles and smirks have shown up and become more prevalent for

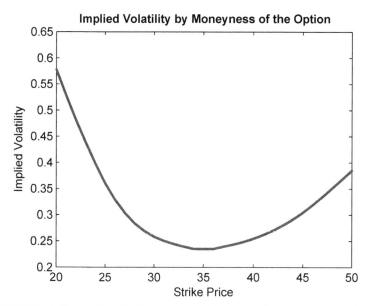

FIGURE 7.9: Implied volatility obtained from the Black-Scholes option pricing formula for plain-vanilla European call option with stock price, $S_t =\$35$, $\sigma = 23\%$, $T - t = 1/2$ year, and short-term interest rate of $r = 2\%$.

equity options after the US stock market crash of October 1987. Today implied volatility is so important that traders and brokers often quote options by their implied volatilities rather than dollar prices. Implied volatilities may be used as an input into the Black-Scholes model to calculate the price of a slightly different options series with different exercise prices or maturities. Implied volatilities produce an entire *volatility surface*, once the options series by times to maturity and strikes are both accounted for. Finally, implied volatility is not directly observed, instead it must be inferred based on a model for option pricing. This can be a source of operational risk.

7.2.3 Sensitivity to the Parameters

Options, and in general all derivatives, derive their value from the value of their underlying instruments and depend on the few parameters that define the option contract. The underlying instrument, in turn, depends on a set of parameters. As we studied in the previous section, volatility and other parameters defining an option can change. Therefore, the hedging, speculative or arbitrage objectives the option is being used to serve would be affected by these changes. Sensitivity of the price of an option with respect to its key parameters helps measure the impact of the changes of these parameters on the hedging, speculative or arbitrage objectives. These sensitivities are depicted

Variable Change / Option	c	p	C	P
S ↑	↑	↓	↑	↓
K ↑	↓	↑	↓	↑
σ ↑	↑	↑	↑	↑
T ↑	-	-	↑	↑
r ↑	↑	↓	↑	↓

↑ : **positive relation**
↓ : **inverse relation**

FIGURE 7.10: The chart marks the dependence of European and American vanilla call and put option prices on parameters that determine the price.

by some symbols of the Greek alphabet, and therefore in short are called *the Greeks*.

Figure 7.10 displays the nature of dependence of plain-vanilla European and American call and put options on the relevant parameters. The first column indicates the parameter going up, and each cell displays by the arrow if the price of that option goes up (up arrow) or down (down arrow). The value of the underlying going up, shown in the first row, results in call options' values going up, while put options' values go down. This is obvious from the pay-off structure of call and put options, respectively. The reverse characteristic is seen for the strike price going up, again seen from the pay-off structure of the options.

Increase in volatility of the underlying stock uniformly increases the value of all the options. The intuition behind the relationship between option price and short-term interest rate lies in the fact that in the risk-neutral world, the stock evolves with a drift of r. If the interest rate increases, it essentially has the effect of the stock price increasing. Therefore, relationship between option price and interest-rate matches the relationship between option price and stock price in Figure 7.10. The precise dependence can be measured in terms of the Greek, $\rho = \frac{dC}{dr}$, which is obtained in the case of the Black-Scholes option price formula below.

As the time to maturity, T, increases, it offers more opportunity to the American style option for a higher pay-off. Therefore the American option price has a positive relation with time to maturity, when time to maturity decreases, the American option price also decreases. This is usually also true for the corresponding European-style options, but not always since European options are only exercised at a specific time, rather than in the entire period leading up to maturity.

We describe each Greek of an option in detail, along with providing exact formulas in the case of the Black-Scholes European call option pricing formula. The formulas for Greeks for European put options in the Black-Scholes framework can be similarly obtained.

Delta or Δ is the rate of change of the option price with respect to the underlying.

$$\Delta = \frac{dC}{dS} = \Phi(d_1), \tag{7.67}$$

where $\Phi()$ is the cumulative distribution function for standard normal distribution, $N(0,1)$, and d_1 was defined in Eqn. (7.65).

Gamma or Γ is the rate of change of Delta or Δ with respect to the price of the underlying stock. Gamma is the largest for near-the-money options. This can be seen in Figure 7.7, where the option price curve has the highest curvature around at-the-money range. Gamma can be explicitly computed for the Black-Scholes option price as follows.

$$\Gamma = \frac{d(dC)}{dS^2} = \frac{\phi(d_1)}{(S_0\sigma\sqrt{T})}, \tag{7.68}$$

where $\phi()$ is the probability density function of standard normal distribution, $N(0,1)$.

Vega is the all important measure of change in value of the option due to increase (or decrease) of volatility of the underlying stock. It tends to be largest for near in-the-money options, and can be computed explicitly for Black-Scholes option price formula as follows.

$$Vega = \frac{dC}{d\sigma} = S_0\sqrt{T}\phi(d_1). \tag{7.69}$$

Theta or Θ of a derivative is the rate of change of the value of the derivative with the passage of time or as the time to maturity shortens. The theta of a call or a put option is usually negative. This is always true for American put and call options, that is, as the time to maturity nears, the value of an American call or put option decreases. Therefore, as time passes with the price of the underlying asset and its volatility remaining the same, the value of a long position in the call or put option declines. Theta (Θ) for European call option can be computed from the Black-Scholes option price formula as follows.

$$\Theta = -\frac{S_0\phi(d_1)\sigma}{2\sqrt{T}} - rK\exp(-rT)\Phi(d_2). \tag{7.70}$$

Rho or ρ is the rate of change of the value of an option with respect to the short-term interest rate, and is computed as follows.

$$\rho = \frac{dC}{dr} = KT\exp(-rT)\Phi(d_2). \tag{7.71}$$

When the Greeks cannot be computed explicitly by taking the desired derivative (or differentiation) of a closed-form pricing formula, they should be estimated using simulation. Following is a sample algorithm for computing the vega of a derivative. A similar algorithm can be developed for estimation of all the remaining four Greeks.

For i=1:n

$\quad$ Generate $Z_{i,1} \sim N(0,1)$;

$\quad$ $S_{i,1}(T) = S_0 \exp((r - \frac{\sigma^2}{2})T + \sigma\sqrt{T}Z_{i,1})$;

$\quad$ Generate $Z_{i,2} \sim N(0,1)$;

$\quad$ $S_{i,2}(T) = S_0 \exp((r - \frac{(\sigma+\delta\sigma)^2}{2})T + (\sigma + \delta\sigma)\sqrt{T}Z_{i,2})$;

$\quad$ $C_{i,1} = e^{-rT} \max(S_{i,1}(T) - K, 0)$;

$\quad$ $C_{i,2} = e^{-rT} \max(S_{i,2}(T) - K, 0)$;

End

$\widehat{Vega}(n) = \dfrac{\frac{\sum_{i=1}^{n} C_{i,1}}{n} - \frac{\sum_{i=1}^{n} C_{i,2}}{n}}{\delta\sigma}$.

Beyond estimating the point estimate of the Greeks, confidence interval for the Greeks can be developed at the desired confidence level, $(1 - \alpha)100\%$. The sampling distribution for difference in mean estimator, $(\bar{X}_1 - \bar{X}_2)$, under the case of known and unknown standard deviation would be useful for creating these confidence intervals. This is because the form of estimator, $\frac{\frac{\sum_{i=1}^{n} C_{i,1}}{n} - \frac{\sum_{i=1}^{n} C_{i,2}}{n}}{\delta\sigma}$ for $\widehat{Vega}(n)$, resembles the difference in mean estimator.

All of the above sensitivities are relevant for a portfolio of options on the same underlying stock. Portfolio Greeks can be obtained by computing individual instrument's Greeks and combining together by portfolio weight for each instrument. Therefore, if a portfolio of options on an underlying stock is defined as, $\Pi_t = \sum_{i=1}^{N} n_i X_i(t, S_t)$, where $X_i(t, S_t)$ is the value of a derivative instrument in the portfolio, then portfolio Delta, Δ_Π is obtained as follows.

$$\Delta_\Pi = \sum_{i=1}^{N} n_i \Delta_{X_i}(t, S_t). \tag{7.72}$$

Similarly, all the other Greeks for the portfolio may be obtained. At the portfolio level, being able to define portfolio Greeks also provides the opportunity to craft portfolios that are delta-neutral, gamma-neutral, vega-neutral, etc. This implies that the portfolio weights are chosen so that the portfolio Greeks are zero. One or more Greeks being zero implies that the portfolio is desensitized to (small) changes in the specific parameter. Re-balancing the portfolio weights, mostly by small adjustments, can maintain delta-, gamma-, and vega-neutrality.

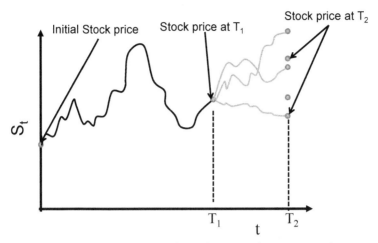

FIGURE 7.11: Trajectories for valuation of a compound option.

7.2.4 Exotic Options

Options with pay-offs determined in a more complex manner than by just comparing the price of the underlying with a strike are called exotic options. These non-vanilla options have been mostly over-the-counter bi-lateral contracts, which has allowed a high degree of innovation in these products. Pay-offs of exotic options may also depend on the entire path of the underlying stock during the 'life' of the option. For this property of the pay-off structure of these options, they earn the name 'path-dependent options.' Some examples of path-dependent options include Asian options, barrier options, and lookback options. Actually to think of it, American options are also path-dependent options, since they are exercised along the path of the life of the option, whenever it is considered beneficial to exercise. However, American options are mostly not referred to as path-dependent options.

We consider a few exotic options here, as a sampling of vast variety of existing exotic options, in order to demonstrate the fact that the option pricing and risk management consideration goes far beyond the plain-vanilla European or American options we have studied thus far. For evaluating the price of these exotic options, either analytical formulas may be obtainable, or binomial tree framework may be good in some cases. In other cases, simulation proves to be a useful alternative, where better accuracy may be obtained for more realistic models. For instance, path-dependent options like the barrier options are only approximately implemented in the binomial or even trinomial tree, but by using simulation they can be more accurately implemented.

We describe some of the simple exotic options, that are not path-dependent, defined on a single underlying stock.

Binary Option: This option provides a set pay-off, Q, based on whether the

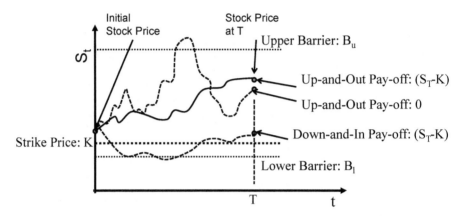

FIGURE 7.12: Pay-offs of an up-and-out barrier call option and a down-and-in barrier call option.

underlying asset is below or above a strike price. Therefore, the pay-off of a binary option is described as, $h(S_T) = Q\mathcal{I}_{\{S_T \geq K\}}$.

Compound Option: A compound option is an option on an option, which means it gives the buyer of the compound option the right to buy an underlying option to buy or sell the underlying stock at a future date. Therefore, the buyer of a compound option has the right to buy a call or a put option at a time T_1 for K_1 that will in its own turn mature at time $T_2 > T_1$. The underlying option has its strike set at K_2, which is the price the underlying stock may be bought or sold at depending on whether it is a call or put option. The pay-off of a compound option is, $h(S_{T_1}) = \max(c(T_1, S(T_1)) - K_1, 0)$ in the case of a compound option on a call, and $h(S_{T_1}) = \max(p(T_1, S(T_1)) - K_1, 0)$ in the case of a compound option on a put. Pricing of a compound option, when utilizing simulation, requires nested simulation of stock evolution as shown in Figure 7.11. For each stock price realization at T_1, in order to determine the pay-off of the compound option, the underlying option must be priced. This is accomplished by continuing to simulate stock evolution until the underlying option's maturity, T_2.

Chooser: In a way this is a compound option. It gives the buyer the right to choose to buy either a call or a put option for K_1 at time T_1 that may be exercised at T_2 for a strike K_2. The pay-off of a chooser option, therefore, is $h(S_{T_1}) = \max(c(T_1, S(T_1)) - K_1, p(T_1, S(T_1)) - K_1, 0)$. More complex chooser options can be created by a different combination of strikes and maturities.

The following options have weak or strong dependence on the path of the underlying stock leading up to the maturity of the option. The precise rules of pay-off and the way a buyer may utilize these options differ by the risk management objectives of the buyer.

Shout Option: In this option, the owner (buyer) of the option gets one chance to 'shout,' then the pay-off gets determined as European pay-off or the intrinsic value at the time of the shout, whichever is higher. In other words, the option gives the buyer the right to reset the rights features of the option. A shout call option, therefore, will have the following pay-off, $h(S_T) = \max(S_T - K, S_\tau - K, 0)$, where τ is the time of the shout. Since the buyer gets to choose when to shout, the option borrows some features of an American option, where one must determine the optimal exercise policy. In the case of a shout option, the buyer must determine the optimal shout strategy.

Bermudan Option: A Bermudan option resembles an American option, however early exercise is restricted to certain dates during the life of the option. Due to this restriction, a Bermudan option would be less costly than its American counterpart, however the mechanism for pricing the option would resemble that of pricing an American option.

Lookback Option: A lookback option earns that name because it determines its pay-off by looking back at what levels the stock price had realized during the life of the option. The pay-off of this option is determined from the maximum or minimum value of the stock price during the life of the option. There may be a fixed strike price utilized against this minimum or maximum value of the stock. Alternatively, the minimum or maximum value of the stock may serve as the strike, relative to the terminal value of the stock. Therefore, a sample pay-off of a lookback option could be, $h(S_T) = \max(\max_{0 \le t \le T}\{S_t\} - K, 0)$ or $h(S_T) = \max(S_T - \min_{0 \le t \le T}\{S_t\}, 0)$.

Barrier Option: This option utilizes a barrier above or below the current stock price to define the pay-off of the option. The barrier (up or down) is used to define either the activation or deactivation of the option. Therefore, an up-and-out barrier option will yield a pay-off only if the stock does not hit the up-barrier, while an up-and-in barrier option will yield a pay-off only if the stock does hit the barrier during the life of the option. Similarly, one can define down-and-in or down-and-out barrier options. Moreover, the pay-off can be call or put type pay-off defined in terms of a strike price, K. Figure 7.12 provides a display of a few trajectories indicating which ones yield positive versus zero pay-offs.

Asian Option: Asian options determine their pay-off in terms of averaging the stock price through the life of the option. These could be discrete-time averages or continuous-time averages, and averages could be arithmetic or geometric. For instance, the arithmetic continuous-time average of stock through the life of the option is given by,

$$\bar{S}_C = \frac{1}{T} \int_0^T S_t dt, \tag{7.73}$$

while arithmetic discrete-time average over chosen times of observations, $\{t_0, t_1, \ldots, t_N < T\}$, is given by,

$$\bar{S}_D = \frac{1}{N} \sum_{i=1}^{N} S_{t_i}. \tag{7.74}$$

Similarly, the geometric average can be defined for the life of the option as,

$$\bar{S}_G = \exp(\frac{1}{T} \int_0^T \ln(S_t) dt). \tag{7.75}$$

As is in the case of a lookback option, the average of the stock price can serve either as a strike or be compared to a fixed strike price, K, for determining the pay-off of the Asian option.

Options may also be constructed as a bundle of other derivatives or on a bundle of underlying stocks. We present some examples of these options.

Ladder: Ladders may be described as a mix of lookback and barrier type options. The 'ladder' is constructed by a sequence of increasing barriers, as each one is hit, a critical level of pay-off is ascertained. The fact that the pay-off remembers the maximum (or minimum) level of stock levels through the life of the option invokes resemblance with a lookback option.

Range Forward: A range forward is best described as a bundle of long-position in plain-vanilla European call and a short-position in a put option with different strikes but the same maturity. Usually the strike of the call is above the strike of the put option. The combined effect of the call and put package is the ability to buy the underlying for a range of price lying between the lower and the upper strike price. Therefore, a range forward resembles a forward contract, with the difference that a range forward will need a premium at the initiation of the contract.

Exchange: An exchange option essentially depends on several underlying assets, rather than a single underlying, essentially allowing exchange of assets. In the case of stocks, an exchange option can be designed to buy certain shares of one stock in exchange for certain shares of a different stock. There is no fixed strike price involved. If S_{1t} and S_{2t} are the two stocks on which an exchange must be defined, the pay-off of the option can be constructed as, $h(S_{1T}, S_{2T}) = \max(q_1 S_{1T} - q_2 S_{2T}, 0)$. This pay-off implies that the buyer of the option will have the right to buy q_1 units of the first stock in exchange for q_2 units of the second stock whenever this transaction is in her favor.

Two-Color Rainbow: Two-color or multi-color rainbow option is an option that depends on two or more underlying assets. The option pay-off is defined in terms of two or all the underlying stocks moving together in the intended direction. Therefore, a rainbow option is considered an option on the correlation between underlying stocks.

Basket: A basket option's pay-off is defined in terms of the value of a portfolio or a basket of assets/stocks. Therefore, the underlying instrument for a basket option is a weighted sum of different stocks, for instance an index option or an option on a stock portfolio.

Where relevant, the above options can be of American or European style, which implies that they can be exercised at any time during the life of the option or only at maturity.

7.2.5 American Options

American options are designed for complete flexibility on when the buyer exercises her right to utilize the option. For this reason, an American option is worth at least as much as the corresponding European option. Therefore, $C(t, S_t) \geq c(t, S_t)$, and $P(t, S_t) \geq p(t, S_t)$. However, in order to determine the precise value of an American option, we need to determine the optimal exercise policy alongside pricing the option. There is some chance that an American option will be exercised early, before the option expires. Whenever an American option is exercised early, its value would differ from its European counterpart. In the case of an American call option defined on a non-dividend paying stock, it can be shown that an early exercise is suboptimal.

In Figure 7.13, we display the algorithm to determine the price of an American option by utilizing the binomial tree model. The algorithm begins by first finding the terminal value of the option at maturity of the option. Following this, the algorithm progresses with backward recursion. At each step or node, one should determine whether an early exercise is better than the continuation value. If an early exercise is a better option, the node is marked as an early exercise node, and its value is that of an early exercise or the intrinsic value of the node. Once the backward recursion folds over to the initial time, the price of the American option is the initial node value.

The binomial tree approach is a popular way to compute the price of an American option due to its ease of implementation. The theme of the binomial tree based approach can be extended to continuous-time stock price evolution models for obtaining the price of an American option. The following sketch of an algorithm can be adopted to develop simulation based pricing of an American option under continuous-time evolution of the underlying stock price. Assume an American option matures at time, T, with a strike price, K, and the underlying stock evolves by a general model, such as in Eqn. (7.43), in the risk-neutral world.

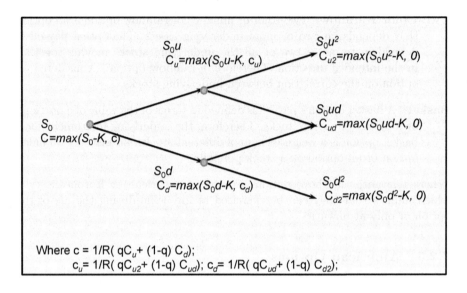

FIGURE 7.13: Pictorial display of algorithm to determine the price of an American option using the binomial tree model.

1. Generate a large sample of paths for the stock price evolution using any time-discretization, and analytical or simulation-based approach to solving the stock evolution model of Eqn. (7.43).

2. Discretize the time to maturity into any chosen granularity that matches the sample paths generated for stock evolution.

3. Begin the backward recursion starting with determining the intrinsic value of each terminal point (node) of each sample path, $V_T = \max(S_T - K, 0)$.

   (a) For each time point and each scenario of the stock price along the sample paths, determine the intrinsic value of the American option, $V_i = \max(S_t - K, 0)$, and the continuation value, $V_c = e^{-r\Delta t}V_{t+1}$, where Δt is the time discretization being used and V_{t+1} is the value of the node one time step in the future on the same sample path.

   (b) Fit a quadratic, $(a + bS_t + cS_t^2)$, or any other higher degree polynomial, least squares interpolation between the continuation values, V_c, obtained for all nodes on sample paths at time, t.

   (c) Compute the fitted continuation value, $V_c' = \hat{a} + \hat{b}S_t + \hat{c}S_t^2$ for each node on the sample paths at time, t. This step is done to achieve better accuracy in estimating continuation value by incorporating information from all nodes at time, t, instead of just considering the discounted value from a single node on each sample path.

(d) Compare and assign the value of each node as the higher of the fitted continuation value and intrinsic value, $V_t = \max(V'_c, V_i)$.

(e) Advance the recursion back to the previous time step.

4. Once the backward recursion reaches the initial time, $t = 0$, the value of the initial node, V_0, is the price of the option. Therefore, $C(0, S_0) = V_0$, and all the exercise points along all the sample paths define the optimal exercise policy.

In the case of European options, once the European call option was priced, the put-call parity could be used to price the corresponding put option. In the case of American options, the put-call parity reduces to the following inequalities,

$$S_t - K < C(t, S_t) - P(t, S_t) < S_t - Ke^{-r(T-t)}. \tag{7.76}$$

Therefore, once the price of an American call option is determined, it will only provide a range for the price of the corresponding American put option. In order to determine the precise price, the above algorithm will need to be adopted to compute the price of the American put option.

7.2.6 Generalizing the Models in Black-Scholes-Merton

In Section 7.2.2, we entertained a long discussion regarding implied volatility and the insights it can offer us. Volatility is by far the most important factor in derivatives pricing. The notion of implied volatility is considered very useful to assess suitability of historic volatility estimates in predicting future volatility. By using market price of options traded in a liquid options market, implied volatility offers crucial information about the market's expectation of future volatility.

Insight obtained from the shape of the volatility curve, as shown in Figure 7.9, or the entire volatility surface, also shows the extent to which the lognormal distribution of the underlying stock evolution model captures the true tail risk of the stock price. If deep in-the-money and/or deep out-of-the-money options have much more enhanced implied volatility than the near at-the-money options, this suggests that the lognormal distribution is failing to capture the tail risk in the future stock price. Moreover, asymmetries in the volatility curve, such as in a volatility smirk with a skew to the left, are indicative of investors' worry about sudden large downward movement of the stock, or more adversely, a market crash.

In this section, we present a few enhancements of the Black-Scholes geometric Brownian motion model for stock price evolution, with an intent to expand the modeling toolkit that addresses the above challenges in predicting future stock price characteristics more accurately. Our intent here is only to provide an overview that serves as pointers to directions for further exploration to the reader. For more detailed development of these more advanced models, please refer to other sources [45, 81, 27, 18, 25, 77].

7.2.6.1 Constant Elasticity of Variance (CEV) Model

A simple generalization of the Black-Scholes model is obtained by modifying the diffusion term in the stochastic differential equation, as follows.

$$dS_t = \mu S_t dt + \sigma S_t^\alpha dW_t, \tag{7.77}$$

where either $\alpha > 1$ or $0 < \alpha < 1$. Of course, if $\alpha = 1$, then the model reduces to the geometric Brownian motion model. This modification of the diffusion term gives us the opportunity to explore some variety of behavior in stock volatility. Note that now the coefficient σ does not retain the same interpretation as the geometric Brownian motion model, i.e., it is no longer the volatility of the log-return of the stock price.

When $0 < \alpha < 1$, volatility of the stock increases as the stock price decreases, resulting in a **heavy left-tail** and **less heavy right-tail**. This lends a skewness to the log-return distribution. As stated above, it captures the asymmetry in the volatility curve when the implied volatility for different strikes is different displaying a volatility smirk. Similarly, a choice of $\alpha > 1$ models a stock whose volatility increases with increase in stock price. This will result in a **heavier right-tail than left-tail**. Therefore, this model modification is suitable to capture asymmetric volatility smiles or volatility smirks.

The price of a European call option on the stock, or any other European style option on the stock, is a function, $c(t, S_t)$. Applying the Ito formula, the price of the option must evolve by the following equation,

$$dc(t, S_t) = (\frac{\partial c}{\partial t} + \mu S_t \frac{\partial c}{\partial x} + \frac{1}{2}\sigma^2 S_t^{2\alpha} \frac{\partial^2 c}{\partial x^2})dt + \sigma S_t^\alpha \frac{\partial c}{\partial x}dW_t. \tag{7.78}$$

The rest of the derivation for option pricing follows as developed for the general case in Section 7.2.1. The solution for the price of the option is obtained as,

$$c(t, S_t) = E[e^{-r(T-t)}(S_T - K)_+], \tag{7.79}$$

where the stock evolves in the risk-neutral world as follows.

$$dS_t = rS_t dt + \sigma S_t^\alpha dW_t. \tag{7.80}$$

The price of the option can be obtained by simulation as described in Section 7.2.1.2

Although we know how the option would be priced under the CEV model, its actual value would be obtainable only if the parameters of the model are determined. We first look at model calibration under the real-world measure. Clearly, if the model is a better representation of the stock evolution than the geometric Brownian motion model, the model can also be used for developing optimal investment or hedge strategies. In these cases, the stock evolves by its real-world model of Eqn. (7.77). Therefore, we calibrate the real-world CEV model by first applying the Euler scheme to the model in Eqn. (7.77).

$$\Delta S_t \approx \mu S_t \Delta t + \sigma S_t^\alpha \Delta W_t, \tag{7.81}$$

which is rearranged to obtain,

$$\frac{\Delta S_t - \mu S_t \Delta t}{S_t^\alpha} \approx \sigma \Delta W_t. \tag{7.82}$$

At this stage, estimates of the parameters, α, μ and σ, can be obtained by applying the quasi-maximum likelihood method. The quasi-maximum likelihood method will utilize the fact, $\sigma \Delta W_t \sim N(0, \sigma \sqrt{\Delta t})$. The method of moments will utilize the first two moments of the same distribution. However one drawback of the method of moments in this case is that the parameter α will need to be determined by some other means, since the distribution of $N(0, \sigma \sqrt{\Delta t})$ is completely determined by its first two moments. The method of moment essentially implies solving the following two moment equations,

$$E[\frac{\Delta S_t - \mu S_t \Delta t}{S_t^\alpha}] \approx 0, \tag{7.83}$$

$$Var(\frac{\Delta S_t - \mu S_t \Delta t}{S_t^\alpha}) \approx \sigma \sqrt{\Delta t}. \tag{7.84}$$

We solve for μ and σ in the above two equations, by first solving for $\hat{\mu}$ from Eqn. (7.83), followed by substituting this value in Eqn. (7.84) to obtain $\hat{\sigma}$.

We have computed the price of an option under the CEV model using simulation and calibrated the model using the method of moments with the goal of following the simplest approach. In actuality, analytical solution of the CEV model is obtainable. Moreover, the price of the European call and put option can also be obtained in closed-form, albeit as a more complicated formula than the Black-Scholes option pricing formula. Both of these can be utilized for calibrating the model in the real world, as well as calibrating the model under the risk-neutral probability measure.

Suppose the closed-form price of European call and put option under the CEV model are $c_{cev}(t, S_t; K_i, T_i)$ and $p_{cev}(t, S_t; K_i, T_i)$, respectively, for a range of strike prices, $\{K_i\}$, and maturities, $\{T_i\}$. If data for the market price of these options is available, $c_{mkt}(t, S_t; K_i, T_i)$ and $p_{mkt}(t, S_t; K_i, T_i)$, then the parameters, α and σ, can be obtained by minimizing the sum of squared difference between CEV model-based price and market price of the options [41], as follows.

$$\min_{\alpha, \sigma} \sum_i (c_{cev}(t, S_t; K_i, T_i) - c_{mkt}(t, S_t; K_i, T_i))^2 \tag{7.85}$$

$$+ (p_{cev}(t, S_t; K_i, T_i) - p_{mkt}(t, S_t; K_i, T_i))^2.$$

The calibrated risk-neutral CEV model can then be used to price other options defined on the underlying stock.

7.2.6.2 Model for Several Correlated Stocks

In Section 6.4.1, we had presented a multidimensional asset price evolution model, given in Eqn. (6.37). We will now apply the model to the evolution of a pair of correlated stock prices. The model can then be easily generalized to a more general case of d-dimensional correlated stock price evolution. In the following model, the two stocks, S_{1t} and S_{2t}, individually evolve by the geometric Brownian motion model, however the two Wiener processes driving the two stock prices, W_{1t} and W_{2t}, are correlated. The model is summarized as follows.

$$
\begin{aligned}
dS_{1t} &= \mu_1 S_{1t} dt + \sigma_1 S_{1t} dW_{1t}, & (7.86) \\
dS_{2t} &= \mu_2 S_{2t} dt + \sigma_2 S_{2t} dW_{2t}, & (7.87) \\
cov(dW_{1t}, dW_{2t}) &= E[dW_{1t} dW_{2t}] = \rho dt. & (7.88)
\end{aligned}
$$

From the above model, we first note that individually the two stocks follow the geometric Brownian motion process, however the correlation between the two driving Wiener processes, $corr(dW_{1t}, dW_{2t}) = \rho$. We can alternatively choose any other model for the two stocks, such as model one or both using a CEV model if there is justification for this choice, keeping the correlation structure between the two driving Wiener processes the same.

Suppose we want to price an option defined on the two stock, $f(t, S_{1t}, S_{2t})$, then by applying the two-dimensional Ito formula, the price of the option will satisfy the following stochastic differential equation.

$$
\begin{aligned}
df(t, S_{1t}, S_{2t}) &= \left(\frac{\partial f}{\partial t} + \mu_1 S_{1t} \frac{\partial f}{\partial x_1} + \frac{1}{2}\sigma_1^2 S_{1t}^2 \frac{\partial^2 f}{\partial x_1^2} \right. & (7.89) \\
&+ \left. \mu_2 S_{2t} \frac{\partial f}{\partial x_2} + \frac{1}{2}\sigma_2^2 S_{2t}^2 \frac{\partial^2 f}{\partial x_2^2} + \frac{1}{2}\rho\sigma_1\sigma_2 S_{1t} S_{2t} \frac{\partial^2 f}{\partial x_1 \partial x_2} \right) dt \\
&+ \sigma_1 S_{1t} \frac{\partial f}{\partial x_1} dW_{1t} + \sigma_2 S_{2t} \frac{\partial f}{\partial x_2} dW_{2t}.
\end{aligned}
$$

In order to create a risk-free portfolio towards pricing the two-stock option, we would need to invest in the two-stock option and both the stocks, as follows.

$$
\Pi(t, S_{1t}, S_{2t}, f(t, S_{1t}, S_{2t})) = w_1 f(t, S_{1t}, S_{2t}) + w_2 S_{1t} + w_3 S_{2t}, \quad (7.90)
$$

where in order to eliminate the risky terms arising from the two Wiener processes, (dW_{1t}, dW_{2t}), we will opt for portfolio weights as follows.

$$
\begin{aligned}
f(t, S_{1t}, S_{2t}) \quad &: \quad -1, & (7.91) \\
S_{1t} \quad &: \quad \frac{\partial f}{\partial x_1}, & (7.92) \\
S_{2t} \quad &: \quad \frac{\partial f}{\partial x_2}. & (7.93)
\end{aligned}
$$

The above choice of portfolio weight makes the portfolio risk-free, therefore it should evolve by the following equation to eliminate arbitrage.

$$
d\Pi(t, S_{1t}, S_{2t}, f(t, S_{1t}, S_{2t})) = r\Pi(t, S_{1t}, S_{2t}, f(t, S_{1t}, S_{2t})) dt. \quad (7.94)
$$

As done in the case of an option based on a single stock, we match the two definitions for the evolution of the replicating portfolio, to obtain the following partial differential equation for the price of the two-stock option.

$$\frac{\partial f}{\partial t} + \frac{1}{2}\sigma_1^2 S_{1t}^2 \frac{\partial^2 f}{\partial x_1^2} \quad + \quad \frac{1}{2}\sigma_2^2 S_{2t}^2 \frac{\partial^2 f}{\partial x_2^2} + \frac{1}{2}\rho\sigma_1\sigma_2 S_{1t} S_{2t} \frac{\partial^2 f}{\partial x_1 \partial x_2} \quad (7.95)$$

$$+ rS_{1t}\frac{\partial f}{\partial x_1} \quad + \quad rS_{2t}\frac{\partial f}{\partial x_2} = rf(t, S_{1t}, S_{2t}),$$

along with the end condition, $f(T, S_{1T}, S_{2T}) = h(S_{1T}, S_{2T})$. We invoke the two-dimensional version of the Feynmann-Kac theorem to obtain the price of the option to be given as,

$$f(t, S_{1t}, S_{2t}) = E[e^{-r(T-t)}h(S_{1T}, S_{2T})], \quad (7.96)$$

with the stocks evolving in the risk-neutral world as follows.

$$dS_{1t} = rS_{1t}dt + \sigma_1 S_{1t}dW_{1t}, \quad (7.97)$$

$$dS_{2t} = rS_{2t}dt + \sigma_2 S_{2t}dW_{2t}, \quad (7.98)$$

$$cov(dW_{1t}, dW_{2t}) = E[dW_{1t}dW_{2t}] = \rho dt. \quad (7.99)$$

Calibration of the individual stock parameters, μ_1, μ_2, σ_1, and σ_2, would be as before, depending on whether this calibration is required in the real-world or the risk-neutral world. The remaining parameter, correlation coefficient ρ, will be estimated as follows.

$$\hat{\rho} = corr(\ln(\frac{S_{1(t+\Delta t)}}{S_{1t}}), \ln(\frac{S_{2(t+\Delta t)}}{S_{2t}})) \sim corr(\frac{\Delta S_{1t}}{S_{1t}}, \frac{\Delta S_{2t}}{S_{2t}}). \quad (7.100)$$

For a higher dimensional set of correlated stocks, both the derivation for option price formula, as well as calibration, will proceed as developed above. For option price, the results for higher-dimensional Ito formula and Feynmann-Kac theorem will be utilized, where writing in matrix notation will simplify the presentation. For calibration, pair-wise correlation between Wiener processes will proceed as above.

Finally, in order to simulate trajectories of a pair of correlated stocks, or in fact a d-dimensional set of correlated stocks, we will need to refer to Chapter 4. In Section 4.4.3.3, the method for generating bi-variate and multi-variate normal random variates was developed, starting from a set of independent standard normal random variates. In our case here, increments in the Wiener process $(\Delta W_{1t}, \Delta W_{2t})$ are normally distributed with mean $(0,0)$, variance $(\Delta t, \Delta t)$, and correlation coefficient $corr(\Delta W_{1t}, \Delta W_{2t}) = \rho$. The algorithm in Section 4.4.3.3 can be applied to these settings for generating correlated Wiener increment for simulating trajectories of the pair (or set) of stocks.

7.2.6.3 Extensions in Option Pricing - Stochastic Volatility

The motivation for this section was to develop models that more closely capture the empirically observed characteristics of stock price evolution.

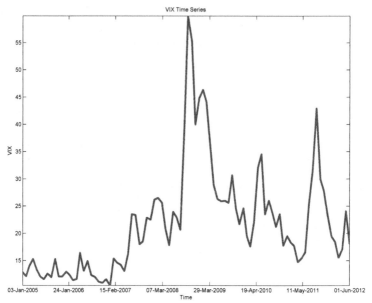

FIGURE 7.14: Monthly observations of VIX index from January 2005 through mid-2012. The variability in the stock market is captured in this index through the financial crises of 2008 and euro crisis evolving through 2011-2012.

Black-Scholes option pricing formula assumes the volatility of the stock to be constant. In reality, stock prices do not seem to satisfy this assumption. In Figure 7.14, we display the time-varying volatility of the stock market (VIX index) over a period of seven years. The plot makes the point about volatility not being constant rather evident.

VIX is the ticker symbol and shortened name of CBOE (Chicago Board Options Exchange) Market Volatility Index. VIX is a weighted sum of implied volatility for a range of options on the S&P 500 stock index. It is quoted in percentage points and translates, roughly, to the expected movement in the S&P 500 index over the upcoming 30-day period, which is then annualized. In 2004, futures on the VIX index began trading, followed by VIX options in 2006.

In our discussion of implied volatility in Section 7.2.2, we observed that if the implied volatility is plotted for a range of strike prices or a range of *moneyness*, the plot is not a straight line. If the assumptions of Black-Scholes framework were true, the plot would be a straight line. But for the case of most stocks this is a convex curve, with a skew to the left, earning the name **volatility smirk**. The leftward skew of the volatility curve for stocks is often termed the *leverage effect*, implying that investors become extra nervous when stock price goes too low for a debt financed firm.

In practice, stock returns show skewness, excess kurtosis, serial correlation and time-varying volatility. Focusing on the time-varying volatility, we now develop an extension to the Black-Scholes framework, where we make the volatility of a stock not only time-varying, but also stochastic. This extension brings with it new complexities, such as, first, volatility of a single stock is not spanned by traded assets in the economy, and second, risk related with the volatility cannot be eliminated by usual arbitrage methods. Finally, in order to address the 'incompleteness' of the market for volatility risk, market price of risk due to volatility appears in the option price derivation. We provide an overview of these enhancements.

We consider the following extension of the Black-Scholes stock evolution model, where the volatility of the stock price, $\sigma_t = \sqrt{v_t}$, evolves by a separate model given as follows.

$$dS_t = \mu S_t dt + \sqrt{v_t} S_t dW_{1t}, \tag{7.101}$$
$$dv_t = -\gamma(v_t - \bar{v})dt + \eta\sqrt{v_t}dW_{2t}, \tag{7.102}$$

where W_{1t}, W_{2t} are taken as independent Wiener processes, or as a further generalization, they can be modeled as correlated Wiener processes, as in the two stock case in Section 7.2.6.2. The equation for stock evolution, Eqn. (7.101), is similar to the usual Black-Scholes stock price equation, with the volatility, $\sqrt{v_t}$, being time-dependent and stochastic. The stochastic volatility model, Eqn. (7.102), is a typical square-root process model of Cox, Ingersoll, and Ross (CIR), seen again in Chapter 8 for modeling interest rates. The above stochastic volatility model is the popular Heston model [39].

In the Heston model, the drift term for the stock variance, v_t, has a specific form, $-\gamma(v_t - \bar{v})$, which lends mean-reversion to the volatility. This implies that when the volatility becomes too high (or too low), it has a tendency to revert to its long-run mean, $\bar{v}$. The rate of mean reversion is captured by the parameter, γ. The model picks after the empirical evidence in volatility being mean-reverting, also visible in the VIX time series in Figure 7.14. The constant, η, is the volatility of the volatility of the stock.

We shift our attention to pricing options under stochastic volatility models for stock evolution. The price of an option now depends on two variables, due to two sources of risk from two Wiener processes. Therefore, let's say the price of an option is, $f(t, S_t, v_t)$. By applying the two-dimensional Ito's formula we obtain that the price of the option will satisfy the following stochastic differential equation.

$$df(t, S_t, v_t) = (\frac{\partial f}{\partial t} + \mu S_t \frac{\partial f}{\partial x_1} + \frac{1}{2}v_t S_t^2 \frac{\partial^2 f}{\partial x_1^2} \tag{7.103}$$
$$- \gamma(v_t - \bar{v})\frac{\partial f}{\partial x_2} + \frac{1}{2}\eta^2 v_t \frac{\partial^2 f}{\partial x_2^2} + \frac{1}{2}\eta v_t S_t \frac{\partial^2 f}{\partial x_1 \partial x_2})dt$$
$$+ \sqrt{v_t}S_t \frac{\partial f}{\partial x_1}dW_{1t} + \eta\sqrt{v_t}\frac{\partial f}{\partial x_2}dW_{2t}.$$

In order to create a risk-free portfolio now, we will need the option, the stock, as well as another asset whose value depends on the variance, v_t. We call this asset, V_t. The portfolio is constructed as follows.

$$\Pi(t, S_t, v_t, f(t, S_t, v_t)) = w_1 f(t, S_t, v_t) + w_2 S_t + w_3 V_t, \qquad (7.104)$$

where in order to make this portfolio risk-free, we need to eliminate the risky terms arising from both the Wiener processes, (dW_{1t}, dW_{2t}). We will opt for the following portfolio weights to achieve this.

$$
\begin{aligned}
f(t, S_t, v_t) \quad &: \quad -1, \\
V_t \quad &: \quad \frac{\frac{\partial f}{\partial x_2}}{\frac{\partial V}{\partial x_2}}, \\
S_t \quad &: \quad \frac{\partial f}{\partial x_1} - [\frac{\frac{\partial f}{\partial x_2}}{\frac{\partial V}{\partial x_2}}]\frac{\partial V}{\partial x_1}.
\end{aligned}
\qquad (7.105)
$$

One can test that the above choice of portfolio weights makes the portfolio risk-free, therefore it should evolve by the following equation to eliminate arbitrage.

$$d\Pi(t, S_t, v_t, f(t, S_t, v_t)) = r\Pi(t, S_t, v_t, f(t, S_t, v_t))dt. \qquad (7.106)$$

As done in the case of simple option pricing derivation, we match the two definitions for the evolution of the replicating portfolio. However, this time we separate the terms depending on $f(t, S_t, v_t)$ and V_t as follows.

$$
\frac{\frac{\partial f}{\partial t} + \frac{1}{2}v_t S_t^2 \frac{\partial^2 f}{\partial x_1^2} + \frac{1}{2}\eta v_t S_t \frac{\partial^2 f}{\partial x_1 \partial x_2} + \frac{1}{2}\eta^2 v_t \frac{\partial^2 f}{\partial x_2^2} + rS_t \frac{\partial f}{\partial x_1} - rf}{\frac{\partial f}{\partial x_2}}
$$
$$
= \frac{\frac{\partial V}{\partial t} + \frac{1}{2}v_t S_t^2 \frac{\partial^2 V}{\partial x_2^2} + \frac{1}{2}\eta v_t S_t \frac{\partial^2 V}{\partial x_1 \partial x_2} + \frac{1}{2}\eta^2 v_t \frac{\partial^2 f}{\partial x_2^2} + rS_t \frac{\partial V}{\partial x_1} - rV}{\frac{\partial V}{\partial x_2}}. \qquad (7.107)
$$

The key observation we make here is that the left- and right-hand sides of Eqn. (7.107) depend on arbitrarily picked option, $f(t, S_t, v_t)$ and asset, V_t. Therefore, the two sides of the equation must depend only on terms that are independent of $f(t, S_t, v_t)$ and $V(t, S_t, v_t)$. Without loss of generality, we choose this term to have the form, $\phi(t, S_t, v_t) = \gamma(v_t - \bar{v}) - \eta\sqrt{v_t}\lambda(t, S_t, v_t)$, where $\lambda(t, S_t, v_t)$ is a yet-to-be-defined term. Therefore, the option price under stochastic volatility should satisfy the following partial differential equation.

$$
\frac{\partial f}{\partial t} + \frac{1}{2}v_t S_t^2 \frac{\partial^2 f}{\partial x_1^2} + \frac{1}{2}\eta v_t S_t \frac{\partial^2 f}{\partial x_1 \partial x_2} + \frac{1}{2}\eta^2 v_t \frac{\partial^2 f}{\partial x_2^2} \qquad (7.108)
$$
$$
+ rS_t \frac{\partial f}{\partial x_1} + (\gamma(v_t - \bar{v}) - \eta\sqrt{v_t}\lambda(t, S_t, v_t))\frac{\partial f}{\partial x_2} - rf(t, S_t, v_t) = 0,
$$

along with the end condition at option's maturity, T, given by, $f(T, S_T, v_T) = h(S_T)$. We can show that the solution of the above equation is,

$$f(t, S_t, v_t) = E[e^{-r(T-t)}h(S_T)], \qquad (7.109)$$

where the stock evolves in the risk-neutral measure by the following equations,

$$dS_t = rS_t dt + \sqrt{v_t} S_t dW_{1t}, \qquad (7.110)$$

$$dv_t = (\gamma(v_t - \bar{v}) - \eta\sqrt{v_t}\lambda(t, S_t, v_t))dt + \eta\sqrt{v_t}dW_{2t}. \qquad (7.111)$$

The quantity we introduced in Eqn. (7.109), $\lambda(t, S_t, v_t)$, is called **market price of risk** for the non-tradable risk factor, v_t. This is defined as the ratio of difference of mean of the variable and risk-free return to the volatility of the variable. Therefore, if μ_v is the drift and σ_v is the diffusion of the non-tradable risk factor, then market price of this risk is defined as, $\lambda = \frac{\mu_v - r}{\sigma_v}$. Let's see why this term is called market price of risk. If we consider the portfolio we constructed in the simple (non-stochastic volatility) case, we would have, $\Pi_1 = -f + \frac{\partial f}{\partial x_1} S_t$. If we apply the Ito's lemma to determine the change in the value of this portfolio, we obtain the following by utilizing the fact that the option price, $f(t, S_t, v_t)$, satisfies Eqn. (7.109).

$$d\Pi_1(t, S_t, v_t) - r\Pi_1 dt = \eta\sqrt{v_t}\frac{\partial f}{\partial x_2}(\lambda(t, S_t, v_t)dt + dW_{2t}). \qquad (7.112)$$

This derivation shows that the portfolio which became risk-free in the simple, non-stochastic volatility case, now must offer some excess return. The excess return per unit of volatility, dW_{2t}, is given by $\lambda(t, S_t, v_t)$, hence it is called the market price of volatility risk.

Finally, there is the issue of calibrating the model, both in real-world and risk-neutral measure. There are now additional parameters to estimate beyond those in the non-stochastic volatility cases. Specifically, the initial value of volatility (v_0), market price of volatility risk (λ), long-run volatility ($\bar{v}$), volatility of volatility (η), and the mean reversion rate (γ).

We apply the Euler scheme to discretize the model in Eqns. (7.101) and (7.102), or alternatively to calibrate the model in the risk-neutral world discretize the model of Eqns. (7.110) and (7.111), on a time discretization, $\{t_0, t_1, \ldots, t_N\}$, to obtain the following.

$$S_{t_{k+1}} = S_{t_k} + \mu S_{t_k} \Delta t + \sqrt{v_{t_k}} S_{t_k} \Delta W_{1t_k}, \qquad (7.113)$$

$$v_{t_{k+1}} = v_{t_k} - \gamma(v_{t_k} - \bar{v})\Delta t + \eta\sqrt{v_{t_k}}\Delta W_{2t_k}. \qquad (7.114)$$

On re-writing the above two equations as follows,

$$S_{t_{k+1}} = S_{t_k}(1 + \mu\Delta t + \sqrt{v_{t_k}}\Delta W_{1t_k}), \qquad (7.115)$$

$$v_{t_{k+1}} = \gamma\bar{v}\Delta t + v_{t_k}(1 - \gamma\Delta t) + \eta\sqrt{v_{t_k}}\Delta W_{2t_k}. \qquad (7.116)$$

we observe their resemblance to the well-known and widely popular GARCH(1,1) econometric model of stochastic volatility.

$$S_{k+1} = S_k(1 + \tilde{\mu} + \sigma_k\epsilon_{1k}), \qquad (7.117)$$

$$\sigma_{k+1}^2 = \omega + \beta\sigma_k^2 + \alpha\epsilon_{2k}. \qquad (7.118)$$

If we had used higher-order discretization schemes for the original continuous-time model, they would be equivalent to the more general GARCH(p,q) models. In order to calibrate the GARCH models, therefore their equivalent continuous-time models of stochastic volatility, we refer the reader to extensive literature on estimating parameters of Generalized Autoregressive Conditional Heteroskedasticity (GARCH) models [13, 76]. Stock return data would be utilized for calibrating the model in the real-world, while for risk-neutral world, implied volatility data based on market prices of options will be necessary. Once parameters are estimated, stock price trajectories can be obtained by simulation using the same discretizations as given in Eqns. (7.113) and (7.114) or higher order ones.

7.2.6.4   Large Sudden Changes in Prices - Jump Diffusion Model

Skewness and excess kurtosis in stock returns arise from one other empirical characteristic, sudden large changes either in the stock price or volatility, or both. This is the last direction of enhancement we consider in modeling stock price evolution. For incorporating sudden large changes, we utilize the continuous-time Markov chain process studied in Chapter 5. We specifically utilize the Poisson process described in Section 5.3.2.

Poisson process, N_t, as stated above, is a continuous-time Markov chain. It is a counting process with events happening at inter-arrival times that are independent, exponentially distributed. The Poisson process has an intensity, λ, for any $t > 0$, and for any time t, the probability distribution is,

$$P(N_t = n) = e^{\lambda t} \frac{(\lambda t)^n}{n!}. \tag{7.119}$$

Therefore, we can also show that at any time, t, the expected number of events occurred is $E[N_t] = \lambda t$, while the variance is, $Var(N_t) = \lambda t$. Poisson process has independent and stationary increments. Independent increments implies, if $s > 0$, $N_{t+s} - N_t$ is independent of $\mathcal{F}_t$, where $\mathcal{F}_t = \sigma(N_s, s \leq t)$ is the filtration generated by the Poisson process. Poisson process has stationary increments since $N_{t+s} - N_t$ has the same distribution as $N_s - N_0$ for all s, t.

The jump-diffusion process is a mixture of jump process, such as the Poisson process indicating the arrival of a jump, and diffusion process. The process mostly evolves diffusively, interrupted by jumps causing discontinuities in the trajectory of the process. The Poisson process captures the random occurrence of events, the actual impact, i.e., size of the jump must be additionally described. We define a sequence, U_j, of independent, identically distributed random variables taking values in the interval, $[-1, +\infty]$, to model the size of the jumps. Therefore, j^{th} jump arriving by the Poisson process is of size U_j. Therefore, the cumulative effect of jumps up to time t is, $Y_t = \sum U_j \mathcal{I}_{j \leq N_t}$, where $\mathcal{I}$ is an indicator function. The Y_t process is called a compound Poisson process.

If τ_i are random times when jumps occur, in the time interval $[\tau_j, \tau_{j+1})$, the jump-diffusion process follows the Black-Scholes model, $dS_t = \mu S_t dt + \sigma S_t dW_t$, or any other advanced diffusion model. At time τ_j, the impact of jump on the stock prices is given by $\Delta S_{\tau_j} = S_{\tau_j} - S_{\tau_j^-} = S_{\tau_j^-} U_j$, where $S_{\tau_j^-}$ refers to the left-limit of the diffusive evolution as $t \uparrow \tau_j$. Thus, $S_{\tau_j} = S_{\tau_j^-}(1 + U_j)$.

Combining the joint impact of diffusion, arrival of jumps and the size of jumps the stock evolution is given as follows.

$$S_t = S_0 \prod_{j=1}^{N_t} (1 + U_j) e^{(\mu - \sigma^2/2)t + \sigma W_t}, \tag{7.120}$$

using the convention $\prod_{j=1}^{0} = 1$. The process thus constructed is a **right-continuous** process, adapted to a filtration constructed by union of filtration of each component of the process. And finally, on each interval $[0, t]$, it has only finitely many discontinuities. We have constructed one example of a jump-diffusion process; other varieties can be constructed by picking different choices for each component of the process.

Pricing of derivatives under jump-diffusion processes becomes more involved. Constructing a replicating portfolio is achievable, however the replication would break down should there be a jump in the stock price at the time of maturity. In order to deal with this complication, the derivation becomes a little more involved. The derivation follows the objective of minimizing the risk of mismatch of the replicating portfolio due to jump occurring at option maturity. It can be proven that an option defined on the stock evolving by the above jump diffusion process will be priced as [90],

$$f(0, S_0) = E_0[e^{-rT} h(S_T)], \tag{7.121}$$

$$\text{where } S_T = S_0 \prod_{j=1}^{N_T} (1 + U_j) e^{(r - \lambda m - \sigma^2/2)T + \sigma W_T}. \tag{7.122}$$

As usual, $h(S_T)$ represents the pay-off of the option at maturity, T, and $m = E[U_i] - 1$. Among other generalizations, one that may be useful is when λ, the rate of arrival of jumps, is also made time-dependent, i.e., $\lambda(t)$. We will use this feature in Chapter 9 for default risk modeling.

In Section 5.3.2, we had described one method for simulating trajectories of the Poisson process. This can be combined with outcomes of the diffusive component to simulate the trajectories of the jump-diffusion process. Alternatively, one can also utilize the fact that $\Delta N_t \sim Po(\lambda \Delta t)$ for generating Poisson trajectories, especially when one is attempting to match Poisson trajectories on a time discretization created for the diffusive component. This may be needed, for instance, for pricing path-dependent options. The latter approach will also be required when the rate of arrival of jumps is time-dependent, $\lambda(t)$.

We have seen several examples of extensions beyond the Black-Scholes model for stock evolution in this section. The more general class of processes that have been applied to problems in finance, with Poisson process, Wiener process, and jump-diffusion processes being special cases, is Lévy processes.

A stochastic process that almost surely starts at zero, has independent and stationary increments, and whose trajectories are right-continuous with left-limits is a Lévy process. This is clearly a very general definition, and also includes processes that have infinitely many small jumps in a finite interval, $(0, t]$. For exploring this topic further, the reader may refer to a dedicated book on this topic [77].

7.3 Equity Hedging Strategies

Stocks are a pure investment asset class. While firms issue shares of their stock for financing their investments in exchange for a residual claim on the firm's assets and earnings, investors seek the upside potential of obtaining high returns from their stock investment. Stocks expose investors to varying levels of high risk, therefore depending on their goals of investment, investors must determine their preferred equity risk-return profile and ways by which this would be achieved. Inclusion of equity derivatives in the development of investment strategies can help crave out precise strategies to achieve the investment goals of the investor. We will develop equity hedging strategies keeping this as one of the motivating contexts. Exposure to equity derivatives is not only done to reduce risk, but they may also be utilized to enhance the upside potential or for speculation.

Writing options, whether vanilla ones or the large range of exotic equity options, is very risky. The downside of the long position in options is losing the initial premium, while the upside may be unlimited. Therefore, options are considered highly leveraged instruments. The reverse is true for the option writer. The upside of writing an option is limited, but the downside can be huge. Option writers and market makers for equity options must perform very active risk management of their options portfolios. Some of the strategies we will discuss are geared to address this challenge.

In the development of this chapter thus far, we have focused heavily on equity options, both vanilla and exotic ones, on single stocks. There is a broader kind of equity derivatives that should be included in the discussion at this point. Instruments beyond equity options we have not seen so far include equity futures, equity swaps, index options, equity basket derivatives, and fund derivatives. As discussed in the context of stochastic volatility models in Section 7.2.6.3, over the years equity volatility has emerged as another asset class. The variety of instruments in this category includes VIX futures and options, which we mentioned earlier, variance swaps on indices as well as single stocks, variance derivatives, such as conditional variance, options on variance, and correlation.

Some of the strategies we will construct are static in nature, implying once the positions are established in relevant instruments with desired features, the positions are maintained until the derivatives mature. Static strategies are cost effective and low-maintenance, therefore are attractive to consider whenever it is possible to achieve the hedging objectives using static strategies. In other cases, the hedge positions must be modified in response to changing conditions of the risk factors. Clearly, such strategies would require active monitoring, taking periodic action to implement the strategy and to achieve the goals of the strategy. Periodic action can also result in incurring high transaction costs. Finally, the fundamental principle for pricing derivatives is to eliminate arbitrage opportunities. No-arbitrage conditions are theoretical motivations for interrelation of derivative prices and price of the underlying stocks or stock indices. In practice, arbitrage opportunities can emerge in the equity and equity derivative prices, and strategies that are designed to take advantage of these opportunities are called arbitrage strategies.

In order to create the best risk management response to equity risk, using equity derivatives, we must utilize frameworks that help determine optimal hedge decisions in all the contexts we have discussed above. In Section 7.3.2, we will pick some of the discussed contexts as illustration to develop an optimal hedge framework that can be applied to other cases.

7.3.1   Static Hedging Strategies

Consider a long position in a stock: for the duration of time the investor intends to maintain this position, she is exposed to the risk of the stock price dropping, say relative to her purchase price, S_0. A protective put strategy adds a long position in a plain vanilla European put option with the desired maturity, depending on the availability of maturities of put options and the investor's planning horizon to hold the stock. The strike of the put option depends on the level of protection sought by the investor, keeping in mind that the higher the strike price, the higher the cost of the hedge. The resulting pay-off of this strategy is shown in the left panel of Figure 7.15. The combination of a long position in a stock and put option resembles a long position in a call on the underlying asset with the same strike as the put option, as expected from the put-call parity.

Moreover, the strategy may be applied to an equity portfolio or to a stock index by utilizing a basket put option or an index put option. In these cases, one must also address the issue of basis risk, i.e., the protective put option not exactly matching the underlying portfolio, thus not resulting in a perfect protective put hedge. Purchasing put options against the assets held in a portfolio is synonymous with taking out an insurance on those assets. The reverse of a protective put or a short protective put strategy is shown in the right panel of Figure 7.15.

If a market maker has written a call option, it is exposed to the downside risk of the option being exercised, which can result in an unlimited loss. One way the option writer can protect itself of this loss is to take a long position in the underlying stock. This hedge will come at the cost of the current price

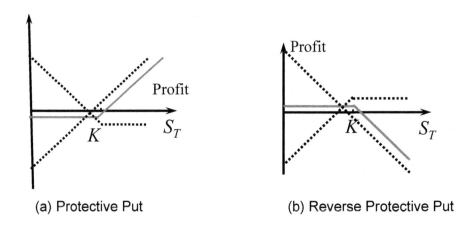

(a) Protective Put   (b) Reverse Protective Put

FIGURE 7.15: (a) Profit and individual positions of a protective put. (b) Profit and individual positions of a reverse protective put.

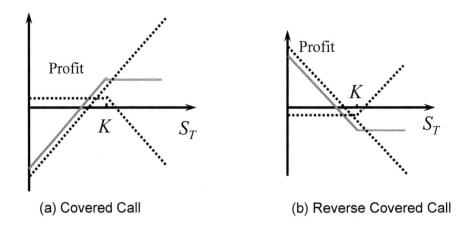

(a) Covered Call   (b) Reverse Covered Call

FIGURE 7.16: (a) Profit and individual positions of a covered call. (b) Profit and individual positions of a reverse covered call.

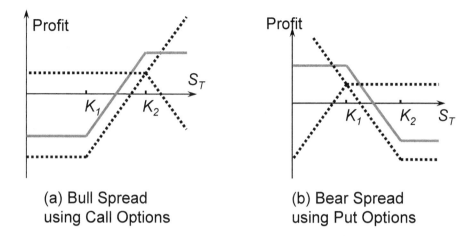

(a) Bull Spread
using Call Options

(b) Bear Spread
using Put Options

FIGURE 7.17: (a) Profit and individual positions of a bull spread using call options. (b) Profit and individual positions of a bear spread using put options.

of the underlying stock, but will protect the option writer from having to buy it at a potentially higher price should the option be exercised at maturity. Figure 7.16 displays the pay-off of the covered call strategy in the left panel, and reverse of a covered call in the right panel. Again as expected, the combination of a short call and long stock, $(-c + S)$, takes the form of a short put, as the put-call parity would indicate.

Options theory has proven invaluable to portfolio and risk management, since the rudimentary pay-off function of a call or put option on an underlying stock can be combined to create a staggering variety of pay-off structures. One may be reminded that piece-wise linear functions as basis functions can be used to approximate functions of any support and shape, provided put and call options with the desired strike prices are available at affordable prices. Therefore, portfolio managers can dynamically tailor investment positions to reflect changing expectations, market conditions, and client needs.

Purchasing and selling combinations of calls and puts can help investors maneuver in volatile and uncertain markets. If an investor or portfolio manager has a bullish or bearish view on a stock, portfolio or index, she can act on this view to try to benefit should it come true. A call option on the underlying is the natural bullish instrument for the prospects of the underlying, offering reward with much leverage. This would be a naked call strategy, however depending on how the investor picks the strike price of the call, it can turn out to be quite costly.

As an alternative to a naked call, the investor can construct a bull spread, which is one of the most popular spread strategies. In a bull spread, the long call with a strike at K_1 is combined with a short call on the same underlying and maturity, but at a higher strike, $K_2 (> K_1)$. The result is the cost of the combination reduces relative to a naked call with strike, K_1, along with of course a reduction in the upside potential. We show the bull spread constructed using call options in Figure 7.17 in the left panel. In fact, a bull spread can also be created using put options, however this construction is not shown in the figure. For a bull spread using put options, the investor will long a put option with strike K_1 and short a put with strike $K_2(> K_1)$. Given, $p(t, S_t; K_1) < p(t, S_t; K_2)$, whereas $c(t, S_t; K_1) > c(t, S_t; K_2)$, a bull spread using calls is more costly than the bull spread constructed using put options.

Similarly for an investor with a bearish view on the market, an index or an individual stock, she can either long a put option on the underlying, or construct a bear spread. The bear spread can be constructed by a long position on a put with strike (K_2) combined with a short position in a put option of the same maturity but a lower strike (K_1). Therefore the cost of the bear spread using put options is $p(t, S_t; K_2) - p(t, S_t; K_1)$, which is positive and can be lowered by either increasing the lower of the two strikes, K_1, or decreasing the upper strike price, K_2. Either of these moves will lower the profit from the strategy. The pay-off of a bear spread using puts is shown in the right panel of Figure 7.17. Just as in the case of bull spread, bear spread can be constructed using call options.

We have considered some examples of hedge strategies, followed by a couple of speculative, higher-return seeking strategies. We now develop a strategy to take advantage of an arbitrage opportunity. As per the put-call parity, a call option, a put option on the same underlying stock, with the same strike and maturity should satisfy the following relationship.

$$c(t, S_t; K) + Ke^{-r(T-t)} = p(t, S_t; K) + S_t. \tag{7.123}$$

Suppose the above relationship is seen to not hold for a time t, then the following strategy will result in arbitrage profit. Suppose we have, $c(t, S_t; K) - p(t, S_t; K) > S_t - Ke^{-r(T-t)}$, then we borrow funds in the amount, $Ke^{-r(T-t)}$, at the risk-free rate, r. With these funds we would buy the stock at S_t, short a call option and long a put option leading to a net cashflow, $c(t, S_t; K) - p(t, S_t; K) - S_t + Ke^{-r(T-t)} > 0$. At maturity, either the call or the put will be exercised. If the put is exercised, this implies $K > S_T$, the proceeds from selling the stock at price K is used to settle the risk-free loan. Thus, there is a positive cashflow at t for no later obligation. On the other hand, if the call option is exercised, this implies $K < S_T$, again the proceeds from selling the stock at price K is used to settle the risk-free loan. We have again created arbitrage profit.

Let's now consider the reverse case to be true. Suppose the put-call parity is violated at a time t, with $c(t, S_t; K) - p(t, S_t; K) < S_t - Ke^{-r(T-t)}$. In this case, we short the stock and use the proceeds to long a call option, short a put option and invest $Ke^{-r(t-t)}$ of the rest in the risk-free bond, leading to a net cashflow, $-c(t, S_t; K) + p(t, S_t; K) + S_t - Ke^{-r(T-t)} > 0$ at t. At maturity, either the call or the put will be exercised. If the put is exercised, this implies $K > S_T$, the proceeds from the risk-free investment are used to buy the stock at price K. The stock is used to settle the short position in

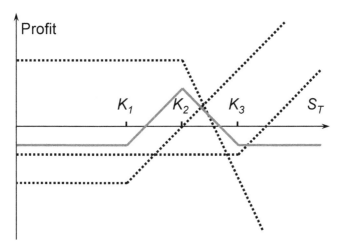

FIGURE 7.18: (a) Profit and individual positions of a butterfly spread.

the stock. Thus, there is zero net cashflow at T. On the other hand, if the call option is exercised, this implies $K < S_T$, we would still buy the stock at K and use it to settle the short position in the stock, thus again making arbitrage profit.

Bull and bear spreads were constructed with a view that the stock, portfolio or market was going to either go up or down. What if for a stock, a sector or the market, it is believed that it is going to stay where it is for a period of time. We construct a more complex spread of four call options of the same maturity, long one call of strike (K_1), short two calls of a higher strike (K_2), and long one call of a yet higher strike (K_3). The combined pay-off of these positions is shown in Figure 7.18. The strategy pays off if the stock remains in the neighborhood of its current price, and not if the stock wanders far away in either direction. This is the butterfly spread.

Volatility is increasingly becoming an asset class in itself, for indices as well as for single stocks. We now consider an aggressive strategy that bets on the volatility of a stock or an index. The strategy is called a straddle and is constructed by a long position in a call and a put of the same strike and maturity. The resulting pay-off is shown in the left panel of Figure 7.19. This strategy pays off only if the stock moves far enough away from its current value, no matter in which direction, hence a bet on the volatility. A straddle may be quite costly, therefore on a similar theme we can construct a less costly strategy called a strangle. The difference in a strangle is that we employ two strike prices, K_1 and K_2, with $K_1 < K_2$. The put option has a strike K_1 and call option has a strike of K_2.

Straddle and strangle can also be mixed with a directional bias, which results in a strip and a strap, as shown in Figure 7.20. A strip has a bias to a higher likelihood of a downward move, thus taking a 2:1 long position in a put and call option of the same strike and maturity. Similarly, a strap has a

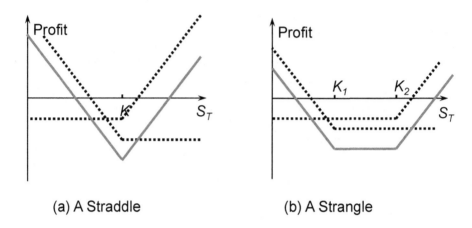

FIGURE 7.19: (a) Profit and individual positions of a straddle. (b) Profit and individual positions of a strangle.

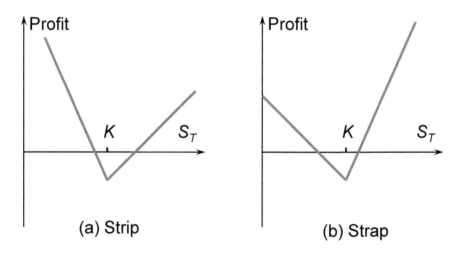

FIGURE 7.20: (a) Profit and individual positions of a strip. (b) Profit and individual positions of a strap.

bias to a higher likelihood for the stock to move up, thus taking a 2:1 long position in a call and put option of the same strike and maturity.

We have so far constructed strategies for single stocks, portfolios or indices using only vanilla European call and put options. Strategies using exotic options can also be constructed as and where deemed appropriate. Exotic options are more complex and have so far been traded over-the-counter, hence there are additional issues to consider when engaging in a risk management strategy using exotic options. The investor should completely understand the terms of the exotic option contract, and ensure that the exotic option will help meet their hedging needs better than the vanilla options. This is true because by virtue of being over-the-counter contracts, exotic options carry additional counterparty risk.

Instead of a put option, in order to develop a protective strategy for a long position in a stock or a portfolio, one can seek a knock-out put option or a compound option with the right to buy a put option in the future with set strike price and time to maturity. In both cases, the hedge could be as effective, but end up being much less costly. A chooser option strategy can achieve the goal of strategies like a straddle, strangle, strip, or strap, but in much more cost-effective terms. Similarly, lookback options, Asian options, spread options, and basket options can be tied into risk management strategies; since all these options have a way to reduce or access risk to construct the desired risk-return profile.

We have provided an overview of some important strategies to cover the grounds of themes by which equity hedging strategies are developed. The reader should consult other sources for further building their knowledge on this topic [17, 41].

7.3.2 Optimal Hedge Problem

All the hedging strategies constructed in the previous section were developed based on a theme for the objective of the strategy. However, in developing these strategies we didn't seek the best setting for the strategy to achieve the objective. All the strategies, in fact, had key parameters that could be used to fine-tune the strategy for reaching the strategy's objectives optimally. For instance, the protective put strategy must determine the strike price of the put option to construct a cost-effective insurance against downside risk of the stock.

We now develop a framework to facilitate this fine-tuning, as well as development of new strategies that offer optimal hedges. Since equity is purely an investment asset, the key quantity that summarizes performance is the investment return. In contexts other than equity risk, the key quantity that may define risk management objectives could be cashflow of a project, annual firm-level cashflow, or at the highest level, the value of the firm or the shareholder value of the firm. Along with the appropriate quantity that defines the risk management objective, one must also choose a suitable risk measure. In some cases expected shortfall, semi-variance, or other measures of tail risk or downside risk may be considered relevant, whereas in others volatility of return may be the guiding risk measure to develop the optimal hedge strategy.

If V is the quantity that defines performance of a firm or portfolio given its risk exposures, R, and $\rho(V)$ is the chosen risk measure, then our objective is to find a hedge instrument, $X(\theta)$, and its weight h, so that risk is minimized. We can formulate this problem in its general form as follows,

$$\min_{\theta,h} \rho(V(R, hX(\theta))), \qquad (7.124)$$

where the solution to the optimization problem, (θ^*, h^*), would be called the *optimal hedge ratio* or the *optimal hedge weight*.

As a specific case, let's consider the return of a portfolio, $V = R_\Pi$, being the quantity of interest, where the risk measure chosen is the volatility of return. If an index futures contract is being considered to construct the hedge, with return of the futures position being, R_f, then we seek the minimum variance hedge h^* that minimizes the volatility of the hedged portfolio, Π_H, given as follows.

$$\min_h var(R_{\Pi_H}) = \min_h var(R_\Pi) + h^2 var(R_f) - 2h\rho_{\Pi f}\sigma_{R_\Pi}\sigma_f, \qquad (7.125)$$

where $\rho_{R_\Pi f}$ is the correlation between the portfolio and the hedge instrument - index futures contract, and σ_{R_Π}, σ_f are standard deviation of portfolio return and index futures, respectively. In order to minimize the risk, we take the first derivative, equate it to zero and solve the equation to obtain the following solution.

$$h^* = \rho_{R_\Pi f}\frac{\sigma_{R_\Pi}}{\sigma_f}. \qquad (7.126)$$

Taking second derivative confirms this to be the optimal hedge; we call it the *minimum variance hedge*. The above static hedge framework can be cast as a dynamic hedge problem by allowing all state and decision variables to be time-dependent. In the next section, we develop some dynamic hedging strategies.

7.3.3   Dynamic Hedging Strategies

The covered call static hedge, discussed in Section 7.3.1, protects an option writer of a call option, by eliminating the unlimited downside risk of the naked call option. The advantage of a static strategy of this kind is that once implemented, it doesn't require any further monitoring and control. However, as seen in Figure 7.16, the covered call exposes the option writer to a significant amount of downside risk of the stock. We now consider a dynamic hedge to avoid the disadvantage of the static hedge.

The stop loss strategy is the simplest dynamic hedge which takes an action

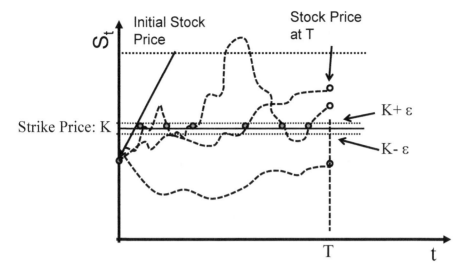

FIGURE 7.21: The points of time along the life of an option when trades must be made to cover the naked short call position. A margin around the strike, K, is created of width 2ϵ to avoid rapid trades when the option is near at-the-money range.

only if the need for a cover is anticipated. In the case of a short position in a call option, this happens when stock prices goes above the strike price, or when the option becomes in-the-money. When the option becomes in-the-money, the writer buys the underlying stock to cover the written call option, and holds the position until the intrinsic value of the option remains positive. If the option matures in the money, the writer is covered due to the long position in the stock; however, if the option's intrinsic value becomes zero during the life of the option, the writer unwinds the long stock position. Therefore, the writer is long in the stock whenever $S_t > K$, and holds a naked call when $S_t < K$. Figure 7.21 shows the response of the strategy for different sample paths of the stock until maturity of the option.

One practical issue with the strategy is that if the stock price hovers in the vicinity of the strike price, the writer may end up buying and selling the stock in quick succession, raking up the transaction costs from these trades. In order to avoid this undesirable aspect of the strategy, we create a band of width 2ϵ around K, and don't execute the buy and sell decision for the stock until stock price goes above $K + \epsilon$ or falls below $K - \epsilon$, respectively. The band of execution is also shown in Figure 7.21. Despite this adjustment, the strategy still has the flaw of swinging in and out of the entire stock position, which can prove to be quite costly.

For the derivation of the arbitrage-free price of an option, we constructed a portfolio, $\Pi(t, S_t) = -c(t, S_t) + \frac{\partial c}{\partial x} S_t$, which was instantaneously risk-free. We take advantage of this fact in constructing an improvement on the stop loss

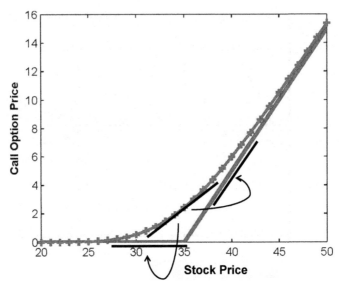

FIGURE 7.22: Delta hedge strategy takes advantage of the fact that the slope of the option price curve will converge to the terminal pay-off level as option reaches its maturity.

strategy. For a written call option, we take a long position in the stock with a weight, $\Delta = \frac{\partial c}{\partial x}$, which is the first Greek of the option (delta). This is called delta hedging. This portfolio is instantaneously risk-free, but as the stock price changes, the delta of the option also changes. Hence, the hedge position should be adjusted. With responsive adjustments, as shown in Figure 7.22, as the option approaches maturity, the hedge weight converges to where it should be.

The frequency of adjusting the hedge or the size of adjustment to the hedge needed when using delta hedging can be reduced if we add another hedge instrument, say another option on the same underlying stock, to neutralize the second order sensitivity of option price, $c(t, S_t)$ to the underlying, which is the gamma of the option. This is called delta-gamma hedging. We would simultaneously solve the following pair of equations for (w_1, w_2).

$$\frac{\partial c}{\partial x} + w_1 + w_2 \frac{\partial c_2}{\partial x} = 0, \qquad (7.127)$$

$$\frac{\partial^2 c}{\partial x^2} + w_2 \frac{\partial^2 c_2}{\partial x^2} = 0, \qquad (7.128)$$

where $c_2(t, S_t)$ is a second option defined on S_t chosen to construct the gamma hedging along with using the underlying stock S_t for the hedge.

Additionally, the volatility underlying the stock may itself be time-varying, as discussed in the stochastic volatility model in Section 7.2.6.3. In order to neutralize the impact of changing volatility on the option price, a second option on the underlying would be added to construct the hedge. This is called delta-vega hedging. We simultaneously solve the following two equations, instead of Eqns. (7.127) and (7.128), to obtain (w_1, w_2).

$$\frac{\partial c}{\partial x} + w_1 + w_2 \frac{\partial c_2}{\partial x} = 0, \tag{7.129}$$

$$\frac{\partial c}{\partial \sigma} + w_2 \frac{\partial c_2}{\partial \sigma} = 0, \tag{7.130}$$

where $c_2(t, S_t)$ is the second option on the underlying included to construct the vega hedge along with using the underlying stock S_t for the hedge.

Finally, in order to simultaneously neutralize the option price for the changing volatility and second order changes in stock price, a third instrument or a third option on the underlying can be added to construct the hedge. We would now solve the following three equations simultaneously for (w_1, w_2, w_3).

$$\frac{\partial c}{\partial x} + w_1 + w_2 \frac{\partial c_2}{\partial x} + w_3 \frac{\partial c_3}{\partial x} = 0, \tag{7.131}$$

$$\frac{\partial^2 c}{\partial x^2} + w_2 \frac{\partial^2 c_2}{\partial x^2} + w_3 \frac{\partial^2 c_3}{\partial x^2} = 0, \tag{7.132}$$

$$\frac{\partial c}{\partial \sigma} + w_2 \frac{\partial c_2}{\partial \sigma} + w_3 \frac{\partial c_3}{\partial \sigma} = 0, \tag{7.133}$$

where $c_3(t, S_t)$ is the third option on the underlying added to the hedge to achieve a delta-gamma-vega hedging.

Usually the above hedges will not need to be constructed individually for every single option exposure, since all the above sensitivities are relevant for a portfolio of options defined on the same underlying, Π_t (refer to Eqn. (7.72)). Portfolio Greeks can be obtained by weighted sum of individual instruments' Greeks, with weights being the portfolio weights for each instrument. Therefore, delta, gamma, vega, gamma-vega, etc. hedging can be done for the entire portfolio by adding additional options as above, and simultaneously solving the following equations for (w_1, w_2, w_3).

$$\frac{\partial \Pi_t}{\partial x} + w_1 + w_2 \frac{\partial c_2}{\partial x} + w_3 \frac{\partial c_3}{\partial x} = 0, \tag{7.134}$$

$$\frac{\partial^2 \Pi_t}{\partial x^2} + w_2 \frac{\partial^2 c_2}{\partial x^2} + w_3 \frac{\partial^2 c_3}{\partial x^2} = 0, \tag{7.135}$$

$$\frac{\partial \Pi_t}{\partial \sigma} + w_2 \frac{\partial c_2}{\partial \sigma} + w_3 \frac{\partial c_3}{\partial \sigma} = 0, \tag{7.136}$$

where $c_2(t, S_t)$ and $c_3(t, S_t)$ are options on the same underlying added to construct gamma-vega neutrality in the portfolio, Π_t.

7.4   MATLAB Tools for Equity and Portfolios

MATLAB mathematical software has a vast array of functions for working with equity analysis, equity derivatives, and optimization methodologies in its Financial and Optimization Toolboxes, respectively. We list a few of these functions here. The reader is advised to look up the extensive help documentation available with MATLAB to see the details of these and other related functions. At the bottom of each function description in MATLAB help documentation, look for 'See Also' to explore other related functions. Resources such as MATLAB Primer [20] are also useful.

Portfolio optimization: `quadprog`, `fmincon`, `plotFrontier`, `estimateFrontier`, `estimatePortReturn`, `estimatePortRisk`

Equity analysis: `corrcoeff`, `cov`, `priceandvol`, `movavg`

Equity options: `blsprice`, `blsimpv`, `binprice`, `opprofit`

Greeks: `blsdelta`, `blsgamma`, `blsvega`, `blsrho`, `blstheta`

GARCH: `ugarch`, `ugarchllf`, `ugarchpred`, `ugarchsim`

7.5   Summary

Equity market risk is the most important component of risk in the market risk segment, both in terms of volume and broad impact on firms and investors. Therefore, in this chapter we focused on the risk management of equity risk. In the context of equity risk, we introduced principles of portfolio management and risk diversification. Derivatives are used extensively for the management of equity risk. We developed the option pricing techniques for a variety of options, as well as methods for estimating sensitivity of these prices for the key parameters. Knowing the price and price sensitivity of the equity derivatives is important, both for the buyer and the writer of the option, in order to determine its impact on their respective overall hedging strategy. We also considered pricing of options under more advanced models, which are well-known improvements over the standard Black-Scholes-Merton framework for pricing derivatives. Finally, we developed rigorous frameworks for developing optimal hedging strategies, starting with assessing some standard examples of hedging strategies and their objectives.

7.6 Questions and Exercises

Review Questions

1. What is the implication of role of equity in the capital structure of a firm?

2. How does fundamental analysis of equity differ from technical analysis?

3. What is the trade-off between objective, strictly model-driven quantitative analysis, and intuition investors may have developed for the markets, sectors, industries, and specific firms?

4. What is the goal of mitigation or diversification of risk?

5. How are the mean and variance of portfolio return constructed in terms of mean and variance of individual stock returns?

6. What is the impact on the return of a portfolio of two stocks, when there is perfect positive or negative correlation between returns of the two stocks?

7. What is a short-selling constraint in a portfolio optimization problem? How is it defined and specified?

8. What is the efficient frontier?

9. How does mean-variance portfolio optimization differ from mean-mean-absolute-deviation portfolio optimization?

10. When should one use tail measures of risk in portfolio optimization?

11. How can simulation analysis be utilized for portfolio decisions?

12. What are the approaches available for addressing parametric uncertainty in portfolio optimization?

13. What are a plain-vanilla European call and put option? For a chosen strike price, K and maturity, T, draw the terminal pay-off function of a short and long position in a call and a put option.

14. What are an at-the-money, in-the-money, and out-of-the-money options?

15. What is arbitrage?

16. What is a binomial tree, how can it be used to model stock evolution?

17. What are risk-neutral probabilities?

18. What is implied volatility? What is the significance of this quantity?

19. What are the Greeks of an option? Give some examples and their exact definition.

20. How can one compute the Greeks of a portfolio of options on the same underlying?

21. What are delta-, gamma-, vega-neutrality of a portfolio of options?

22. What are exotic options? Give some examples.

23. What is a compound option? How does it differ from a chooser option?

24. How is a shout option different from it European or American counterpart? How will its price compare?

25. What are an up-and-out and a down-and-in barrier option?

26. What are the different varieties of Asian options?

27. What is a rainbow option? Why is it considered an option on correlation between underlying stocks?

28. How does the put-call parity change when applied to American call and put options?

29. When is the Constant Elasticity of Variance (CEV) model useful for modeling stock price evolution?

30. When is a stock price evolution model described and calibrated under the real-world measure versus risk-neutral measure?

31. What is VIX index?

32. What is a volatility smirk? What is a smirk indicative of?

33. What is a stochastic volatility model for stock evolution? When is this model enhancement important?

34. What is a jump-diffusion process? When is such a model useful for stock evolution?

35. What are the protective put and covered call static hedge strategies?

36. What are a bear spread and a bull spread? When is each spread constructed?

37. What is the butterfly spread, how and when is it constructed?

38. What are a straddle and a strangle? What is the cost of these two strategies?

39. What are a strip and a strap? How do these differ from a straddle?

40. What is a minimum variance hedge? What is optimal hedge ratio?

41. What is a stop loss strategy? What is the drawback of this dynamic hedging strategy?

42. What is delta hedging? How is delta-gamma hedging an improvement over delta hedging?

43. What is delta-vega hedging? When would this hedge be required?

44. What is delta-gamma-vega hedging?

Exercises.

1. Consider three stocks with the following summary information regarding their annual returns. The mean annual return of three stocks is estimated to be, $\vec{\mu} = [0.09; 0.05; 0.16]$; the annual standard deviation of returns is $\vec{\sigma} = [0.10; 0.06; 0.25]$, and the correlation matrix is given as follows.

$$\rho = \begin{pmatrix} 1 & 0.3 & 0.1 \\ 0.3 & 1 & -0.05 \\ 0.1 & -0.05 & 1 \end{pmatrix} \qquad (7.137)$$

Assuming a planning horizon of one year, construct and analyze the following portfolios in MATLAB.

   (a) Construct the minimum variance optimal portfolio under no short-selling constraints.

   (b) Choose a desired target mean return, r_{th}, which should serve as a lower bound for the optimal portfolio return. Compute the minimum variance portfolio under no short-selling constraint. Relax the no short-selling constraint and re-optimize your portfolio. How do your portfolio weights change?

2. Consider a continuous-time stock price evolution model for three stocks of the form,

$$dS_{it} = \mu_i S_{it} dt + \sigma_i S_{it} dW_{it}, \qquad (7.138)$$

for $i = 1, 2, 3$, where initial stock price is, $\vec{S_0} = [19; 53; 26]$, $\vec{\mu} = [0.09; 0.05; 0.16]$ and $\vec{\sigma} = [0.10; 0.06; 0.25]$. The three correlated Wiener processes are described by the following correlation matrix.

$$\rho = \begin{pmatrix} 1 & 0.3 & 0.1 \\ 0.3 & 1 & -0.05 \\ 0.1 & -0.05 & 1 \end{pmatrix} \qquad (7.139)$$

Assume a monthly trading strategy for an annual planning horizon, and a chosen terminal wealth performance measure, $U(W_T) = E[u(W_T)]$, where $u(x) = \frac{x^\gamma - 1}{\gamma}$ is a constant relative risk aversion utility. Analyze different investment strategies for these three stocks in MATLAB using simulation for different choices of coefficient of relative risk aversion, γ. Pick a $\gamma < 0$ for degree of high risk aversion and a $\gamma > 0$ for low risk aversion. Explore how the optimal strategy may be constructed in each case using simulation.

3. Consider a stock price evolving by the following model,

$$dS_t = 0.19 S_t dt + 1.2 S_t^{0.8} dW_t, \qquad (7.140)$$

where the current price of the stock is, $S_0 = \$20$.

(a) Determine the price of a European vanilla call option with strike price $K = \$18$ and time to maturity, $T = 0.25$ years. The short-term risk-free interest rate is 2.3%. If put-call parity holds, what is the price of the corresponding European put option?

(b) Assume that the accurate representation of the real-world stock prices is by the model in Eqn. (7.140). Irrespective of this fact, if we want to represent stock evolution by the Black-Scholes model, what parameters would you describe the Black-Scholes stock price evolution by that are consistent with the above model? What will be the price of the two options in the Black-Scholes world? How different are the prices of the two options?

4. For the stock evolving by the model in Eqn. (7.140), define a call-on-call and a call-on-put compound option with a maturity of $T = 0.1$ years. Price the compound option, assuming the short-term risk-free interest rate is 2.3%. Define and determine the price of a chooser option with same maturity as the compound option, and compare the prices of the two options.

5. A discrete-average Asian option is defined on the stock evolving by the model given in Eqn. (7.140) with the current price of the stock being, $S_0 = \$20$. The Asian call option pay-off is determined by the arithmetic average of weekly closing price of the stock, with a maturity of $T = 0.25$ years and strike price $K = 20$. Estimate the price of the option using simulation. Assume that the short-term risk-free interest rate is 2.3%.

6. Define an up-and-out Barrier call option and down-and-in Barrier put option on the stock evolving by the model in Eqn. (7.140). Estimate price of these options, and compare the price with the corresponding vanilla call and put options. Assume that the short-term risk-free interest rate is 2.3%.

7. Consider a stock evolving by the following Black-Scholes model

$$dS_{2t} = 0.10 S_{2t} dt + 0.24 S_{2t} dW_{2t}, \, S_{20} = \$45. \qquad (7.141)$$

Define an exchange option between this stock and the stock defined in Eqn. (7.140) with initial price of $S_0 = \$20$. Estimate the price of the exchange option, assuming that the correlation between the two driving Wiener processes is, $\rho = -0.15$ and the short-term risk-free interest rate is 2.3%.

8. Price an American put option with strike price $K = \$28$ and time to maturity, $T = 0.25$ years, using simulation analysis for the stock evolving by the model in Eqn. (7.140). Approximate this model by a recombining binomial tree representation and estimate the price of the American put option using the binomial tree. Compare the prices obtained by the two approaches.

9. Obtain the market price of options defined on your favorite stock for range of strikes and maturities. Plot implied volatility curves and surfaces, and comment on the properties of these plots regarding assumptions of constant volatility.

10. Define a bull and a bear spread for the stock defined to be evolving by Eqn. (7.141). Compute the cost of the spreads and summarize the distribution of the profit from the spread by a histogram, and in terms of mean and variance.

11. Consider a European vanilla call option written on the stock evolving by the model in Eqn. (7.140) with strike price $K = \$18$ and time to maturity, $T = 0.25$ years. Implement a stop loss strategy for this position in order to estimate the cost of this dynamic hedge strategy.

12. Consider the Barrier options defined in Problem 6. Implement a delta hedge and a delta-gamma hedge strategy for these options for the same frequency of adjustment to the hedge. Report some statistics for the comparison of the performance of the two hedges.

13. Consider the following portfolio of written options on the stock evolving by the model in Eqn. (7.140), reported by number of option contracts (100 shares/contract), maturity, strike, and plain vanilla European call versus put.

    (a) $n_1 = 1000$; $T_1 = 0.1$ years, $K_1 = \$22$, plain vanilla European put.

    (b) $n_2 = 2000$; $T_2 = 0.13$ years, $K_2 = \$20.50$, plain vanilla European call.

    (c) $n_3 = 1200$; $T_3 = 0.15$ years, $K_3 = \$17.00$, plain vanilla European call.

    (d) $n_4 = 500$; $T_4 = 0.2$ years, $K_4 = \$20$, plain vanilla European put.

    (e) $n_5 = 200$; $T_5 = 0.25$ years, $K_5 = \$19.80$, plain vanilla European put.

(f) $n_5 = 2000$; $T_6 = 0.4$ years, $K_6 = \$21$, plain vanilla European put.

By adding the underlying stock and additional options of your choice, construct a delta-gamma hedge and delta-vega-gamma hedge for the portfolio.

Chapter 8

Managing Interest Rates and Other Market Risks

In an introductory finance course, the first concept explained is that of *Time Value of Money*. This is an important and fundamental concept at the core of all monetary transactions with a temporal element, and the core mechanism by which monetary resources move in an economy to be allocated for their most beneficial use. The entity extending credit by parting with some idle monetary resources expects to get them back, but expects to be rewarded for the act of parting with its resources. One can consider this reward to be the opportunity cost of the entity extending the credit using its monetary resources. This opportunity cost constitutes the interest rates involved in borrowing and lending in an economy.

This is a simple description for interest rates, albeit giving an impression that an interest rate is a static, universally applied quantity to assess the time value of money. That is, in fact, how it is dealt with in an introductory exposition, and the way we will begin here. However, in the real world there are a multitude of interest rates used depending on the nature of entities at the two ends of contracts, borrower's financial health, duration of borrowing and lending, collateral used, economic conditions, and the monetary policy, to name a few. Changing market and economic conditions also put pressures on interest rates to move, change, and fluctuate. This is the risk we study in this chapter, along with its impact and management.

Entities that concern themselves with interest rates and the underlying risks range from individuals, governments, corporations, investment institutions, banks, and insurance firms. Therefore, interest base for interest rates is broad. Interest rate risk is closely related with two other risks - credit risk and liquidity risk - both of which we will study in later chapters.

Other market risks important in an increasingly globally connected production economy are risks underlying the value of commodities and various currencies firms may buy and sell materials, products, and services in. After developing a detailed discussion of risk management of interest rate risk, in the later part of this chapter, we will also study important features of these two additional market risk segments.

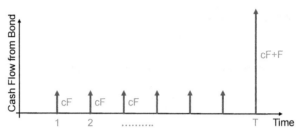

FIGURE 8.1: Cash flow from a bond with maturity, T years, and annual coupon of $c\%$.

8.1   Pricing Fixed Income Instruments

Fixed income instruments get their name from the nature of underlying cash flow due to the buyer over the life of the instrument. In order to receive this 'fixed income' cash flow, an investor buys the instrument for a price. The currently applicable interest rates determine the price the investor will pay for a specific fixed income instrument. The changing interest rates in the market for the fixed income instruments will dictate change in their prices, therefore volatility of interest rates is the fundamental risk underlying fixed income instruments. This underlines the importance of the relation between interest rates and prices of fixed income instruments, as well as their risk management.

Banks, pension funds, insurance companies, etc. take positions in fixed income instruments, not only as assets (long positions), but also as liabilities (short positions). Therefore, interest rate risk can affect these institutions in complex ways, depending on what fixed income instruments constitute their assets and liabilities. In order to characterize these complex risks, interest rate risk is further classified as 'curve risk,' 'basis risk,' 'gap risk,' etc. Valuation of the fixed income instruments is the basis for measurement of these risks, which can help quantify the impact of the risk on the firm's interest.

8.1.1   Bond Pricing

A bond constitutes a loan to the bond-issuer, where the bond-issuer promises to repay the loan principal in a set duration of time, called the 'maturity' of the bond. The issuer may also pay interest in the interim period on the loan, which is called a 'coupon.' The coupon level, expressed as a percentage of the principal, is set at the time of issue of the bond in accordance of the prevalent interest rates. Once set, the coupon level and frequency stays fixed for the life of the bond. Figure 8.1 shows a stylized cash flow diagram for a bond with an annual coupon level of $c\%$, a principal value, F, which is repaid in T years.

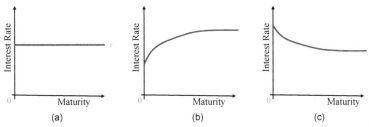

FIGURE 8.2: Different shapes of the term structure of interest rate by maturity. (a) Constant (b) Upward sloping (c) Inverted.

To price this bond at a time point after it is issued, one would need to determine the present value of all the (deterministic) cash flow from the bond using an appropriate discount rate. If the relevant market interest rate is a constant, say R, arising from a flat term-structure of interest rates, as shown in Figure 8.2(a), this serves as the discount rate. The price of the bond can then be stated as,

$$P_0 = \sum_{j=1}^{T} \frac{cF}{(1+R)^j} + \frac{F}{(1+R)^T}. \tag{8.1}$$

A bond's price sensitivity to interest rate is an important input for risk management. If the interest rate goes up or down, in this case a parallel shift of the flat curve in Figure 8.2(a), the price of the bond will change. This dependence is nonlinear, as is clear from the formula in Eqn. (8.1). If the risk underlying interest rate change, ΔR, can be described by a distribution, $f_R(\Delta t)$, its impact on bond price change can be evaluated using the following simulation. **Algorithm:**

Step 1: Generate ΔR with probability density $f_R(\Delta t)$.

Step 2: Compute Price of Bond with new interest rate: $P_1 = \sum_{j=1}^{T} \frac{cF}{(1+R+\Delta R)^j} + \frac{F}{(1+R+\Delta R)^T}$.

Step 3: Repeat until sufficient sample is generated.

Step 4: Compute appropriate summary statistics for bond price change: $\Delta P = P_1 - P_0$

Well known and widely used quantities, albeit deterministic in nature, exist that measure sensitivity of bonds to (small) interest rate changes, such as, duration, modified duration, convexity. We will discuss these in Section 8.2.1.

The other significant complexity in pricing bonds comes from the fact that interest rates are rarely ever the same for all ranges of maturities or timings of cash-flow, as in Figure 8.2(a). Instead the shape of term-structure given in Figure 8.2(b) is the one more commonly observed, which lends it the name, normal shape of the yield curve. Sometimes the inverted shape in Figure 8.2(c) is also observed. For either of these shapes, the term-structure must be qualified by a range of values, $\{R_{t_0}, R_{t_1}, \ldots, R_{t_N}\}$, depicting the applicable rates for different maturities or timing of cash flow, $\{t_0, t_1, \ldots, t_N\}$. If these rates were fixed and known, the price of the bond would become:

$$P_0 = \sum_{j=1}^{T} \frac{cF}{(1 + R_j)^j} + \frac{F}{(1 + R_T)^T}, \qquad (8.2)$$

where $\{R_j\}_{i=1}^{T} \subset \{R_{t_0}, R_{t_1}, \ldots, R_{t_N}\}$.

Now when there are T different interest rates used in the formula to price the bond, it is not possible to simply restate the sensitivity of bond price on interest rate changes, since it would have to capture the change of all combinations for the N applicable rates. However, it is a relief to note that these N rates don't move completely out of sync of each other, the smoothness of their moves relative to each other can be utilized in modeling the term structure of interest rates. We will explore this in the next section. The other option we have is to attempt to summarize the impact of these N rates on the price of the bond using one representative discount rate, the internal rate of return of the bond. We will call this the 'yield to maturity (YTM)' of the bond. It is the internal rate of return for the cash flow of the bond with respect to the quoted price of the bond, and can be obtained by solving the following equation for 'y'.

$$P_0 = \sum_{j=1}^{T} \frac{cF}{(1 + y)^j} + \frac{F}{(1 + y)^T}, \qquad (8.3)$$

where P_0 is the quoted market price of the bond. Using this single-number summary of interest rates for the bond, the algorithm described earlier in this section can be applied to quantitatively measure bond price risk due to yield changes described by a distribution, $f_y(\Delta t)$.

Impact of interest rate risk on bond prices, and other fixed-income instruments, is most amenable to be modeled by considering interest rates to be continuously compounded. This is, in fact, how the modeling of risk in financial models was demonstrated in Chapter 6. Equivalent continuously compounded interest rate, r, can be obtained from discretely compounding rates, R, by the following equation,

$$r = \frac{1}{T} \ln(1 + \frac{R}{m})^{Tm}, \qquad (8.4)$$

where m is the frequency of compounding per annum. For instance, if $m = 2$, the compounding is semi-annual, and therefore in a $T = 3$ year period, there are Tm compounding periods. Using the continuously compounded interest rates, the price of a bond equation becomes:

$$P_0 = \sum_{j=1}^{T} cFe^{-r_j t_j} + Fe^{-r_T T}, \qquad (8.5)$$

where $\{r_j\}_{i=1}^{T} \subset \{r_{t_0}, r_{t_1}, \ldots, r_{t_N}\}$ is the equivalent continuously compounded term structure of interest rates.

Based on the current term structure of interest rates, a crucial implied term structure derivable is that of interest rates applicable for times in the future. These are termed as the **forward rates**, and the corresponding term structure defines the forward curve. The forward curve plays a crucial role for risk management and interest-rate derivatives. Forward rates can be derived from the current spot rates by an arbitrage argument. For instance, the rate applicable one-year later for a duration of one-year can be derived in terms of the current one-year rate and the current two-year rate. By shorting a one-year Treasury-note and buying a two-year Treasury-note of the same face value, the one-year interest rate can be locked for the time one year in the future. The forward curve is summarized by, $f(t, T)$, where t signifies the time in the future when the interest rate will be applicable, and $T - t$ is the duration for which it will be applicable.

When interest rates are stated for discrete compounding, the forward rates can be computed by repeated application of the arbitrage argument for the bond instruments of the appropriate maturities, t and T. For any face value, F, we must have $F(1 + R_t)^t(1 + F(t,T))^{(T-t)} = F(1 + R_T)^T$, to eliminate arbitrage. From this relationship, $F(t,T)$, can be derived as,

$$F(t, T) = \left[\frac{(1 + R_T)^T}{(1 + R_t)^t}\right]^{1/(T-t)}. \tag{8.6}$$

If the compounding is at a higher than annual frequency, or if T or t are not whole multiples of years, the formula will need to be appropriately adjusted. With continuous compounding, the relation to eliminate arbitrage needs to be $Fe^{r_t t}e^{f(t,T)(T-t)} = Fe^{r_T T}$. Therefore, the continuously compounded forward rate becomes,

$$f(t, T) = \frac{r_T T - r_t t}{T - t}. \tag{8.7}$$

If we reorganize the terms in Eqn. (8.7) to obtain the following restatement,

$$f(t, T) = r_T + \frac{(r_T - r_t)t}{T - t} \tag{8.8}$$

we can see how the forward rate $f(t, T)$ relates to the current shape of the term-structure of interest rates. For instance, if the term-structure is upward sloping (i.e., $r_T - r_t > 0$), as in the middle panel of Figure 8.2, the forward curve is above the spot curve (i.e., $f(t, T) > r_T$). Similarly, if the term-structure is downward sloping (i.e., $r_T - r_t < 0$), as in the right panel of the figure, the forward curve falls below the spot curve (i.e., $f(t, T) < r_T$). This relationship is summarized in Figure 8.3

The forward rate and interest rate concepts discussed in this section are utilized to define stochastic interest rate models, and define risk measures and risk management strategies for fixed-income instruments. We begin developing the stochastic interest rate models next.

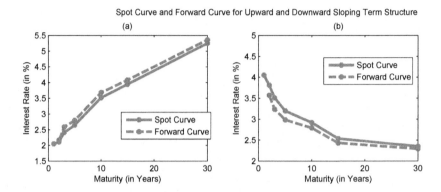

FIGURE 8.3: Relation of the forward curve to the spot curve for different shapes of the term structure of interest rates.

8.1.2   Stochastic Interest Rate Models

There are several theories that describe the theoretical or the empirical properties of the term structure of interest rates. These theories, such as the expectations theory, the market segmentation theory, or the liquidity preference theory, provide the justification for the shape and movement of the term structure of interest rates. Since the price and risk underlying fixed-income instruments is derived from the shape and movement of the term structure of interest rates, the models developed attempt to capture, and are judged on the basis of their ability to capture, these properties. Moreover, the models can be developed based on these theories as they pick on specific economic factor(s) to describe interest rates, or are developed with the sole purpose of fitting the current price data accurately without any specific economic motivation. The former are often called equilibrium models, while the latter are termed no-arbitrage models. No matter the motivation for the model, the models for interest rates are developed in the risk-neutral world, since interest rates are observed through prices of fixed-income instruments.

We will explore stochastic models for term structure of interest rates from both the equilibrium and no-arbitrage motivation following two significant themes. The first approach attempts to capture the term structure dynamics by utilizing models for the evolution of short-term interest rates. The second approach allows higher degrees of freedom by utilizing several factors to capture the entire term-structure dynamics.

8.1.2.1   Short Rate Models

A short rate model attempts to capture the entire term structure of interest rates in terms of short-term interest rates and their evolution in time. In this approach, the short-term interest rate becomes the single factor in the one-factor representation of the term structure. For modeling short-term interest rate dynamics, albeit in the nominal terms, one must pay heed to certain properties of short-term interest rates. For instance, nominal interest rates should not be negative, and if the interest drops to near zero level, it doesn't get stuck at that level. Empirically, when interest rates are too low, they tend to go up, and when they are too high, they tend to adjust downward. This can be captured by, for instance, incorporating mean-reversion in the process dynamics.

The short rate models are often defined in the risk-neutral world, where the driving Wiener process is governed by the risk-neutral measure. This is done so that the term structure of interest rates is consistent with the bond prices that will be used to calibrate the model. A general short rate model may be described by,

$$dr_t = (u(r_t, t) - \lambda(r_t, t)w(r_t, t))dt + w(r_t, t)dW_t, \qquad (8.9)$$

where W_t is a Wiener process in the risk-neutral measure, $u(r_t, t) - \lambda(r_t, t)w(r_t, t)$ is the drift term of specific structure that lends to mean-reversion, and $w(r_t, t)$ is the diffusion coefficient. Having stated the general structure of the single factor model, we will be quick to state a simpler version of the model for ease of handling of the model. In this attempt, we will make the drift term take the form: $u(r_t, t) - \lambda(r_t, t)w(r_t, t) = \eta(t) - \gamma(t)r_t$, and the diffusion term $w(r_t, t) = \sqrt{\alpha(t)r_t + \beta(t)}$. This simplification in the model structure results in simpler bond price formulas obtained based on these model, which in turn helps in the calibration of the models.

In order for these models to show evolution properties that match those of the real-world interest rates, we will need to impose some additional constraints and bounds on the terms, $\alpha(t), \beta(t), \gamma(t), \eta(t)$. These bounds and constraints are geared towards keeping the interest rates positive and bounce back from the lower boundary of (near) zero rather than remain stuck at it. The properties $\alpha(t), \beta(t), \gamma(t)$ and $\eta(t)$ must satisfy are as follows.

- If $\alpha(t) > 0$ and $\beta(t) \leq 0$, then interest rates can be bounded below by a positive number, chosen to be near-zero. If, however, in a specific model we choose $\alpha(t) = 0$, then we must have $\beta(t) > 0$, i.e., $\beta(t)$ must be strictly positive.

- The condition we must have satisfied so that the interest rates don't remain stuck at the lower bound, once they hit the lower bound is that $\eta(t) \geq -\frac{\beta(t)\gamma(t)}{\alpha(t)} + \frac{\alpha(t)}{2}$.

- Finally, we observe that when interest rates are too low, they tend to go up, and when they are too high, they tend to adjust downward. The form of drift term as $\eta(t) - \gamma(t)r_t$ assures the mean-reversion property.

The single factor interest models we will explore are of the form,

$$dr_t = (\eta(t) - \gamma(t)r_t)dt + \sqrt{\alpha(t)r_t + \beta(t)}dW_t, \qquad (8.10)$$

but before we begin looking at specific examples of these models, we demonstrate the derivation of pricing of bonds using these single factor models. This relationship is crucial for the calibration of the interest rate models.

Pricing a bond using stochastic interest rate model

We take the stochastic interest rate process in the real-world measure to evolve by the following equation,

$$dr_t = \mu(r_t, t)dt + \sigma(r_t, t)dW_t, \tag{8.11}$$

where $\mu(r_t, t)$ is the drift term, $\sigma(r_t, t)$ is the diffusion term, and W_t is a Wiener process in the real-world measure. We have reverted to the real-world measure in order to clarify why stochastic interest rate models for pricing bonds are developed and calibrated under the risk-neutral measure. It should also be clarified that, even while the interest rate is now stochastic, if it is an interest rate applicable to a default-free bond, the interest rate is referred to as a risk-free interest rate.

Pricing a bond under stochastic interest rates is harder than pricing an equity option, because there is no underlying asset to hedge the risk away. This is because in the stochastic interest rate case, a bond essentially becomes a derivative contract defined on the interest rate process. When the standard replicating portfolio based arbitrage-free derivatives pricing argument cannot be applied, we construct a portfolio of two bonds, each with an arbitrarily chosen maturity of T_1 and T_2, respectively. We consider two zero-coupon bonds, for simplicity of the derivation, the price of each at time t being $P_1(t; T_1)$ and $P_2(t; T_2)$, respectively. We hold 1 bond contract of the former and $-\Delta$ contracts of the latter. Thus, the value of the portfolio is $\Pi = P_1(t; T_1) - \Delta P_2(t; T_2)$.

Change in the portfolio value, $d\Pi$, can be obtained using Ito's formula as follows.

$$
\begin{aligned}
d\Pi &= (\frac{\partial P_1}{\partial t} + \mu(t, r_t)\frac{\partial P_1}{\partial x} + \frac{1}{2}\sigma(t, r_t)^2\frac{\partial^2 P_1}{\partial x^2})dt \\
&+ \sigma(t, r_t)\frac{\partial P_1}{\partial x}dW_t \\
&- \Delta[(\frac{\partial P_2}{\partial t} + \mu(t, r_t)\frac{\partial P_2}{\partial x} + \frac{1}{2}\sigma(t, r_t)^2\frac{\partial^2 P_2}{\partial x^2})dt \\
&+ \sigma(t, r_t)\frac{\partial P_2}{\partial x}dW_t].
\end{aligned}
\tag{8.12}
$$

For this portfolio to be risk free, we would want to eliminate the dW_t term. This can be achieved by selecting the position in the second bond appropriately. The choice of Δ that would do the trick is,

$$\Delta = \frac{\frac{\partial P_1}{\partial x}}{\frac{\partial P_2}{\partial x}}, \tag{8.13}$$

and with this choice of Δ, the portfolio value should instantaneously increase at the risk-free interest rate, r_t, i.e., $d\Pi = r_t \Pi dt$. This and the one in Eqn. (8.13) are representations of change in the same portfolio. We match the dt terms in the two representations and shuffle the terms so that all terms involving the first bond, P_1, that matures at T_1 are on one side of the equation, while all the terms involving the second bond that matures at T_2, are on the other side, as follows.

$$\frac{\frac{\partial P_1}{\partial t} + \frac{1}{2}\sigma(t, r_t)^2 \frac{\partial^2 P_1}{\partial x^2}) - r_t P_1(t, r_t)}{\frac{\partial P_1}{\partial x}} \tag{8.14}$$

$$= \frac{\frac{\partial P_2}{\partial t} + \frac{1}{2}\sigma(t, r_t)^2 \frac{\partial^2 P_2}{\partial x^2} - r_t P_2(t, r_t)}{\frac{\partial P_2}{\partial x}}.$$

The key observation we must make at this time is that these maturities were arbitrarily picked, therefore the two sides of the equation should be a value that is independent of bond maturity, say $a(r_t, t)$. Without loss of generality, we will choose a special form for $a(r_t, t) = \sigma(r_t, t)\lambda(r_t, t) - \mu(r_t, t)$, where $\lambda(r_t, t)$ is a yet-to-be-defined term. Therefore, either of the bonds, and for that matter any zero-coupon bond based on this interest rate term structure, should satisfy the following equation.

$$\frac{\partial P}{\partial t} + \frac{1}{2}\sigma^2 \frac{\partial^2 P}{\partial r^2} + (\mu - \lambda\sigma)\frac{\partial P}{\partial r} - rP = 0. \tag{8.15}$$

The end condition is $P(r_T, T) = F$, where T is the bond's maturity. We can show that the solution of the above equation, along with the end-condition, is,

$$P(t, r_t) = E[e^{-\int_t^T r(s)ds} F], \tag{8.16}$$

where the short-term interest rate, $r(t)$, evolves driven by a Wiener process, W_t, in the risk-neutral measure by the following equation,

$$dr_t = (\mu(r_t, t) - \sigma(r_t, t)\lambda(r_t, t))dt + \sigma(r_t, t)dW_t. \tag{8.17}$$

The fact that the price of the contract is the expected discounted pay off at the termination of the contract using the risk-free rate for discounting confirms that the valuation is under the risk-neutral measure. What remains to be shown is the role and meaning of the term, $\lambda(r_t, t)$, that was introduced in the derivation. For this purpose, we will describe the change in the price of the bond, $P(t, r_t)$, in the real-world measure by applying Ito's formula. In the real-world, interest rate was taken to evolve by equation, $dr_t = \mu(r_t, t)dt + \sigma(r_t, t)dW_t$, where W_t is a Wiener process under real-world probability measure. Change in price of a bond is described by,

$$dP = (\frac{\partial P}{\partial t} + \mu\frac{\partial P}{\partial r} + \frac{1}{2}\sigma^2 \frac{\partial^2 P}{\partial r^2})dt + \sigma(r_t, t)\frac{\partial P}{\partial r}dW_t \tag{8.18}$$

From the earlier derivation, price of a bond also satisfies Eqn. (8.15), which we substitute in the above Eqn. (8.18) to obtain,

$$dP - rPdt = \sigma(r_t, t)\frac{\partial P}{\partial r}(dW_t + \lambda dt)). \tag{8.19}$$

Eqn. (8.19) indicates that the excess return of a bond (left-hand side of the equation) in the real world must be dictated by the degree of unhedged risk due to non-tradability of the underlying risk factor. $\lambda(t, r_t)$ is therefore the expected rate of reward per unit risk exposure, W_t, also called the **market price of risk**. After incorporating this reward into the drift term of the interest rate dynamics, hence transforming the measure to the risk-neutral measure, the price of a bond simply becomes the expected discounted face value by risk-free discounting. The reader should note that this derivation is valuable not just in this context of pricing a zero-coupon bond, but also for other fixed income instruments and interest rate derivatives, as well as in other contexts where an underlying risk factor is not tradable. For instance, if a bond has a fixed given frequency of coupon payments, these can be seen as a bundle of zero-coupon bonds. Price of such a coupon-bearing bond can be determined additively by pricing each known cashflow through the life of the bond. We will consider other fixed income instruments and derivatives later in this chapter. For now, we bring our attention back to specific single-factor interest rate models.

Equilibrium Stochastic Interest Rate Models

There are numerous single-factor interest rate models proposed, developed and utilized in the past decades. The fundamentally desirable property of any model is that it is useful. For usefulness of a model, it should be flexible in capturing a variety of situations, consistent with market prices, yet simple to facilitate fast numerical computations. Equilibrium models approach the model development based on capturing macroeconomic justifications, both from the real sector and the monetary sector, for the dynamics of the interest rates themselves. Once an interest rate model is identified, pricing of bonds and other instruments can be achieved, which in turn allows judging consistency and coherence of the model relative to market prices. An increasing complexity to the models may be brought in with an intention of improving model properties, albeit at the expense of higher computational burden.

As a reminder, we will explore single factor interest models of the following form.

$$dr_t = (\eta(t) - \gamma(t)r_t)dt + \sqrt{\alpha(t)r_t + \beta(t)}dW_t. \tag{8.20}$$

Vasicek Model: In this model, we will modify Eqn. (8.20) by taking $\alpha = 0$ and $\beta > 0$, while all the other parameters are constants. The model thus becomes,

$$dr_t = (\eta - \gamma r_t)dt + \sqrt{\beta}dW_t. \tag{8.21}$$

The model shows mean-reversion due to the structure of the drift term, but interest rates can become negative. The greatest attraction of this model is its tractability in terms for determining closed-form prices of bonds and other interest-rate derivatives. Moreover, the model is able to capture the different shapes of the yield curve observed in the real world, as discussed in Section 8.1.1 (Figure 8.2).

Cox Ingersoll Ross (CIR) Model: In the CIR model, we set $\beta = 0$ and $\alpha > 0$. The model becomes,

$$dr_t = (\eta - \gamma r_t)dt + \sqrt{\alpha r_t}dW_t. \tag{8.22}$$

This model retains the mean-reversion property, and if we ensure that $\eta > \frac{\alpha}{2}$, then the short-term interest rate stays positive. This model is more complex than the Vasicek model, thus becoming relatively less tractable, but is more realistic. The lower tractability comes from a square-root diffusion term, and the difficulty with obtaining closed-form solutions.

Both the Vasicek and the CIR models are examples of **affine** short-term interest rate models, implying that the zero-coupon rate implied by these models is affine function of the short-term rate. Other models in this category include Merton (1973), Pearson and Sun (1994), while other popular equilibrium short rate models include Dothan (1978), Brennan and Schwartz (1980).

No Arbitrage Stochastic Interest Rate Models

While equilibrium models approach modeling interest rate dynamics on the basis of fundamentals of the economy, no-arbitrage models take a greedy approach to design models that capture the current prices and yield curve accurately. Their name indicates this objective, in that the model's attempt to capture prices of all bonds consistently so that there is no arbitrage possibility. We will present some examples of single-factor no-arbitrage models, where in some cases their multi-factor extensions are also available, since a single-factor model lacks the degrees of freedom to truly capture the entire yield curve consistently.

Ho and Lee Model: This is the simplest of models in this category, where the model is created so that a zero-coupon bond price is of the form, $e^{A(t,T)-r(T-t)}$, where T is the maturity of the bond. $A(t,T)$ is the term that has more complex time-dependence to allow the model to fit different shapes of the yield curve. The model for interest rate evolution is therefore one with $\alpha = \gamma = 0$, $\beta > 0$ constant, and $\eta(t)$ as a function of time lends the flexible time-dependence to $A(t,T)$ function. The model is,

$$dr_t = \eta(t)dt + \sqrt{\beta}dW_t. \tag{8.23}$$

This model shows good yield-curve fitting capability among the one-factor models.

Hull and White Model: This single-factor model is an extension of the

Vasicek model, where the drift term, $\eta(t)$, is made time-dependent in order to better fit the yield curve. Therefore, the model is,

$$dr_t = (\eta(t) - \gamma r_t)dt + \sqrt{\beta}dW_t. \tag{8.24}$$

This model is also an example of an affine interest rate model, with zero-coupon bond prices working out to be of the form, $e^{A(t,T)-rB(t,T)}$. The functions $A(t,T)$ and $B(t,T)$ are a lot more complex than in the Ho-Lee model case.

Other single-factor no-arbitrage models include Black and Karasinski (1991), Black, Derman, and Toy (1990). These are both improvements on the above two models, in order to avoid the short-term interest becoming negative. They achieve this by instead modeling the evolution of natural-log of interest rate, $\ln r$, although this brings in considerable additional complexity.

8.1.2.2   Multi-Factor Interest Rate Models

Single-factor models face significant challenges in capturing the complex dynamics of the term structure of interest rates. Multi-factor improvements to both equilibrium and no-arbitrage models are considered. The **Brennan and Schwartz** (1982) model identified the two factors to model as short-term and long-term interest rates, and described the evolution of this pair to capture the term-structure of interest rates. The long-term rate is taken to be the consol rate $l(t)$, and the joint evolution is captured by the following model,

$$dr_t = (\eta_1 - \gamma_1(l_t - r_t))dt + \alpha_1 r_t dW_{1t}, \tag{8.25}$$
$$dl_t = (\eta_2 - \gamma_2 r_t - \lambda l_t)dt + \alpha_2 l_t dW_{2t}, \tag{8.26}$$

where W_{1t} and W_{2t} are correlated Wiener processes with instantaneous correlation coefficient, ρ.

Other models, such as, Fong and Vasicek (1991) and Longstaff and Schwartz (1992) can be described as stochastic volatility short-rate models, where the volatility of the short-rate is itself seen to evolve by a stochastic process, driven by a second Wiener process. Whereas Chen (1996) and Balduzzi-Das-Foresi-Sundaram (1996) models describe the interest rate dynamics by three factors, short-term interest rate, its volatility, and its long-run mean.

In the no-arbitrage category, the **Hull and White** (1994) model is a two-factor model that is similar to the Brennan and Schwartz (1982) model, however it follows the single-factor Hull and White model in its development by keeping the drift term, $\eta(t)$, to be time dependent. The second factor is not consol rate, as in the Brennan and Schwartz model, but instead an adjustment to the mean-reversion level of the short rate. This adjustment factor has its mean reversion level of zero. Therefore, the two-factor Hull and White model can be summarized as,

$$dr_t = (\eta(t) + u_t - \gamma_1 r_t))dt + \beta_1 dW_{1t}, \tag{8.27}$$
$$du_t = -\gamma_2 u_t dt + \beta_2 dW_{2t}, \tag{8.28}$$

where W_{1t} and W_{2t} are correlated Wiener processes with instantaneous correlation coefficient, ρ.

Heath Jarrow Morton (HJM) Model: This last model we discuss takes an approach that can be said to be infinite dimensional, since it models the entire yield curve and its evolution. The HJM model describes the dynamics of the instantaneous forward rate, $F(t,T)$, for time t and all future time, T, and in the most general case states these dynamics so that all the previous models discussed so far are special cases of this model. The instantaneous forward rate is modeled to evolve by,

$$dF(t,T) = \mu(t,T)dt + \sigma(t,T)dW_t, \tag{8.29}$$

where $\mu(t,T)$ and $\sigma(t,T)$ are general drift and diffusion coefficients, and W_t is a Wiener process. Given $F(t,t) = r(t)$, i.e., the instantaneous forward rate at time t is the spot rate, specific choices of the drift and diffusion coefficient can yield the one-factor models discussed. In the risk-neutral world, the instantaneous forward rate is related to the price of a zero-coupon bond, $P(t,T)$, by the following relation,

$$P(t,T) = Fe^{-\int_t^T F(t,s)ds}, \tag{8.30}$$

where F is the face value of the zero-coupon bond. Although we don't demonstrate it, this relationship between zero-coupon bond price and forward rate entails that the drift and the diffusion coefficient in Eqn. (8.29) are related by the following relation,

$$\mu(t,T) = \sigma(t,T) \int_t^T \sigma(t,s)ds. \tag{8.31}$$

It is also important to note, in the most general case, the HJM model results in the spot interest rate, $r(t)$, to be non-Markovian. This implies that $r(t)$ at any time is not *memoryless*, it depends on the path taken by the interest rates in the past. This property makes it considerably more difficult to work with this model, hence simulation methods must be utilized, which can also be slow. Simplifications, such as by considering the most significant explanatory factors by, say, principal component analysis, are developed, or simple cases of the volatility term, $\sigma(t,T)$, are considered that bring the Markovian property back in effect. The latter approach will result in models similar to those considered earlier. We will consider simulation techniques in the next section, for the various interest rate models developed here.

8.1.2.3 Other Fixed-Income Instruments

So far in our discussion of fixed income instruments, we have focused on zero-coupon bonds to develop the modeling approach for interest rates in the simplest context. We now expand the space of instruments where the models developed so far will find use. First, different governments issue bonds in different markets (domestic, Eurobond, Sovereign bonds) with different maturities (from days to tens of years) and payment structures (coupon type, rate and frequency) to fund their projects. For instance, the US Treasury issues T-Bills, Notes, and Bonds, the German government issues Bunds, the UK has Gilts, and Japan has Japan Government Bonds (JGBs), etc.

Besides national treasuries and central banks, state and local governments, such as counties, districts, cities and towns issue bonds, called municipal bonds, to raise funds for public projects. Separate Trading of Registered Interest and Principal, in short STRIPs, are zero-coupon bonds created by stripping government bonds of the G7 countries. Among bond issuers, agencies that issue bonds to fulfill specific public purpose, such as home loans or student loans, are also significant contributors to bond markets. Finally, corporations fund their investment projects by raising funds by issuing corporate bonds.

Beyond the vanilla zero-coupon or coupon-bearing bonds, there are a variety of bond types defined by how the payments underlying the bond are determined. Floating rate notes (FRNs) are bonds with coupon rates that can change with time, either determined by a short-term market reference interest rate, like 3-month LIBOR, or federal funds rate. Inflation-adjusted bonds are bonds that offer coupons and principal adjusted for future inflation rates. There are also bonds that are designed to be paid off sooner than their term, called callable bonds, and bonds that can be converted into another type of security, such as equity of a corporation, called convertible bonds.

At the far end of the maturity range spectrum is the very short term debt instruments, which constitute the money market instruments. These maturities can range from a few hours, for example overnight, to several days and months, but less than a year. Governments, central banks, commercial banks, corporations issue instruments in this market to serve their needs for funds, as well as in case of central banks, to control the money supply in the economy. Treasury bills or T-Bills are short-term debt instruments that mature in one year or less, and pay no coupons. T-Bills are often used as collateral for borrowing funds for short durations, such as in the overnight money markets. This borrowing rate is called a **Repo** rate, short for repurchase agreement rate. It is a repurchase agreement since the buyer of a repurchase agreement sells a security for funds needed with a promise to buy the security back at a (near) future time for a set price. The counterparty of this transaction is engaged in a reverse-repo. Repo rates and federal funds rates are key rates that cascade through the whole spectrum of interest rates in the economy, affecting the demand and supply of funds.

Banks also issue certificates of deposits (CDs) with weeks, months to years maturities to finance their lending activity. CDs can have a fixed or floating interest rate, and counterparties may be commercial or retail customers. Similarly for short-term borrowing, corporations issue commercial paper. These run in maturity of less than a week to almost a year. Firms issue commercial papers to either raise short-term funds or as bridge financing until favorable long-term debt issuance opportunity arrives.

We have provided an overview of interest rate models and fixed-income instruments in this section. The interested reader should look at some of the references listed to expand on the limited description provided here of the models, their extensions and the variety of instruments. Some suggestions include, Wilmott [90], Martellini et al. [59], and Sundaresan [84].

8.1.3   Simulation of Interest Rate Models

The generic short-term interest rate model we studied in the last section is,

$$dr_t = (\eta(t) - \gamma(t)r_t)dt + \sqrt{\alpha(t)r_t + \beta(t)}dW_t. \qquad (8.32)$$

The simplest approach to simulate this process for any pricing or risk management objective is by discretizing the model using the Euler scheme. If we need to simulate the interest rate process on time interval, $[0, T]$, the Euler scheme on a discretization of time, $\{t_i : i = 1 \ldots N, t_1 = 0, t_N = T\}$, is obtained as,

$$r(t_{i+1}) = r(t_i) + (\eta(t_i) - \gamma(t_i)r(t_i))\Delta t_i + \sqrt{\alpha(t_i)r(t_i) + \beta(t_i)}\Delta W(t_i), (8.33)$$

where $\Delta W(t_i)$ are increments of the Wiener process in time steps Δt_i. In specific cases, one can do better than the Euler scheme in terms of improving upon the discretization error entailed by the scheme by taking advantage of known properties of the model. For instance, when $\alpha(t) = 0$, as is the case for Vasicek, Ho, and Lee, and Hull and White models, the short rate process becomes a Gaussian process. The solution of these models takes the form,

$$r_t = e^{\Lambda(t)}r_0 + \int_0^t e^{(\Lambda(t)-\Lambda(s))}\eta(s)ds + \int_0^t e^{(\Lambda(t)-\Lambda(s))}\sqrt{\beta(s)}dW_s, \quad (8.34)$$

where $\Lambda(t) = \int_0^t -\gamma(s)ds$. This solution can be mapped onto the time discretization, $\{t_i : i = 1 \ldots N, t_1 = 0, t_N = T\}$, for more accurate results from a simulation-based solution. We display the solution on a time discretization for a simpler case where $\gamma(t)$ and $\beta(t)$ are constants.

$$\begin{aligned} r(t_{i+1}) &= e^{-\gamma(t_{i+1}-t_i)}r(t_i) + \gamma \int_{t_i}^{t_{i+1}} e^{-\gamma(t_{i+1}-s)}\eta(s)ds \qquad (8.35) \\ &+ \sqrt{\beta \int_{t_i}^{t_{i+1}} e^{-2\gamma(t_{i+1}-s)}ds} Z_{i+1}, \end{aligned}$$

where Z_i is drawn from standard normal distribution, $N(0,1)$. In the yet simpler case, where $\eta(s)$ is a constant, Eqn. (8.36) becomes,

$$\begin{aligned} r(t_{i+1}) &= e^{-\gamma(t_{i+1}-t_i)}r(t_i) + \eta(1 - e^{-\gamma(t_{i+1}-t_i)}) \qquad (8.36) \\ &+ \sqrt{\frac{\beta}{2\gamma}(1 - e^{-2\gamma(t_{i+1}-t_i)})} Z_{i+1}. \end{aligned}$$

In Eqn. (8.32), if $\alpha(t)$ is a non-zero constant, the model is a general square-root diffusion model. We encountered it in Chapter 7 in the context of stochastic volatility models. In order to improve upon the application of the Euler scheme to solve this model, we utilize the knowledge of distributional properties of the short rate under this model. The conditional distribution of $r(t)$, given $r(s)$ for $s < t$, is shown to have noncentral Chi-square distribution. More specifically, for a time discretization, $\{t_i : i = 1 \ldots N, t_1 = 0, t_N = T\}$,

$$r(t_{i+1}) \; \frac{\alpha(1 - e^{-\gamma(t_{i+1}-t_i)})}{4\gamma} \chi_d^2 \left(\frac{4\gamma e^{-\gamma(t_{i+1}-t_i)}}{\alpha(1 - e^{-\gamma(t_{i+1}-t_i)})} r(t_i) \right), t_{i+1} > t_i, \quad (8.37)$$

where $d = \frac{4\eta}{\alpha}$ is the degrees of freedom of the χ_d^2 (Chi-square) distribution.

We have provided a brief discussion of simulation of a set of interest rate models. For a comprehensive discussion on this topic, the reader is encouraged to refer to Glasserman [30].

8.2   Interest-Rate Risk Management

Fixed income instruments and their derivatives are affected by the movements in the term structure of interest rates. As such the yield curve can be thought to have infinite degrees of freedom. However, the movements in the yield curve can be captured to a high degree of accuracy by a few principal properties, namely the level, slope, and curvature. This is demonstrated by a principal component analysis, a methodology designed to extract the most important explanatory factors to summarize variations in high-dimensional random variables. With this understanding, assessing risk in fixed income instruments is reasonably well-served by assessing the impact of changes in level, slope, and curvature in the yield curve on the instruments. We begin with defining measures that capture these sensitivities.

The primary risk that affects valuation of fixed income instruments is the interest rate structure and its dynamics. However, it should be noted that there are at least two other important factors that affect fixed income instrument valuation. Bond and all the other instruments discussed in Section 8.1.2.3 are primarily fixed income instruments that are fundamentally affected by credit risk. Credit risk is realized when either the interest (coupon) or the principal underlying the instrument is not paid or is not paid on time. While default is the extreme case, where the counterparty is unable or unwilling to fulfill its obligation, credit risk is also realized with changes in the credit quality of a counterparty, since this also significantly affects the value of the fixed income instruments. For instance, an institution or a government may get downgraded by a rating agency. We will study this risk in Chapter 9.

Participants in the fixed income market are often participating in the market for their long-term objectives. For instance, insurance companies must support their long-term liabilities by investing in long-term bonds. This may imply that their portfolio of assets is not traded actively. This results in illiquid markets for a large segment of fixed income instruments. This is often incorporated in the price of the bonds as liquidity risk premium. However, changes in liquidity of fixed income instruments can result in price fluctuations. Liquidity risk is also realized when an institution is unable to convert the value of a fixed income instrument to cash due to lack of depth in the market. Since interest rates affect fixed income instruments of different maturities and class differently, this can result in funding liquidity risk, where a firm is unable to raise necessary cash to meet its immediate cash needs. We will study liquidity risk in the context of asset liability management in Chapter 10.

For the rest of this section, we will focus on the impact of interest rate risk. We will first develop some measures for assessing the impact of interest rate risk. This is followed by considering some portfolio level issues for fixed income instruments, since more often than not, risk management must be done not for single instrument exposures. Finally, we will study some interest rate derivatives and strategies for managing interest rate risk by using these instruments.

8.2.1 Interest-Rate Sensitivity in Fixed-Income Instruments

A one-quantity summary of the entire term-structure of interest rates simplifies the effort of measuring the impact of interest rate risk. We will begin with defining some measures based on this summary, and explore the extent of effectiveness of this one-quantity summary. In fixed-income products, the simplest risk measure traders use is the 'DV01' measure. This refers to change in value of a security, DV or Delta V, after a change in yield or change in interest rate of 1 basis point or 0.01%. It is a bond valuation calculation showing the dollar value of a one basis point decrease (or increase) in interest rates. It shows the change in a bond's value compared to a decrease (or increase) in the bond's yield.

A more rigorous definition of measuring sensitivity of price of a bond with respect to interest rates is defined by the bond's Duration. The one-quantity summary of the entire term-structure of interest rates, as is relevant for the bond in question, is captured in the yield to maturity (YTM) of the bond. This was discussed at length in Section 8.1.1, defined in Eqn. (8.3). Duration is a measure consistent with DV01 in measuring sensitivity of bond price to interest rates. Duration is defined as the weighted average of the dates (in years) of each cash flow of the bond, where weights are the present value of the cash payment divided by the sum of the weights, which is the price of the bond. Therefore, duration is computed by the following formula,

$$D = \sum_{j=1}^{T} \frac{j \frac{cF}{(1+y)^j}}{P_0} + \frac{T \frac{F}{(1+y)^T}}{P_0}, \tag{8.38}$$

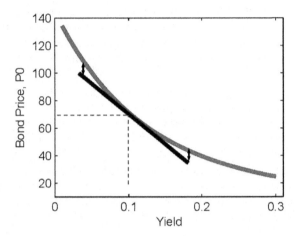

FIGURE 8.4: Bond price as a function of increasing yield.

where T is the bond maturity and the annual coupon rate is $c\%$. From the formula, it is clear that duration of a coupon-bearing bond will always be less than the maturity of the bond. It will be exactly equal to the maturity of the bond if the bond bears no coupons. For any set coupon rate and frequency of coupon payment, the longer the maturity of a bond, the higher its duration. Based on duration, we define modified duration as, $D^* = \frac{D}{(1+y)}$, which is a better indicator of bond sensitivity to interest rate risk since,

$$\frac{-1}{P_0}\frac{dP}{dy} = D^*. \tag{8.39}$$

Therefore, the higher the duration, the greater the sensitivity of the bond price to change in yield. While modified duration is a measure of relative sensitivity of bond price to interest rate changes, albeit summarized in a single number - YTM, 'DV01' is the dollar sensitivity of the bond price to interest rate changes, also measured in terms of YTM. For this reason, 'DV01' is also often termed as $Duration.

Eqn. (8.39) is accurate for an infinitesimally small change in the yield. However, yield changes can and will be greater than infinitesimally small! For larger than infinitesimally small changes, we use the symbol Δy, and the '=' relation in Eqn. (8.39) must be replaced with '$\simeq$'. As the Δy becomes larger, the modified duration becomes increasingly less accurate in determining the change in price of the bond.

To understand this better, we must look at the functional dependence of the price of a bond, P_0 on the yield level. In Figure 8.4, we draw the function given in Eqn. (8.3). The curve is far from linear, therefore the accuracy of the first derivative term (the slope) in determining the change in price of a bond is going to be limited as Δy gets larger. The price-yield relationship for a bond is nonlinear, therefore duration is only the first-order approximation of impact of change in yield on the price of a bond, or in other words, it gives a good approximation of sensitivity of price to yield only in small variations of the yield.

The convex curve defining the dependence of price of bond (y-axis) on yield (x-axis) in Figure 8.4, must be captured by a measure beyond the linear approximation measure of modified duration. A second order adjustment to more accurately predict change in value of a bond due to larger changes in yield is needed. This is motivated from the following Taylor expansion of price of bond around y,

$$P(y + \Delta y) = P_0 + \frac{1}{1!}\frac{dP}{dy}\Delta y + \frac{1}{2!}\frac{d^2P}{dy^2}\Delta y^2 + \frac{1}{3!}\frac{d^3P}{dy^3}\Delta y^3 + O(\Delta y^4). \quad (8.40)$$

We consider the convexity, C, of bond price, defined by,

$$C = \frac{1}{P_0}\frac{d^2P}{dy^2}. \quad (8.41)$$

More strictly speaking, this is the relative convexity of the bond. Similar to 'DV01' or \$Duration, there is also a notion of \$Convexity, defined simply as, $\$C = \frac{d^2P}{dy^2}$. Both measures of convexity, absolute or relative, are positive-valued measures. This is also seen in the shape of the curve in Figure 8.4. Increase in maturity, all else kept constant, increases the convexity of a bond. The coupon rate of a bond has an inverse relationship with convexity, whereas a direct relationship with \$Convexity.

Putting together modified duration and convexity in the Taylor expansion shows that the two measures together provide a much better measure of change in bond price due to interest rate changes. Instead of these relative sensitivity measures, the absolute ones, namely 'DV01' and '\$Convexity,' can also be used.

$$P(y + \Delta y) = P_0 - \frac{1}{1!}P_0 D^* \Delta y + \frac{1}{2!}P_0 C \Delta y^2 + O(\Delta y^3). \quad (8.42)$$

The order of accuracy is $O(\Delta y^3)$, which implies that a 1bp (basis point) change in yield ($\Delta y = 10^{-4}$) will result in the price to be off by $O(10^{-12})$. This is good accuracy, however for large changes in interest rates, the size of unaccounted change in bond price may still be quite large, also depending on the level of exposure. Therefore, more accuracy is desired. Moreover, we have summarized changes in interest rates by a single measure of yield to maturity, not acknowledging the exact source of the change in yield. That is, it is oblivious of whether change in level, slope, or curvature of the term structure of interest rates is causing the change in yield. This is an additional inaccuracy in measuring the impact of interest rate risk.

In this section, our discussion of interest rate risk has been in terms of discretely compounding interest rates, R_{t_j}, as used in Eqn. (8.2). Continuously compounding interest rates for different maturities, as may be obtained from the short rate models of Section 8.1.2, can be transformed into discretely compounding equivalent by Eqn. (8.4). Interest rates for different maturities do not always move in the same direction and magnitude. This is the cause for changes in slope and curvature of the yield curve. Moreover, long-term interest rates are known to be less volatile than short-term rates. We will now like to capture the changes in the term structure of interest rates, $\{R_{t_0}, R_{t_1}, \ldots, R_{t_N}\}$, in terms of a much lower number of factors. By doing so, we would like to develop improvements in the measures for interest rate risk underlying fixed income instruments.

Let's say there are M number of factors chosen to capture the changes, ΔR_{t_j}, in interest rates, where $M << N$. As stated earlier, the three principal components capturing level, slope, and curvature of the term-structure are capable of describing a high percentage of changes in interest rates. Therefore, we would like to describe the changes as,

$$\Delta R_{t_j} = \sum_{i=1}^{M} f_{ji} F_i + \epsilon_j, \text{ for } j = 1 \ldots N, \tag{8.43}$$

where F_i's are the factors and ϵ_j is the residual from the factors describing the change in interest rate. In this set-up, to determine the change in price of a fixed income instrument due to interest rate changes, multi-variable calculus and Taylor expansion is required, since price of a bond, for instance, is given by,

$$P_0 = \sum_{j=1}^{T} \frac{cF}{(1 + R_{t_j})^j} + \frac{F}{(1 + R_{t_T})^T}, \tag{8.44}$$

where each of R_{t_j} is a variable. Moreover, change in each of R_{t_j} is captured in terms of factors F_i, as per Eqn. (8.43). The multi-variable Taylor expansion will give,

$$\Delta P = \frac{1}{1!} \sum_{i=1}^{N} \frac{\partial P}{\partial R_{t_i}} \Delta R_{t_i} + \frac{1}{2!} \sum_{i=1}^{N} \sum_{j=1}^{N} \frac{\partial^2 P}{\partial R_{t_i} \partial R_{t_j}} \Delta R_{t_i} \Delta R_{t_j} + O(\Delta R^3), \tag{8.45}$$

where $\Delta R = \max_{\{i=1 \ldots N\}} \Delta R_{t_i}$. Changes in bond price will be obtained from Eqn. (8.45), in similar flavor as applying the combined effect from duration and convexity, by applying the dependence on key factors, $\Delta R_{t_j} = \sum_{i=1}^{M} f_{ji} F_i$.

A more comprehensive risk measure, such as Value-at-Risk (VaR), Conditional Value-at-Risk (CVaR), does not rely on first-order or second-order sensitivity of prices to risk factors. Instead, they seek out the forward distribution of prices, and summarize the distribution in a risk measure. Both VaR and CVaR are tail risk measures, discussed in several earlier chapters, such as Chapters 2 and 7. The forward distribution of the price can be obtained by simulating (or analytically deriving from) the appropriate short rate model, such as those discussed in Section 8.1.2. The price of a fixed income instrument

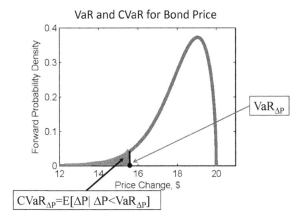

FIGURE 8.5: Value at Risk (VaR) and Conditional Value at Risk (CVaR) display for bond price.

for a future time, $t + \Delta t$, providing the forward distribution can be obtained from,

$$P(t + \Delta t, r(t + \Delta t, \omega)) \;=\; E[e^{-\int_{t+\Delta t}^{T} r(s)ds} F | r(t + \Delta t, \omega)]. \quad (8.46)$$

The VaR or CVaR at a certain confidence level, say $\alpha = 99\%$, indicates how low the price of the bond can fall with $(1 - \alpha)$ probability. While VaR is a percentile measure, CVaR is the expected value of the price (change) given it falls below the α-percentile. Figure 8.5 provides a visual display of these measures for the forward distribution of the bond's price (change). VaR and CVaR can be applied to a single fixed income instrument, but they are most commonly applied at the portfolio level, where interactions of all risk factors pertinent to the constituents of the portfolio must be evaluated. We evaluate the impact of interest rate risk in the portfolio context next.

8.2.1.1 Bond Portfolio Immunization

In almost all contexts, it is not single bonds whose risk we would want to manage. Instead, it is several bonds put together either as a well-defined portfolio under a set investment strategy or to meet a given future liability. Therefore, the sensitivity and other risk measures developed in this section must be extended to portfolios of fixed income instruments. The price sensitivity (changes) of instruments' prices from the same yield curve can be aggregated from individual sensitivities by calculating the weighted-average duration of the instruments held in the portfolio. This helps define portfolio duration, given as

$$D_{\Pi} = -\frac{1}{\Pi}\frac{d\Pi}{dy} = -\frac{1}{\Pi}\left(w_1\frac{dP_1}{dy} + w_2\frac{dP_2}{dy}\right) \simeq \frac{1}{\Pi}(w_1 D_1^* P_1 + w_2 D_2^* P_2), \quad (8.47)$$

where D_1^* and D_2^* are modified duration of bond 1 and bond 2, respectively, and w_1, w_2 are portfolio weights for the two bonds. Similarly the notion of convexity can be extended to bond portfolios, as follows,

$$C_\Pi = \frac{1}{\Pi}\frac{d^2\Pi}{dy^2} = \frac{1}{\Pi}(w_1\frac{d^2P_1}{dy^2} + w_2\frac{d^2P_2}{dy^2}) \simeq \frac{1}{\Pi}(w_1C_1P_1 + w_2C_2P_2), \quad (8.48)$$

where C_1 and C_2 are convexity of the two bonds, respectively. Therefore both modified duration and convexity of the portfolio are weighted sum of individual bond's modified duration and convexity in the portfolio.

The notion of portfolio modified duration and convexity is helpful in crafting certain risk characteristics of the portfolio. These desired risk characteristics could be arising for a certain investment goal or designed to meet certain future liability cash flow. The idea is that the designed bond portfolio has similar or identical response to interest rate risk as the future liability would have, or as the investment goal dictates. This matching 'immunizes' the bond portfolio to interest rate risk as measured by the chosen measures as far as the future liability or the chosen investment goal is concerned. This is termed bond portfolio immunization.

If w_1, w_2 and w_3 are portfolio weights for three (for illustration purposes) candidate bonds included in an investment portfolio, they should be chosen so that,

$$w_1P_1 + w_2P_2 + w_3P_3 = \Pi_{target}, \quad (8.49)$$

$$w_1D_1^*P_1 + w_2D_2^*P_2 + w_3D_3^*P_3 = D_{target}\Pi_{target}, \quad (8.50)$$

$$w_1C_1P_1 + w_2C_2P_2 + w_3C_3P_3 = C_{target}\Pi_{target}, \quad (8.51)$$

where Π_{target}, D_{target}, and C_{target} are the present value of future liability, its duration and convexity, respectively. In practice, Eqns. (8.49)-(8.51) can be extended to an arbitrarily large number, N, of bonds. In this case, the three conditions in Eqns. (8.49)-(8.51) will leave $N-3$ degrees of freedom for portfolio weight determination, which must be set by some other criteria.

As for a single bond, for a general portfolio of fixed income instruments the most general and robust method for risk measurement is in terms of forward distributions and its summary characteristics, such as by VaR or CVaR methodology. This will require models for future term-structures of interest rate for all relevant interest rates for fixed income instruments in the portfolio. In Section 8.1.2.3, we had discussed a variety of fixed income instruments. To determine the forward distribution of prices of these instruments in the portfolio, not only is the spot rate model for the relevant interest rates needed, but also the correlation structure must be described between the set of interest rates in consideration. Modeling correlation was discussed in Chapter 5, however a detailed discussion of modeling correlation in this specific context is beyond the scope of this book.

Based on the interest rate models, the forward distribution for the change in portfolio value can be determined using Eqn. (8.46) for each instrument in the portfolio and portfolio weights relevant for the time window $(t, t + \Delta t)$. Once the forward distribution of change in value of the portfolio is determined, we obtain the 1-day or 10-day VaR or CVaR, as indicated in Figure 8.5, by appropriate choice of Δt. In summary, VaR or CVaR combines both the duration and convexity adjustment into change in the value of a portfolio. Albeit with a much higher modeling and computational effort required.

8.2.2 Interest-Rate Derivatives

In support of the 'avoid-diversify-transfer-keep' paradigm for risk management, we have thus far developed models and risk measures to support the avoid, diversify and keep decisions for interest rate risk, with the greatest focus of methodological development on the diversify component. We now move our attention to the transfer mode. There are numerous instruments used for hedging and transferring interest rate risk. This list includes forwards, futures, swaps, bond options, exotic options, caps, floors, captions, floortions, and swaptions, to name a few. In this section, we will provide an overview of properties of these instruments, since these are key derivative instruments used by investors, corporations, and financial institutions to manage interest-rate risk.

Some of these instruments are exchange traded, but many are over-the-counter (OTC) contracts. As always, the exchange-traded derivatives are mostly of the simpler kind, but the OTC ones can be highly customized, sometimes to a rather high degree of complexity. Due to the bilateral nature of the OTC contracts, they tend to be less liquid and their execution is backed only by the capital of the provider or the dealer, hence the key players in the OTC derivatives market are financial institutions with good credit standing. The worldwide government debt is enormous! And that is an understatement. Something that is creating a significant unease among the Western economies through the 2008-2011 financial crisis.

Coupled with corporate bonds and bank loan portfolios, which are all fixed income instruments, it makes for a huge pool of assets and liabilities that are sensitive to changes in various interest rates. Therefore, it is not surprising that the OTC market for interest-rate derivatives was $450 trillion in notional value in 2010, and growing. In fact, the OTC market for derivatives is dominated by interest rate products. See Figure 8.6 for comparison of interest rate OTC derivatives broken down by currencies. From these facts, it is clear that interest-rate derivatives make for an essential tool for managing interest-rate risk.

Forward and Futures Contracts. A forward contract on an instrument is a contract to purchase or sell specific units of the instrument at a specific price and at a specific time in the future. These are over-the-counter contracts. Forward contracts are binding contracts, the buyer has to buy/sell the underlying at the agreed-upon price, and the seller is obliged to sell/buy the

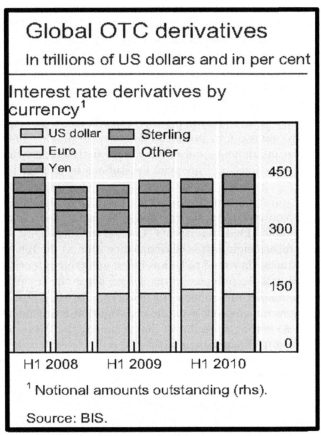

FIGURE 8.6: Volume of over-the-counter (OTC) interest rate derivatives in 2008-2010 period (Courtesy *Bank for International Settlements (BIS) Report*).

underlying. Besides interest rate instruments, forward contracts are commonly used for foreign currency.

Initial payment associated with a forward contract is zero, except that sometimes a security deposit is required of both parties. No cash changes hands before settlement date. Forward price, F, is the price that is applied to the sale/purchase of the underlying at delivery. This price is negotiated so the initial value of the contract is zero, i.e., the value of the contract is zero when it is initiated. Forward contract is also defined directly on interest rates, such as forward rate agreements (FRAs). Interest-rate forwards are cash settled, i.e., one party pays the other the difference between the set forward rate and its spot value at maturity date. Forward contracts on bonds allow selling and buying bonds at set prices. Pricing of forward contract, i.e., determining an arbitrage-free forward price, can be done in the same framework as used for pricing the bond, by changing the end-condition in Eqn. (8.15).

Futures are forwards, but these contracts are traded in an exchange. A margin account is maintained with the broker for each trading party, and positions are marked-to-market every day. Daily installments, positive or negative, that correspond to the change in daily market value of the futures price is paid for the daily settlement. The total of daily installments and payment made at maturity equals the futures price set when the contract was initiated.

Swaps. Swaps are over-the-counter agreements between two parties to exchange the cash flows of two different securities throughout the life of the contract. These can be seen as a series of forward contracts and, as with forwards, the contract is binding on both sides of the transaction. Interest-rate swaps are by far the most popular hedging instrument, used by treasurers for asset-liability management, and by portfolio managers to reduce or extend duration of an open position.

A common form of interest-rate swap is fixed-to-floating interest-rate swap, where the 'fixed' side (or leg) pays a fixed interest rate on a notional amount, and the 'floating' side pays a floating rate on the notional amount. The reference rate on the floating side could be LIBOR or any other reference rate agreed upon by the parties of the contract. A spread is added to the reference rate reflecting the 'fixed' rate to be exchanged. There is no exchange of the principals, as the principals on both sides of the swap cancel out, both at the inception and at the maturity of the contract.

As in forwards and futures contracts, no up-front fee is payable when a swap is initiated, as all swap transactions are priced initially so that the net present value of both legs of the swap is the same. LIBOR (London InterBank Offered Rate) refers to any of a number of short-term indicative interest rates compiled by the British Bankers Association (BBA) at 11:00 am London time each business day. For a given currency and maturity, LIBOR represents the simple interest rate at which banks are willing to lend to each other. As a practical matter, it is the rate other banks must pay to borrow.

Interest-rate swaps offer a great advantage to a hedge strategy, since either counterparty doesn't have to transact in large volumes of underlying fixed income instruments, and yet, change the interest rate risk characteristic of their portfolio. Therefore, interest-rate swaps are advantageous for corporations or financial institutions in allowing them to change the nature of payments on loans either from fixed to variable or from variable to fixed rates to match their income stream. Swap transactions are often used by corporate treasurers as a way of bridging the gap that tends to exist between the particular needs of a company and the demands of the investors.

For example, Bank 1 and Bank 2 enter into a 3-year interest-rate swap with notional value of \$100 million. Bank 1 will pay Bank 2 each year, at year-end, a sum equal to \$100 million times a fixed interest rate, say 2%, and will receive from Bank 2 a sum equal to \$100 million times the 1-year T-Bill rate plus a spread of, say, 1.5%. In practice, a netting procedure is applied, i.e., only the difference in the fixed and floating rate is paid. Therefore, if the T-Bill rate at the beginning of the year is less than 0.5%, Bank 1 pays Bank 2 the difference between 2% and T-Bill rate + 1.5% times \$100 million. If 1-year T-Bill rate is 0.8%, then Bank 2 will pay Bank 1 the sum of $((0.008 + 0.015) - 0.02) * \$100$ million $= \$300,000$. Therefore, without making large payouts, both banks transform the nature of their interest rate risk exposure.

Beyond the plain vanilla fixed-to-floating or floating-to-floating swaps is a vast variety of non-plain vanilla swaps, such as, basis swap, forward-starting swap, inflation-linked swap, accrediting, amortizing, and roller coaster swaps, etc.

Options. Options on fixed income instruments function exactly as they were discussed to work for stocks in Chapter 7. A bond option gives the buyer the right to buy (call) or sell (put) the underlying for the set strike price at maturity. This is the European-style bond option. Since bond prices decrease with interest rate increase, a stand-alone put option on a bond functions as a bet on the decline in the value of the bond, or equivalently, a bet on an increase in the interest rates. Similarly, a stand-alone call option is a reverse bet. Beyond the European options, there are fixed income exotics, American, Asian-style, and path-dependent options.

As discussed in the previous chapter, put options also allow the holder of an open position in the underlying to insure against a loss of value, exactly like the protective put stock option strategy. Other combinations obtained by the no-arbitrage principle hold. For instance, a forward contract can be constructed by buying a call option and simultaneously selling a put option on the same underlying with exercise price equal to the forward price of the bond. A call option on a bond can be constructed by buying a forward contract and also buying a put option on the same underlying. These are all byproducts of a put-call parity for bond options.

A large number of hedging strategies can be constructed for interest-rate risk by buying-selling call and put options at different exercise prices for different maturities. A straddle is a bet on increased volatility in interest rates; the investor is insuring against major increase or major decrease in price of the underlying. In the case of interest rate risk, an investor sells an anticipated volatility in interest rates by selling a straddle. Traders may often resort to using straddles when an announcement about a change in key interest rate is expected and when the outcome of the announcement is uncertain, or before some other major macroeconomic decision by a government or central bank. As discussed in Chapter 7, volatility can be purchased somewhat more cheaply by buying a strangle instead of a straddle. A strangle will involve buying a put option at a lower exercise price than current value of the underlying and buying a call option at a higher exercise price than current value of the underlying. Options on futures are also an important option category for fixed income instruments, giving the buyer the right to buy or sell a futures contract at the maturity of the option.

Caps, Floors, and Collars. These are best explained in terms of adjustable-rate mortgages (ARMs). The interest rate applicable to ARMs might be based on a floating rate, such as the rate of a 6-month T-Bill, for the upcoming 6 months. The borrower will need to pay that rate plus a spread, say, an additional 2% per annum, for the interest payment for the next 6 months. Often adjustable rate (AR) borrowers are offered 'caps' on interest rates for their long-term loans, so that in case the short-term interest rates rise above a predetermined level, say 5% in this example, the borrower doesn't pay more than the 5% cap plus the pre-set spread. A cap on the floating rate is definitely attractive for interest rate risk management, but it comes at a cost, paid as a fee upfront.

In order to reduce cost of the cap, the borrower may also be offered a 'floor.' The floor sets a minimum interest payment per period when short-term rate declines substantially; the borrower doesn't fully benefit from this. For instance in the above example, it may be set at a T-Bill rate of 2% plus the spread. Therefore, if the T-Bill rate drops to 1.5%, the borrower must still pay 2% plus the spread on his loan. The floor and the cap may be set at such a level that their premiums offset each other; such an arrangement is called a 'zero-cost collar' or a 'zero-cost cylinder.' Both these terms are used in the context of options strategy also. As extension to caps and floors, one can also purchase a 'caption' or a 'floortion' as options to enter a 'cap' or a 'floor' with certain terms at a future time.

In a similar vein, a swaption is an options on swaps, which gives the right to the buyer to enter a swap on or before a specified date at currently determined terms. And there is a wider variety of exotic options, such as barrier caps and floors, cancelable swaps, extendible swaps, moving average caps and floors.

8.2.3   Interest-Rate Hedging Strategies

The interest rate derivatives discussed in Section 8.2.2 can be utilized for the transfer of interest rate risk. Their precise use, however, depends on the objective of risk management. For a single bond, with price P, or a bond portfolio, Π, the goal of risk management could be to neutralize the effect of interest rate risk. The optimal risk management strategy, based on transfer of risk, depends on creating the optimal hedge for the chosen measure of risk. A detailed discussion of optimal hedging strategy was given in the context of equity risk in Section 7.3 of Chapter 7. We will advance this discussion here for the context of fixed income instruments.

For interest rate risk, in this chapter we developed a few new measures of risk for the first- and second-order effect of changes in the term structure of interest rates on price of bonds and portfolio of bonds. Those focused on yield of the bonds were modified duration and convexity, for first- and second-order, respectively. Suppose we intend to hedge a bond portfolio Π with a yield to maturity y_1 using a hedge instrument, such as a swap or a futures contract, F_1, with a yield y_2. We wish to construct a hedged portfolio that is insensitive to small interest-rate variations. In other words, for small changes in yield from parallel shifts of the yield curve, $dy = dy_1 = dy_2$, we want no change in value of the hedged portfolio. This can be achieved by taking a position h^* in the hedge instrument to construct a hedged portfolio, Π_{hedged}, which from Eqn. (8.47) should satisfy,

$$\Delta\Pi_{hedged} \simeq (D_\Pi^*\Pi(y_1) + h_D^* D_{H_1}^* H_1(y_2))\Delta y = 0, \qquad (8.52)$$

where D_Π^* and $D_{H_1}^*$ are the modified durations of the portfolio and the hedge instrument, respectively. The Eqn. (8.52) allows us to obtain the optimal hedge, h_D^*, to be,

$$h_D^* = -\frac{D_\Pi^*\Pi(y_1)}{D_H^*H(y_2)}. \qquad (8.53)$$

We can extend this to second order sensitivity by considering a convexity hedge, but for this we will need a second hedge instrument, $H_2(y_3)$. The hedge weights, h_{1C}^* and h_{2C}^*, for the two hedge instruments, $H_1(y_2)$ and $H_2(y_3)$, can be obtained by simultaneously solving the following two equations.

$$\Delta\Pi_{hedged} \simeq (C_\Pi\Pi(y_1) + h_{1C}^* D_{H_1}^* H_1(y_2) + h_{2C}^* D_{H_1}^* H_1(y_2))\Delta y = 0 \quad (8.54)$$

$$\Delta^2\Pi_{hedged} \simeq (C_\Pi\Pi(y_1) + h_{1C}^* C_{H_1} H_1(y_2) + h_{2C}^* C_{H_1} H_1(y_2))\Delta y^2 = 0 \quad (8.55)$$

By this hedge, $\Delta\Pi_{hedged} = \Delta^2\Pi_{hedged} \simeq 0$, therefore from a Taylor expansion the portfolio will not respond to changes in the yield up to second-order accuracy (refer to Eqn. (8.40)). However, this is assuming that the yield changes due to a parallel shift in the yield curve, or if it is not a parallel shift the net effect on the yield of all bonds in the portfolio is identical. These are reasonable approximations, given the simplicity of obtaining the hedge weights under these assumptions.

For the development of a more accurate hedging strategy, one can choose a set of factors, such as by principal component analysis, and describe the term structure of interest rates, $\{R_{t_0}, R_{t_1}, \ldots, R_{t_N}\}$, in terms of a much lower number of factors. If there are M factors chosen to capture the changes, ΔR_{t_j}, in interest rates, we will describe the changes as,

$$\Delta R_{t_j} = \sum_{i=1}^{M} f_{ji} F_i + \epsilon_j, \text{ for } j = 1 \dots N, \qquad (8.56)$$

where F_i's are the factors and ϵ_j is the residual from the factors describing the changes in interest rates. In this set-up, change in the portfolio value can be described in terms of weights of each bond in the portfolio and change in price of each bond given by Eqn. (8.44). Moreover, change in each of R_{t_j} is captured in terms of factors F_i, as per Eqn. (8.56). The change in value of chosen hedge instruments, $\{H_j; j = 1 \dots M\}$, can also be expressed in terms of the factors, F_i's. The hedge weights, $\{h_j^*; j = 1 \dots M\}$, can be obtained by solving a system of equations obtained by equating $\partial \Pi_{hedge} / \partial R_{t_i}$ to zero. This results in obtaining a first order hedge. For a second order hedge, a much larger set of hedge instruments and system of equations will need to be solved to obtain the optimal hedge. A second order hedge will attempt to equate $\partial^2 \Pi_{hedge} / \partial R_{t_i} \partial R_{t_j}$ to zero, for all i, j, besides setting the first derivatives to zero.

Finally, in some cases hedge instruments may not be accessible or available for the exact term-structure of interest rates on which the bond portfolio is defined. This will require constructing a *cross-hedge*. Cross-hedging is when one attempts to utilize a hedge instrument based on one risk factor as a substitute for hedging risk of an asset or portfolio based on another risk factor. This is the genesis for *basis risk*, where while the hedge was constructed to neutralize risk, since the two risk factors are not identical, therefore not perfectly correlated, mismatches can arise that result in risks not remaining neutralized.

The determination of cross-hedge weight can follow the above approach, with an intermediate crucial step of describing one risk factor in terms of the second. If the hedge instruments are based on a different term-structure of interest rates, $\{\tilde{R}_{t_0}, \tilde{R}_{t_1}, \dots, \tilde{R}_{t_N}\}$, then we would describe each of the factor $\tilde{R}_{t_i}$ in terms of factor R_{t_i}, for instance by a linear regression model, $\tilde{R}_{t_i} = \beta_i R_{t_i} + \epsilon_i$. We proceed by using this relationship between $\tilde{R}_{t_i}$ and R_{t_i} in the above derivation.

Although in the above discussion we have chosen to remain abstract in terms of specific hedge instruments that are and can be used for the hedging objectives, many of the instruments we discussed in Section 8.2.2 fit the bill perfectly. Vanilla swaps, non-vanilla swaps, bond futures, option on futures, and bond options are all candidate hedge instruments one can consider. For each kind of hedge instrument, the modified duration, convexity, and other sensitivity measures will need to be developed, and the hedge weights will then be derived from the above formulations.

In the chapter thus far, we have focused on interest rate risk, measures designed specifically for interest rate risk, and instruments for managing the risk. We next begin considering the third important class of market risk, commodities risk. In the next section, we will begin with a general discussion of commodities and their price risk, followed by features of their risk characteristics and instruments for risk management.

8.3   Managing Commodities Risk

Commodities are broadly defined as physical (with some exceptions, like electricity, bandwidth) products that can be traded in an organized marketplace. These physical products are crucial raw materials for the production economy, and hence the variability in prices of these raw materials is an important risk affecting the production of goods and services. The variability in prices of these raw materials constitutes commodities risk. Risk in commodity prices behaves quite differently than equity, interest-rates or foreign exchange risks, and are often more complex. This is because price of commodities are often affected by specific demand and supply features of the commodity, including the possibility of supply being controlled by a few suppliers. Interest in commodities doesn't arise solely from actual business use of the commodities; investors expose themselves to commodity price risk also as a means to diversify their investments or to simply speculate.

The variabilities in demand and supply for a commodity, as well as specific properties of the commodity, dependence of its availability on factors such as weather, political issues, etc. can magnify volatility in commodity prices. The fundamentals affecting commodity prices can include the extent of market liquidity or illiquidity, and degree of ease of storage and cost of storage, also called cost of carry. Prices may be affected by seasonality in demand and supply of the physical product, besides being affected by the usual business or economic cycle. Due to all these reasons, commodity prices have higher volatility, and can display sudden large changes or jumps.

Commodities can be classified into the following groups, based on some common shared properties.

Hard (nonperishable): Precious metals (gold, palladium, platinum, silver), base metals (aluminium, copper, iron ore, tin, zinc)

Soft (perishable, short shelf life and hard to store): Agricultural products (grains, soybean, salt, coffee, sugar, orange juice, live cattle, lean hog, cotton, dairy)

Energy: Crude oil, ethanol, natural gas, heating oil, gasoline, diesel, coal, electricity

Therefore, soft commodities are goods that are agricultural or farm products that perish in time if not used, and hence have a shorter shelf life. While hard commodities are ones that are extracted through mining, and are nonperishable. The energy commodities, with exception of electricity, can be stored.

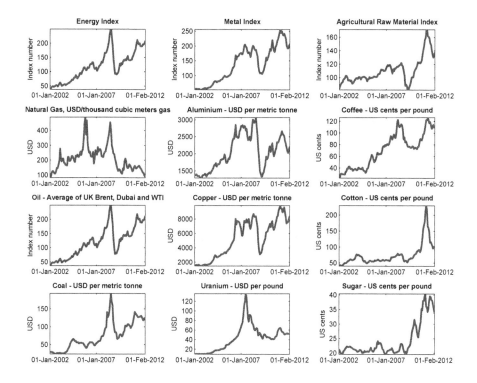

FIGURE 8.7: Prices for some commodities of different type, from January 2002 through 2012.

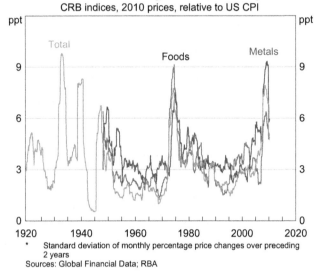

FIGURE 8.8: Level of volatility in commodity indices (Courtesy *Reserve Bank for Australia (RBA) Bulletin*, June 2011).

Therefore, storage and the related cost is a common feature to different extents for all commodities. A display of ten-year evolution of commodities prices is provided in Figure 8.7, with the top panel representing an index for energy, metal, and agricultural commodities, respectively. The high volatility in prices is quite evident, especially enhanced at times for certain commodities, such as crude oil prices towards the end of 2010, uranium price in mid-2007, and a surge in price of cotton in early 2011, each representing several fold increase in price of the commodity.

A high level of volatility of commodities is not a recent phenomenon. Figure 8.8 displays volatility of three commodity indices, metal, agricultural and aggregate, over a much longer time period. It is in fact true that the prices of Figure 8.7 correspond to an enhanced volatility regime, but this is not unprecedented. Other than the generic market price risk of commodities, another type of risk relevant for commodities is basis risk, arising from the differential in price movement of different related commodities, such as among energy commodities, metals, and agricultural products. In Figure 8.7, some evidence for joint movement of some commodities is visible. Additionally, interest rate risk can result in change in the cost of carry, which is an important component for determining commodity and commodity derivative prices.

In the next section, we will develop model enhancements as relevant for commodity price risk, followed by advancing the discussion of risk management of commodity risk. We will introduce a special case for the optimal hedging decisions framework for commodity price risk, followed by providing an overview of derivative instruments applicable for commodity risk.

8.3.1 Modeling Commodity Spot Prices

As discussed above, commodity prices show certain unique characteristics that the models used for commodity risk management should attempt to capture for the analysis to be meaningful. The continuous-time modeling framework developed in Chapters 5 and 6, and extensively utilized in Chapter 7 and this chapter so far, are useful here. Whenever appropriate, discrete-time approximations to the continuous-time models, such as by using binomial or trinomial trees, can be developed. Our focus in the discussion here will be on continuous-time models.

Many commodities are known to show seasonality, arising from either the demand-based or supply-based seasonality. For instance, energy commodities are affected by user needs and behavior in summer's heat and summer travel, while agricultural commodities are affected by production based on crop cycle. Cyclicalities may also arise in different time scales, such as diurnal cyclicality due to day-and-night temperatures or human activities, or in years arising from lag in investment for increase in production capacity for a commodity.

The simplest version of the Black-Scholes model used for modeling stock price evolution in Section 7.2 can be modified to capture the seasonality or cyclicality in commodity prices as follows.

$$dY_t = \mu(t)Y_t dt + \sigma Y_t dW_t, \tag{8.57}$$

where Y_t is the price of a commodity with time-dependent drift, $\mu(t) = \mu_0 + \mu_1 \sin(2\pi\alpha t)$, which is chosen to depict the seasonality or cyclicality in the commodity price. The parameter μ_1 is the amplitude of the cycle and α is the frequency, while σ is the commodity price volatility, as always, taken to be a constant in the above model. In reality, the cyclicality of most commodities may not be so strict as depicted in Eqn. (8.57), instead the price may have a tendency to show seasonality, but may wander significantly away from this pattern. A mean-reverting model, first studied in the context of stochastic volatility in Chapter 7 and later used in this chapter for modeling interest rates, can be utilized for this purpose, as follows.

$$dY_t = -\gamma(Y_t - \mu(t))dt + \sigma Y_t dW_t, \tag{8.58}$$

where $\mu(t)$ now serves the role of long-run mean and γ is the rate of mean-reversion to the long-run mean.

Although in the above two models we have taken the volatility of commodity price to be a constant, in the price trajectories of individual commodities and commodity indices shown in Figures 8.7 and 8.8, there is ample evidence for the volatility to not be constant, and instead be stochastic in its own right. We advance our modeling in response by introducing a stochastic volatility model for commodity prices as follows.

$$\begin{aligned} dY_t &= -\gamma_1(Y_t - \mu(t))dt + \sqrt{v_t}Y_t dW_{1t}, & (8.59) \\ dv_t &= -\gamma_2(v_t - \bar{v})dt + \eta\sqrt{v_t}dW_{2t}, & (8.60) \end{aligned}$$

where W_{1t}, W_{2t} are taken as independent Wiener processes, or as a further generalization, they can be modeled as correlated Wiener processes. The variance of commodity price, v_t, also shows mean-reversion with the rate of mean reversion taken as, γ_2, and long-run mean as, $\bar{v}$, while η is the volatility of the volatility of commodity price. Finally, commodity prices show occasional, sudden large changes due to either a demand shock or a supply shock, or some other fundamental event, such as a political event. A jump-diffusion model first introduced in Chapter 7 can be applied here, keeping the mean-reversion intact in the commodity price evolution as follows.

$$dY_t = -\gamma(Y_t - \mu(t))dt + \sigma Y_t dW_t + \nu dN_t, \qquad (8.61)$$

where N_t is a homogeneous Poisson process with parameter λ and ν is the jump amplitude. Generalization on the Poisson process being non-homogeneous or the volatility being stochastic, as in the stochastic volatility model of Eqn. (8.60), can be introduced in the above model.

As discussed earlier, commodity prices are affected by many factors, which can either be theoretically justified or empirically tested. This explicit dependence on or relation with factors can be explicitly incorporated in factor models, where the factors Z_{it}, for $i = 1 \ldots K$ evolve along with commodity price Y_t as follows.

$$
\begin{aligned}
dY_t &= \mu(t, Y_t, Z_{1t}, \ldots, Z_{Kt})dt + \sigma(t, Y_t, Z_{1t}, \ldots, Z_{Kt})dW_{0t}, & (8.62) \\
dZ_{it} &= \alpha_i(Z_{1t}, \ldots, Z_{Kt})dt \\
&\quad + \eta_i(Z_{1t}, \ldots, Z_{Kt})dW_{it}, \text{for } i = 1 \ldots K, & (8.63)
\end{aligned}
$$

where $[W_{0t}, W_{1t}, \ldots, W_{Kt}]$ are $K + 1$ correlated Wiener processes, with correlation coefficient, $corr(dW_{it}, dW_{jt}) = \rho_{ij}$. The K factors can be stochastic convenience yield, which we will define in the context of commodity futures, long-term trends of commodity price, price of other commodities, such as oil price, macroeconomic variables, such as real interest rate, etc.

In some cases, the factors are not modeled as a diffusion as in Eqn. (8.63), but instead are described as a continuous-time Markov chain in order to depict regime switching for the commodity price evolution. A regime-switching model is used to describe a change in fundamentals, namely regime, that makes the rules for evolution of the commodity price change. If Z_{1t} is a continuous-time Markov chain with M states, each state of the chain is a regime. The commodity price evolution by any of the models discussed thus far can then be stated with a regime-switching effect [14, 15]. We apply regime-switching to the seasonal mean-reversion model in Eqn. (8.58) to obtain the following regime-switching model.

$$dY_t = -\gamma(Z_{1t})(Y_t - \mu(t, Z_{1t}))dt + \sigma(Z_{1t})Y_t dW_t, \qquad (8.64)$$

where $\gamma(Z_{1t}), \mu(t, Z_{1t}), \sigma(Z_{1t})$ are all taken to be functions of the regime-indicator state variable, Z_{1t}. As the state of the regime-indicator variable changes, these crucial terms of the model change form or level. Therefore, the commodity evolves by different rules in different regimes.

In this section, we have presented a sequence of models with increasing complexity to capture the properties of various commodities. Our model descriptions here have been in the real-world measure, treating commodities as an asset class for investment. For risk transfer and derivative pricing, we leave the discussion of the models in risk-neutral measure for Section 8.3.2. The variety of commodities, as listed earlier in this section, is large. It is beyond the scope of this book to discuss specific models for all the commodities. However, we consider some specific examples in the energy commodities, and the newer segments of electricity and weather risk models next.

8.3.1.1 Energy, Electricity, and Weather Risk

Every firm, small or large, and every household is exposed to energy, electricity, and weather risk, irrespective of whether they actively manage them or not. In this section we consider some specific examples for these commodities. Crude oil or petroleum, one of the fundamental sources of energy and other highly useful byproducts, is drilled and pumped from the ground at various regions in the world, such as the Middle East, West Africa, the Americas, and Asia. It is shipped in tankers to oil refineries, where through a complex series of distillation processes common distillates, such as gasoline, jet fuel, kerosene, diesel fuel, and a range of other useful byproducts are obtained. Oil refinery profits are tied directly to spread between the price of crude oil and the prices of products that result from refining crude oil, such as gasoline, diesel fuel, jet fuel, and heating oil. Therefore, models for both crude oil and 'crack' spread, price difference of crude oil and its distillates, are valuable for risk management.

Almost all the models discussed in the previous section have been considered for modeling evolution of crude oil price. The Schwartz [78] model described price of crude oil, Y_t, to evolve by the following model.

$$dY_t = \alpha(\bar{L} - \ln(Y_t))Y_t dt + \sigma Y_t dW_t. \tag{8.65}$$

In a more recent study, however, geometric Brownian motion with jumps was found to be the best model for crude oil price compared to the other commonly used continuous-time stochastic processes [43].

Natural gas is an important energy resource used as a less expensive energy source than electricity in retail settings, as well as for power generation and heating. Natural gas accounts for about a quarter of total energy consumption in North America. After the deregulation of the natural gas industry in 1978, the natural gas market has evolved into a dynamic, highly competitive market with significant price volatility. New York Mercantile Exchange (NYMEX) was the first to launch natural gas futures. Standardized contracts of relatively small size, fungibility, performance requirements, with lack of requirement of physical delivery have led producers, distributors, processors, utilities, consumers, and speculators to participate in this market.

There are a number of fundamental factors that drive the price of natural gas, such as extraction, storage, transport, weather, technological advance, new reserves discovery, etc. As a result, gas price shows mean-reversion, where the mean-reversion seems to have correlation with summer heat waves, floods, and other news that develops into supply or demand imbalances. Dissipation of the news or temperatures reverting to normal range results in the reversion to the mean for natural gas price. The long-term mean for gas is determined by cost of production and long-run demand level. Seasonality is also seen in gas prices, arising from the heating use of gas.

We describe one model where the natural gas spot price is taken as a sum of seasonal factor, $f(t)$, and a non-seasonal factor, Z_t, i.e., $Y_t = f(t) + Z_t$, where each component is described as follows.

$$f(t) = \beta_0 t + \sum_{i=1}^{N} \beta_{1i} \cos(\frac{2\pi it}{365}) + \beta_{2i} \sin(\frac{2\pi it}{365}), \qquad (8.66)$$

$$dZ_t = -\gamma_1(Z_t - L_t)dt + \sigma Z_t dW_{1t}, \qquad (8.67)$$

$$dL_t = -\gamma_2(L_t - \bar{L})dt + \eta L_t dW_{2t}, \qquad (8.68)$$

where $f(t)$ is a sinusoidal function fitted to yearly seasonality, and Z_t is perturbation from the seasonal characteristics. Z_t in its own right has a long-run, time-varying, stochastic mean, L_t, and a mean-reversion rate, γ_1. The long-run mean, L_t, is also mean-reverting with reversion rate, γ_2. Therefore, natural gas price in this model evolves by seasonal variation and mean-reverting perturbations around this variation.

Electricity is (mostly) non-storable and follows laws of physics for transportability, which are essential features of electricity in explaining behavior of electricity prices. Moreover, off-peak and on-peak demand for electricity, and their variation by seasons, are important aspects for electricity price risk, risk models, and electricity markets. One expects electricity spot prices to show high dependence on temporal and local supply-demand levels. Disruptions in transmission lines and power outages due to rare and extreme events can result in a steep rise in prices. Electricity prices display mean-reversion, like natural gas prices, and as discussed, are affected by factors such as weather and load. We present one example model for electricity spot prices, Y_t.

$$dY_t = -\gamma_1(Y_t - \mu_t)dt + \sigma Y_t dW_t + \nu_t dN_t, \qquad (8.69)$$

$$\mu_t = \mu_1 \mathcal{I}_{t \in T_{on-peak}} + \mu_2 \mathcal{I}_{t \in T_{off-peak}}, \qquad (8.70)$$

where μ_t is an off-peak versus on-peak mean that electricity prices revert to at the mean reversion rate of γ. Prices are affected by arrival of jumps by the Poisson process, N_t and jump size, ν_t. The jump size is a product of two random variables, one that dictates the sign and other the size, such as a product of a Bernoulli distribution for sign and exponential for size of jump. Therefore, the above electricity price model is a regime-switching, mean-reverting, jump-diffusion model.

Weather derivatives are the newest kinds of derivatives traded over-the-counter and on exchanges. Weather risk constitutes most significantly the temperature, but it also includes frost, hurricanes, snowfall, and rainfall. Over-the-counter derivatives for weather appeared after the changed weather patterns due to El Niño in 1997. Chicago Mercantile Exchange (CME) started trading weather derivatives in 1999. As seen in the above discussion, dependence of electricity and natural gas prices on temperature implies that producers and service providers in these sector will find weather derivatives very advantageous. As such, weather derivatives are also useful for other businesses, such as in entertainment and leisure, agriculture, grocery supermarkets, and even consumers, especially the temperature related derivatives. Hurricane, frost, rainfall or snowfall can be utilized for risk management objectives of insurance providers, as we will discuss in Chapter 11. An additional attractive feature of weather derivatives is that they offer an opportunity to manage not just price risk, but also producers' quantity risk, i.e., the amount of produce they are able to generate due to weather phenomena.

Weather risk, like electricity, is non-storable and non-transportable, and weather, unlike all the commodities considered so far, is not delivered. We describe a model for temperature based weather risk; where most temperature derivatives are designed as Asian, barrier or lookback options. Temperature shows seasonality and daily patterns, and also displays long range dependence or high degree of autocorrelation. Pricing of these derivatives is done based on temperature models, such as the following.

$$dT_t = -\gamma_1(T_t - T_{\mu_t})dt + \sigma_t dW_t, \tag{8.71}$$

$$T_{\mu_t} = b_0 + b_1 t + b_2 \sin(\omega t + \theta), \tag{8.72}$$

where temperature is taken to evolve as a mean-reverting Ornstein-Uhlenbeck process, with the long-run mean T_{μ_t} taking on a very definite sinusoidal form.

8.3.2 Management of Commodity Risk

Applying the avoid-mitigate-transfer-keep framework for risk management to commodity risk requires specific consideration, since as stated before, commodities serve as crucial raw materials for the production economy. A producer or manufacturer exposed to one or more commodity's price risk doesn't have the luxury of too many alternatives for the diversify or mitigate response of risk management. Avoiding the risk is not an option since the commodity risk exposure is crucial to the producer's or manufacturer's business. Therefore, producers or manufacturers may be left with the keep or transfer responses for risk management.

There is a significant participation in commodity markets for investment, as seen in Figure 8.9. Investors exposing themselves to commodities price risk with the purpose of speculation or diversification can utilize the models developed in earlier chapters and in Section 8.3.1 for the purpose of risk management. In this context, the mitigate or diversify objective of risk management

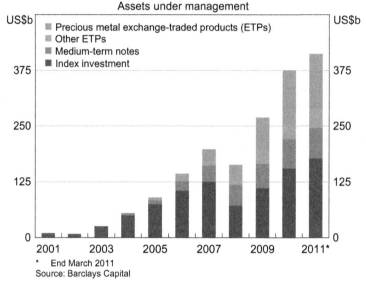

Commodity Investment
Assets under management

FIGURE 8.9: Participation in commodities markets for diversification benefits (Courtesy *Reserve Bank for Australia (RBA) Bulletin*, June 2011).

will be served well by utilizing the framework for portfolio optimization developed in Section 7.1, where commodity indices or specific commodity products may be treated as assets considered for investment from an asset class. This broader participation in the commodities markets also implies that the risk transfer objectives of the participants will be broader than of those engaged in the production economy utilizing these resources. Therefore, a larger type of instruments to serve a variety of risk management objectives may prove useful. Therefore, our focus in this section remains the transfer or keep responses of risk management.

Consider a manufacturer of confectionery goods who needs 100,000 pounds of sugar for use in the next month. The manufacturer can either purchase the sugar in a spot contract at the next month's spot price of sugar, thus exposing itself to sugar price risk. Alternatively, if a forward contract to purchase sugar at $0.35 cents per pound is available for delivery next month, then the manufacturer can enter such contracts to the anticipated level of sugar requirement for next month and lock in the price.

As discussed in Section 8.2.2, a forward contract on a commodity is a contract to purchase or sell a specific amount of the commodity at a specific price and at a specific time in the future. These are over-the-counter (OTC) contracts, and the contract is binding. Pay-off of the contract will be determined by price of sugar next month. If the spot price is $0.40 cents per pound, the contract will be worth $0.05 cents per pound, but if the price of sugar next month is \$0.32 cents per pound, there would be a loss of \$0.03 cent/pound. A producer or manufacturer may still be interested in such a contract since it helps lock in a price, and guarantees the availability of the required raw material at the set price and required time. For this advantage offered by forward contracts, these contracts have been around for thousands of years.

In some cases the future level of need for the crucial raw material may not be known precisely or a perfect substitute forward contract may not be available. In such cases, the minimum variance hedge or optimal hedge framework developed in Section 7.3.2 can be applied, where the producer's or manufacturer's risk management objective would be appropriately captured by variability of cash flow.

Let the forward price, F_0, be the price that is applied at delivery, which is set initially so that the initial value of the contract is zero. A minimum variance hedge, or in general an optimal hedge, is constructed by picking the optimal level, x^*, of forward contracts in order to obtain the lowest variance of cash flow. If the producer is expecting to have Q units of a commodity to sell (or buy) at a future time, T, which is stochastic with known distribution. Let S_T be the stochastic spot price of the underlying commodity at maturity of the forward contract. Therefore, the anticipated cash flow at maturity of the forward contract is,

$$Y = xF_0 + (Q - x)S_T, \tag{8.73}$$

where x units of the underlying commodity will be sold using the forward contracts and $(Q - x)$ units will be sold on the spot market. It should be noted that depending on the choice of x and outcome of Q, $(Q - x)$ can turn out to be negative, implying that the producer will need to buy the commodity on the spot market in order to fulfill the obligation of the forward contracts. The stochastic component of the cash flow in Eqn. (8.73) is,

$$Z = (Q - x)S_T, \tag{8.74}$$

which will remain the focus in the determination of optimal hedge. We define the variance of the cash flow as,

$$
\begin{aligned}
\sigma^2(Y) &= \sigma^2(Z) = E[Z^2] - E[Z]^2 & (8.75) \\
&= E[((Q - x)S_T)^2] - (E[Q]E[S_T] - xE[S_T])^2, & (8.76)
\end{aligned}
$$

where the last term is obtained under the assumption that the quantity produced by the producer, Q, and the spot price at T, S_T, are independent. From Eqn. (8.76), the following steps of simplification yield an equation from which the minimum variance hedge, x^*, is determined.

$$
\begin{aligned}
\sigma^2(Z) &= E[(Q-x)^2 S_T^2] & (8.77) \\
&\quad -(E[Q]^2 E[S_T]^2 + x^2 E[S_T]^2 - 2xE[Q]E[S_T]^2), \\
&= E[Q^2 + x^2 - 2Qx]E[S_T^2] & (8.78) \\
&\quad -(E[Q]^2 E[S_T]^2 + x^2 E[S_T]^2 - 2xE[Q]E[S_T]^2), \\
&= E[Q^2]E[S_T^2] + x^2 E[S_T^2] - 2xE[Q]E[S_T^2] & (8.79) \\
&\quad -(E[Q]^2 E[S_T]^2 + x^2 E[S_T]^2 - 2xE[Q]E[S_T]^2).
\end{aligned}
$$

From Eqn. (8.79), candidate minimum variance hedge is found to be, $x = E[Q]$. An optimality test using the second derivative confirms that this choice of hedge minimizes the variance of cash flow.

8.3.2.1 Commodity Futures and Other Derivatives

Volume of trades of commodity futures in 2011 at 81 worldwide exchanges was at 2.5 billion contracts, which was in fact down from 2.8 billion in 2010. Although this dwarfs before number of equity and interest rate futures and options contracts traded on these exchanges, it remains a sizeable volume. Given the significant participation in commodity markets for investment, as seen in Figure 8.9, as well as variety of commodities relevant for the production economy, this volume is not surprising. We first defined forwards and futures as general risk transfer instruments in Chapter 2, and since then have looked at specific examples of forward and futures contracts for equity and interest rate risks. Forwards and futures are the simplest kind of derivative instruments, as well as given their popularity for commodities, we begin with discussing some aspects of commodity futures, followed by some other commodity derivatives.

We develop a model-free price for a futures contract. If Y_t is the spot price of a commodity and $F(t,T)$ is the futures price of a contract that matures at T, we show that,

$$
F(t,T) = Y_t e^{(r+u)(T-t)}, \tag{8.80}
$$

where u is the storage cost per unit time as a percent of the underlying commodity value and r is the short-term risk-free interest rate. We call $c = r+u$ the cost of carry for the commodity. We arrive at the futures price in Eqn. (8.80) by an arbitrage argument. If $F(t,T) > Y_t e^{c(T-t)}$, we would borrow $Y_t e^{u(T-t)}$ at the risk-free rate for $T - t$ period, we will sell a forward contract maturing at T, buy the underlying commodity for Y_t and keep the rest of the cash for the storage cost of the commodity until T. At time T, the commodity is used to close the short position in the futures contract, with a net positive cash flow. This is an arbitrage profit, since the initial cost was zero. In response, the prices will move so that the arbitrage profit is no longer available.

Let's consider the reverse possibility of $F(t,T) < Y_t e^{c(T-t)}$. In this case, the firms who hold the commodity will sell the commodity and long a futures contract to buy it back at T for a cheaper price, $F(t, T)$. The cash from sale of commodity is invested in the risk-free asset. At time T, from the savings of not having to store the commodity and the return from the risk-free investment, the merchant will have $Y_t e^{c(T-t)}$, which will be used to settle the futures contract. The firms get to bag the difference as arbitrage profit. In order to eliminate the arbitrage opportunity, the prices should move so that Eqn. (8.80) holds.

In reality, this doesn't always happen, since the firms holding the commodity aren't always willing to part with their commodity holdings with the intent to acquire it back later. One reason for this, which is the fundamental property of commodities, is that commodities support the production economy and the firms may need the commodity for their production activities. Therefore, for a consumption asset, we have $F(t, T) \leq Y_t e^{c(T-t)}$. We define a quantity called the *convenience yield* of the consumption asset, y, so that

$$F(t, T) = Y_t e^{(c-y)(T-t)}. \tag{8.81}$$

In practice, convenience yield must be estimated from the observed futures prices. The convenience yield measures the market's expectation regarding future availability of the commodity. The higher the likelihood of shortage of the commodity in future, the higher the convenience yield. If inventories of the commodity are running high, there is a lower chance of shortage, hence the lower convenience yield.

Other types of derivatives for commodities include commodity vanilla European options, exotic options, commodity swaps, and options on futures or futures options. For pricing these derivatives, since the underlying commodity is often not liquid, non-storable, and is often bulky and difficult to trade and store, the price derivation and the corresponding portfolio replication is done based on commodity futures as the underlying contract. The advantage of this is commodity futures are liquid, taking short or long position is equally straightforward. If the commodity spot price evolved by any of the models discussed in Section 8.3.1, and given the futures price is determined by Eqn. (8.81), we can apply Ito's formula to determine the change in the futures price with time. If we want to price a commodity option, $c(t, Y_t)$, the replicating risk-free portfolio, $\Pi(t, Y_t)$, can be constructed as in Section 7.2.1, using the option and the futures contract, $f(t, T)$. The rest of the derivation will follow as we have seen in several previous contexts, including in Section 7.2.1.

Commodity options can be used to develop hedging strategies, as discussed in the context of equity risk in Section 7.3.1, or can be applied to develop an optimal hedge or minimum variance hedge. The objective of the optimal hedge problem can be to minimize the return or cash flow from commodity risk exposure, where the former was developed in the context of equity risk in Section 7.3.2 and the latter was developed earlier in this section.

8.4 Managing Foreign Exchange Risk

Financial firms function in a variety of global markets, in addition to making products available to their customers that expose them to currency risk. Currency or foreign exchange risk arises from open or imperfectly hedged positions in a particular currency. Increasingly many non-financial firms must transact in different currencies due to global operations or markets for their goods. This currency risk exposure arises as a natural consequence of business operations, not due to explicitly taking a position in a currency. For non-financial firms, currency risk can have a pretty drastic impact in terms of sweeping away profits. As a result, it can place a firm at a significant competitive disadvantage, generate huge operating losses, and in the end, inhibit the firm's growth and investment.

Purchasing power parity, a general equilibrium hypothesis, claims that the exchange rate should be such that the purchasing power in one currency matches the purchasing power in the other currency. This might suggest that exchange rates should more or less remain constant, and adjust only periodically and gradually. This is far from today's reality. There is also a well-known, macroeconomically-motivated relationship between currency exchange rates, domestic interest rates and foreign interest rates, known as the international Fisher effect. It is based on a hypothesis in international finance that the difference in the nominal interest rates between two countries determines how their currency exchange rate will change. Therefore, interest rates and markets' perception of future interest rates have a direct impact on currency exchange rates.

If the interest rate in one country is high, the demand for that currency increases, which in turn makes the currency stronger relative to other currencies. The reverse is true for currency with low interest rates. An appreciating currency can lead to higher inflation and a riskier economy. The standard risk-return linkage holds. Therefore, currencies are like any other investment asset, albeit with a stronger macroeconomic and monetary policy underpinnings. When a currency thus becomes more risky, risk-averse investors sell off their positions, thus resulting in the currency's fall in value.

In Figure 8.10, we display a sample of nine exchange rates for key developed and emerging economies from all continents over the duration of several months. Even in this short duration snapshot from October 2011, one observes all exchange rates to fluctuate, some significantly more than others. In this period, the US dollar is fluctuating at par level with the Australian dollar (AUD) and the Canadian dollar (CAD), while through the Greek debt crisis, the euro and British pound seem to be moving with a high level of correlation. After the March 2011 tsunami disaster, the Japanese yen evolved under many pressures, and experienced significant devaluation against the US dollar in this period.

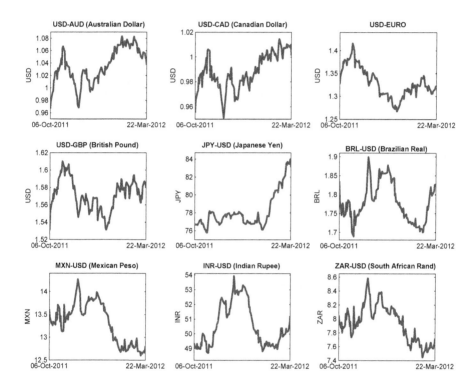

FIGURE 8.10: Sample of key exchanges for developed and emerging economy countries from all continents, as of March 2012.

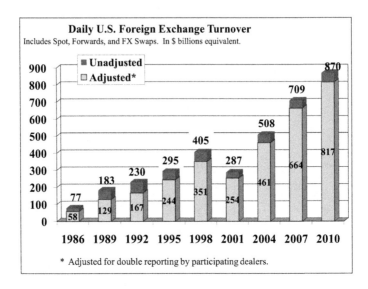

FIGURE 8.11: United States turnover of foreign exchange, all currencies. (Courtesy *Federal Reserve Bank of New York (FRB NY) Report*, April 2010).

If a firm, financial or non-financial, has any extent of international footprint, it inherently has exposure to currency risk. In such a case, the firm can have several motivations to explicitly manage currency risk, just as in the case of commodity risk when this exposure is fundamental to a firm's operations. The motivations can range from attempting to coordinate cash flows and investment, minimize expected taxes, likelihood of financial distress, or simply due to managerial risk aversion. Currency risk may be explicitly managed also due to incentives for smoothing earnings, obtaining competitive pricing advantages for the firm's products.

Figure 8.11 shows the daily turnover of the US dollar for all currencies. The daily turnover shown for an extended period of approximately 25 years shows a steady upward trend, touching a staggering daily turnover of a trillion US dollars in 2010. In Figure 8.12, the daily turnover is split by currency, both for US dollar and the euro. The US dollar has the highest fraction of its turnover for the euro. Additionally, it is not surprising that the next three highest currencies for US dollar turnover are the British pound, Japanese yen, and the Canadian dollar. Such large daily foreign exchange turnover volumes both indicate and justify active currency risk management.

Among the avoid-mitigate-transfer-keep response of risk management, if a firm has seen a strategic advantage in expanding its operations and/or sales to other countries, then the limitations of implementing the 'avoid' response are evident. The 'diversify' response for risk management must be explored. In international financial investment strategies, however, diversifying across currencies (and economies) is regularly practiced, and in principle is not different

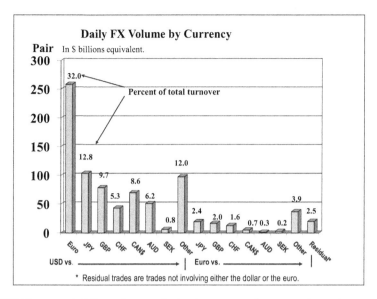

FIGURE 8.12: USD and euro daily foreign exchange volume by currency. (Courtesy *Federal Reserve Bank of New York (FRB NY) Report*, April 2010).

from what we discussed at length in Chapter 7. Therefore, in the discussion here our focus will be on the hedging component of risk management.

8.4.1   Models for Spot and Forward Exchange Rates

Various models have been utilized for the spot exchange rate, many of which we have already applied to equity, interest rate and commodity risks. For instance, mean-reversion is utilized in the models since exchange rates are expected to be at a fundamental level commensurate to the purchasing power parity or international Fisher effect. Additionally, stochastic volatility [62] and jump-diffusion extensions [42] have also been studied for modeling currency risk. In some cases, in order to capture fundamental changes in the monetary policy in a country or other regulatory changes, regime-switching models have also been applied to exchange rate risk [82]. Models for currency risk are needed either to price options contract or to determine hedging strategies. Pricing of currency forwards can be done in a model-free setting, which we describe next.

The relationship between spot exchange rates and exchange rate forwards is crucial, as discussed in the previous section. Let X_t be the spot exchange rate for a currency and $F(t, T)$ be the forward exchange rate for a forward contract maturing at a future time, T. Let's say the domestic short-term risk-free interest rate is r and the foreign short-term risk-free interest rate is r_f. A firm intends to convert one unit of foreign currency into its domestic currency,

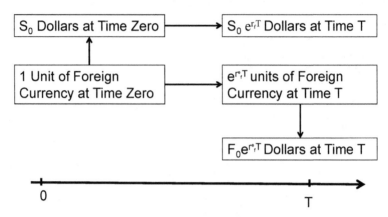

FIGURE 8.13: Spot and forward exchange rates hold a key relationship.

say US dollar, which it anticipates using at a future time, T. The firm has two options. First, it can convert the funds immediately at the current spot exchange rate of X_t, and have the funds appreciate at the domestic risk-free interest rate, r, becoming worth $X_t e^{r(T-t)}$ at time, T.

Alternatively, the firm can enter a long position in a forward contract to buy US dollars at T at the forward rate, $F(t, T)$. Until time T each unit of foreign currency held appreciates at the foreign short-term risk-free interest rate of r_f, becoming worth $e^{r_f(T-t)}$ at time, T. At time T, the firm executes the forward contract to obtain $F(t, T)e^{r_f(T-t)}$ US dollars. Following either alternative, the firm ends up with a certain amount of US dollars in a risk-free manner. If these amounts are different, i.e., $X_t e^{r(T-t)} \neq F(t, T)e^{r_f(T-t)}$, then it opens an arbitrage opportunity. Therefore, a foreign currency is analogous to a security providing a dividend yield, where the continuous dividend yield is the foreign risk-free interest rate, r_f. It follows that the currency forward or currency futures price should satisfy, $F(t, T) = X_t e^{(r-r_f)(T-t)}$. This relation is summarized in Figure 8.13.

8.4.2   Currency Derivatives

We have discussed the role of currency forward contracts in the context of developing models for exchange rates in Section 8.4.1. Figure 8.14 displays the relative significance of currency spot and forward contracts in terms of their daily turnover. It is no surprise that the trajectory of daily spot contract turnover mimics the evolution of daily volume seen in Figure 8.11. The daily forward contract volume shows a steady increase over the 18 year period, without being affected by the modulations of the spot contract volume. However, the volume of currency forward contracts is surpassed by that of currency swaps throughout this period. In this section, we will consider currency swaps and other derivatives that may be used to serve the risk transfer objective of currency risk management.

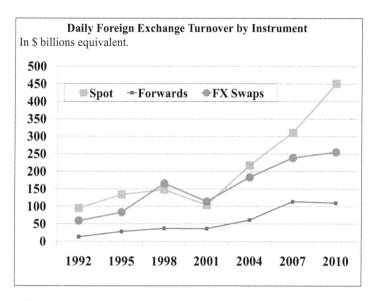

FIGURE 8.14: Daily turnover comparison for foreign exchange spot, forward and swaps. (Courtesy *Federal Reserve Bank of New York (FRB NY) Report*, April 2010).

Currency Swap. Currency swaps provide two counterparties the opportunity to periodically exchange the two underlying currencies for a set period of time at a set exchange rate. A principal is also identified in each currency to define the reference amount in each currency that the swap rate will be applied to determine the amount exchanged periodically. For example, a currency swap may be defined to pay 5% on a British pound principal of £10,000,000 and receive 6% on a US dollar principal of $18,000,000 every year for 5 years. While in an interest rate swap the principal itself is not exchanged, in a currency swap the principal is usually exchanged at the beginning and the end of the swaps life. Therefore, a currency swap can help in converting an investment or a liability in one currency to an investment or a liability in another currency, if the firm is more comfortable to maintain this investment or liability in the second currency. A currency swap can be considered as a bundle of forward contracts, where a common price is set for all the forward contracts in the bundle.

The above example was a fixed-to-fixed currency swap, or a vanilla currency swap. Other currency swaps can be constructed where one or both legs of the swap pay a floating rate. For example, a fixed-to-floating or a floating-to-floating currency swap. In the fixed-to-floating currency swap, the fixed rate is the swap rate, while in the floating-to-floating currency swap the spreads applied to either of the floating legs is the swap rate.

Currency Options. Beyond currency forwards and a variety of swaps, various currency options may also be useful for risk management. Vanilla

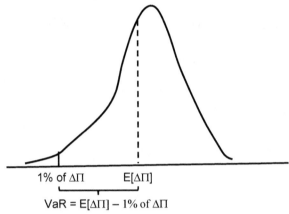

FIGURE 8.15: Distribution of the change in portfolio value, $\Delta\Pi$, in order to compute Market Value-at-Risk.

European call and put are the simplest choice to buy or sell a currency when the exchange rate is favorable. Other options to consider are a barrier option or a compound option. A down-and-in put or an up-and-out put can be more cost-effectively way to sell a currency at a required time. A compound option postpones the decision of having to sell the currency using an option by waiting to see if the currency moves in the favorable direction or not. A gap option is an interesting exotic option, which can serve the role of a 'pay-later' strategy, which implies the option buyer will have a zero premium upfront, but at option maturity will need to give up some of its pay-off. The pay-off of a gap put option is, $g(t, X_t) = (K_1 - X_t)\mathcal{I}_{X_t < K_2}$, where $K_1 < K_2$ is the case when the premium can be zero.

8.5   Value-at-Risk and Stress Testing for Market Risk Management

Our emphasis in this and the previous chapter has been on creating the avoid-mitigate-transfer-keep responses of risk management for the spectrum of market risks. We have developed portfolio analysis and optimization framework for the mitigate objective and hedging framework for the transfer objective. One of the important tasks of risk management is to monitor the performance of risk management decisions, and assess the level of risk kept as an outcome of avoid-mitigate-transfer-keep choices. This corresponds to the feedback loop at the bottom box of Figure 2.1.

Monitoring of risk management strategy is rarely done for single risk exposure or a single instrument. Instead the entire portfolio of assets must be collectively assessed for the impact of all relevant risk exposures. This is important since risk exposures rarely evolve independently of each other. Monitoring may also be developed at two levels, one that utilizes models chosen for each risk exposure and creates a model-based assessment of risk management strategy. The other level 'stresses' the models, so that model risk doesn't lead us into misguided conclusions.

For the first level assessment, a measure of performance must be decided, along with a threshold that allows judging if the strategy is performing satisfactorily according to the performance measure. This can be done with two themes, namely that good aspects of risk management strategy are good enough or bad aspects of the strategy are not too bad. A tail measure chosen as a performance measure would follow the latter theme (see Figure 8.15), while an upside measure will address the first theme. A popular tail measure used, for instance, in the Basel Accord, is the Value-at-Risk (VaR) measure. At a certain confidence level, a certain threshold or risk limit must be chosen, and the risk management strategy is assessed by the VaR measure against the set risk limit.

At the level of a business unit, a portfolio or enterprise-wide, along with a risk-limit, a time duration must also be chosen in which impact of risk is assessed. The value of the portfolio (business unit or enterprise value) is assessed at the start of the time period, Π_0, and at the end of the period, $\Pi_{\Delta T}$. The change in value of the portfolio, $\Delta \Pi = \Pi_{\Delta T} - \Pi_0$ is the key distribution to which the performance measure, in this case VaR, is applied.

For the market risk sensitive instruments in the portfolio, one needs to specify all the relevant risk factors. We also need to specify the dynamics of evolution of each of the risk factors, i.e., stochastic processes, parameters estimated volatilities, correlations, mean-reversion, stochastic volatility, etc. As such, three different levels of sophistication may be applied to estimate portfolio VaR for the period, ΔT, the Historical or Non-parametric approach, Analytic Variance-Covariance or Delta Normal approach, and Monte Carlo Simulation approach.

Historical or Non-parametric Method: This is the simplest of approaches, where no analytic assumptions are necessary for the risk factors. However, 2-3 years of historical data are necessary for meaningful results, since the approach treats the past observations as the population of possibilities the future will be sampled from. With this in mind, the steps followed here are as follows.

1. Using the historical observations, changes in relevant market prices and risk factors are sampled, keeping consistency regarding the time of observation for all the factors. This may need to evaluate price of derivative instruments, based on underlying asset values, if the corresponding historical market price for them is unavailable.

2. Based on the scenarios of change in market price of constituent instru-

ments, and portfolio weights, portfolio under examination is revalued to create a distribution of change in value of the portfolio for which VaR needs to be computed. Each simulated change in the value of the portfolio becomes an observation in its empirical distribution.

3. Construct a histogram of portfolio values and identify the absolute VaR that isolates the α-th percentile of the distribution in the left-end tail. After the computing the empirical mean in change in value of the portfolio, VaR is computed as the difference of mean change in portfolio and absolute VaR.

The major attraction of this approach is that it is completely nonparametric, where no assumptions about distributions of the risk factors are necessary. If the historical data shows fat-tails, the empirical distribution for the future retains it. Assumptions and calibration of variances, correlations, etc. of risk factors are not needed, since historical volatilities and correlations are already reflected in the data set.

The drawback of the approach is its complete dependency on the particular set of historical data and their idiosyncrasies. Since one is fully committing to the notion that the past is a perfectly reliable representation of the future, one must that much more carefully respond to which past is a perfectly reliable representation of the future? Whereas on the other hand, sufficient variety of observations in the population will necessitate including a larger window of past as being relevant. In some cases, data may simply be limited, which will make the population to sample from rather small. Historical method does not allow any innovation in what may happen in the future that is different from what was seen in the past, such as periods of unusual low or high volatility, market crashes, or structural changes in the market.

Analytic Variance-Covariance or Delta Normal Method: This approach assumes that the risk factors and the portfolio values are lognormally distributed, or equivalently, their log-returns or natural logarithm of returns are normally distributed. This assumption is useful since normal distributions are completely characterized by their first two moments, namely mean and variance. One describes the mean and the variance of the portfolio return distribution from the mean, variance and correlation between the risk factors, and the composition of the portfolio. If there are nonlinear instruments, such as derivatives contracts, in the portfolio, their return is also assumed to have a normal distribution and are included in the VaR computation as stated above.

The caution to maintain here is how sure can we be that return distributions of all risk factors are normal. A large amount of evidence suggests that returns exhibit fat-tails. For instance, Figure 8.16 shows the histogram and normal probability plot for two years of equity log-return data for two stocks, both of which show deviations from normal distribution at the tails. This is even more true for nonlinear instruments. In case of fat-tails, there are more observations away from the mean than the normal distribution. Therefore, the

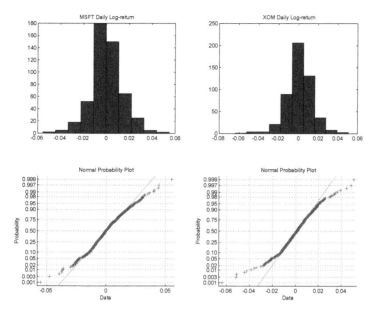

FIGURE 8.16: Histogram and normal probability plot of two years of equity return for Microsoft (MSFT) and Exxon-Mobil (XOM).

analytic variance-covariance method may end up underestimating VaR, since the actual VaR values may fall farther out, perhaps much farther out.

All is not lost however, since even if returns of individual risk factors do not follow normal distribution, it is reasonable to expect that returns of a well-diversified portfolio will still exhibit (near) normal distribution. This is due to the central limit theorem. In practice, therefore, managers can assume that a portfolio has normal distribution of return, provided the portfolio is fairly well-diversified and the risk factors' returns are sufficiently independent from one another, even if they are not themselves normal. However, for nonlinear instruments, which will be naturally correlated with underlying assets they are meant to hedge in the portfolio, fat tails in returns of underlying assets, less diversified lumpy portfolio, highly correlated risk factors should be treated as sources of warning signals, where analytical variance-covariance can be risky to rely on. In such cases, one must resort to using more sophisticated, and realistic, models, which will require the simulation approach to estimate VaR.

Monte Carlo Simulation Method: In the simulation approach, price paths must be constructed by using appropriate simulation techniques for each stochastic process model for risk factors, incorporating correlations and multivariate distributions where necessary. The non-linear contracts, such as derivatives contracts, in the portfolio have to be priced to determine their value in the future for each scenario of underlying asset(s) values. All the models and techniques developed earlier in this chapter and Chapter 7 will be necessary for achieving this task.

Value of the portfolio is determined in each scenario, where each scenario corresponds to possible value of the portfolio at the target horizon. A large number of scenarios creates a good approximation of the distribution of change in value of the portfolio. From this generated distribution, shown in Figure 8.15, we estimate the 1st percentile, or in general α-th percentile, of change in value of portfolio, $\Delta\Pi_{1\%}$, and mean change in value of the portfolio, $E[\Delta\Pi]$. Then $VaR_{99\%} = E[\Delta\Pi] - \Delta\Pi_{1\%}$.

No matter which method is adopted to compute the VaR, if the estimated VaR is higher than the stipulated risk-limit, this is an indication that the risk management strategy is not producing desirable performance. Sources of excess risk must be identified and a change in risk management response must be developed. One must note that there is nothing uniquely special about the VaR measure for the performance assessment of risk management. In fact, as we observed in Section 2.1.3, VaR is neither a coherent nor a convex risk measure in general. Conditional Value-at-Risk, or CVaR, or expected shortfall are also well suited, and must be included as alternative measures to evaluate risk management strategies by.

Distinction is also made for VaR in terms of whether it is evaluating the portfolio under normal conditions of the market or under stressed conditions. For so called 'stressed VaR,' the models are calibrated for periods of stress for the portfolio, and VaR is computed based on these models. The second level of assessment is conducted by scenario analysis and stress testing. Section 2.2 provided a detailed guideline for conducting these analyses. For all the relevant risk factors for a portfolio, different shock levels need to be picked that are in the realm of possibility and imagination, guided by past experience and expert judgement. The guidelines for what is considered a shock include:

- Parallel yield-curve shifts of +/- 100 bp.

- Yield-curve twist of +/- 25 bp.

- Equity index values change of +/- 10%.

- Currency value changes of +/- 6%.

- Volatility changes of +/- 20%.

The goal is to unravel weaknesses in the portfolio not picked by the VaR or stressed VaR assessment. As such, stress testing and scenario analysis is as much an art as a science, therefore the exercise should be structured so that it instigates discussion to benefit from insights and experience from various angles.

The stress tests designed and applied, as well as calibration of those tests, should reflect the nature of the unit being assessed, as well as hedging and trading strategies used by the unit. The stress shocks are applied independently to assess their impact, as well as being combined together to create stress envelopes. The aim is to estimate the expected losses under a range of scenarios. The creative, but justifiable, definition of scenarios is important since in a stressed environment, new interdependencies emerge that were theretofore unseen.

For creating a response to the stress shocks and scenarios, assessment of the time it might take to manage risks under severe market conditions should also be incorporated. Moreover, as the instruments, risk profile, and trading/hedging strategies change, the stress tests should be duly updated to continue to be representative of the unit's condition. Historical stressed periods and events can provide a good guideline for plausibility and sequentiality of interplay within a market segment and interdependencies between markets. Finally, to be useful a good stress testing program should generate insightful visualization and summary of the impact of multitudes of risk factors, shocks, stresses, and scenarios it evaluates.

8.6 MATLAB Tools for Fixed Income, Commodities, and Exchange Rates

MATLAB mathematical software has a vast array of functions for working with financial variables and methodologies in its Financial Toolbox. We list a few of these functions here relevant for interest rate, fixed income instruments, commodity and currency derivatives. The reader is advised to look up the extensive help documentation available with MATLAB to see the details of these and other related functions. At the bottom of each function description in MATLAB help documentation, look for 'See Also' to explore other related functions. Resources such as MATLAB Primer [20] are also useful.

Present value, future value calculations: pvfix, pvvar, fvfix, fvvar

Bond price: bndprice, prtbill

Bond duration: bnddury, bnddurp, bnddur, bndkrdur

Bond convexity: bndconvp, bndconvy

Commodity derivatives: blkprice, blkimpv

Currency options: blsprice, blsdelta, blsgamma, blsvega, blstheta, blsrho, blslambda, blsimpv

8.7 Summary

We began this chapter with the second important market risk segment of interest rate risk. Interest rate risk affects the valuation of a variety of bonds, both government-issued and corporate bonds, but also a host of other instruments collectively called fixed-income instruments and their derivatives. We studied the valuation of bonds and some of the other fixed-income instruments and derivatives. As in the case of equity market risk, the primary motivation behind studying the valuation and sensitivity of instruments to interest rate risk is to support developing hedging strategies. From interest rate risk, we moved to the other two important components of market risk, commodity risk and exchange rate risk. For some firms, one or both of these two market risk components may be very significant to their risk management objectives. We applied the valuation and hedging principles developed in the context of other market risk components to these components, and discussed some examples. We also discussed some unique features of these market risk components and the related challenges. Finally, in this chapter, to summarize the discussion of market risk, we applied the concept of Value-at-Risk to portfolios that are exposed to market risk factors and discussed methods for its computation.

8.8 Questions and Exercises

Review Questions

1. What is interest rate risk? Discuss its most important characteristics.

2. Who are the major participants in the fixed income markets? Discuss the purpose of their participation.

3. What are curve risk, basis risk, and gap risk in the context of interest rate risk?

4. Given the term structure of interest rates, in the simple setting how are bonds priced? What is the yield to maturity of a bond?

5. What are the different possible shapes of the term structure of interest rates? What are the implications of a changing term structure of interest rates?

6. What are continuously compounding interest rates? How are they useful?

7. What are forward rates? How are they computed? How are forward rates used?

8. What is short rate? What theories are stochastic short rate models constructed by?

9. Give examples of equilibrium and no-arbitrage short rate models.

10. What is market price of risk? How is it arrived at, and what does it achieve in the pricing of bonds?

11. What are the multi-factor extensions of single-factor short rate models? Why are these extensions needed?

12. Discuss the types of fixed income instruments, by issuers, maturities, and other terms of the instruments.

13. How are interest rate sensitivities measured? What are duration, modified duration and convexity measures of sensitivity?

14. When are modified duration and convexity measures of interest rate risk not sufficient for risk measurement and management?

15. What are the three factors that capture well the movements of the term structure of interest rates?

16. How can dynamics of term structure of interest rates be captured through movements of key factors?

17. What is bond immunization? What risk measures is the immunization achieved by?

18. What are exchange-traded interest rate derivatives? What are the over-the-counter interest rate derivatives?

19. Discuss the different interest rate derivatives, and their major characteristics.

20. What is floating-to-fixed interest rate swap? What hedging objective can this instrument serve?

21. What are cap, floor, and collar contracts for interest rate risk?

22. How are hedging strategies for interest rate risk constructed? What risk measures are optimal hedges constructed by?

23. How does commodity price risk differ from that of interest rates risk or exchange rate risk?

24. What is the classification of commodities? What are the shared characteristics in this classification?

25. What general characteristics of commodities do different price models capture?

26. Why is mean-reversion or jumps seen in many commodity prices?

27. What are factor models, why are they useful for modeling commodity prices?

28. What is regime-switching and how is it relevant for price evolution of commodities?

29. What are the risk characteristics of each of the following commodities, and how do models capture them?

    (a) Crude oil
    (b) Natural gas
    (c) Electricity
    (d) Weather

30. Why are commodities considered an attractive asset class for investment?

31. What is the minimum variance hedge under cashflow objective?

32. When and why are firms reluctant to sell down their inventory of commodities, even when there is arbitrage opportunity?

33. What is convenience yield? What is cost of carry?

34. Why are firms increasingly exposed to foreign exchange risk?

35. Why should firms manage their currency risk exposure?

36. What is purchasing power parity? What is the international Fisher effect?

37. What are the different type of currency swaps?

38. What is a gap option? What is the advantage of using a gap option?

39. How does Value-at-Risk serve the objective of monitoring a risk management strategy?

40. What is the historical method of computing VaR? What are its advantages and disadvantages?

41. What is the analytic variance-covariance method of computing VaR? What are its advantages and disadvantages?

42. What is stressed VaR?

43. Why is stress testing important for risk management?

Exercises

1. Calibrate the following interest rate models using MATLAB for historical interest rate data.

   Vasicek Model: $dr_t = (\eta - \gamma r_t)dt + \sqrt{\beta}dW_t$
   Ho and Lee Model: $dr_t = \eta(t)dt + \sqrt{\beta}dW_t$
   Cox Ingersoll Ross Model: $dr_t = (\eta - \gamma r_t)dt + \sqrt{\alpha r_t}dW_t$

   Use data for the following interest rates, obtainable from Federal Reserve Economic Research & Data website Selected Interest Rates - H.15.

   - 1-month, Nonfinancial commercial paper
   - 1-month, Financial commercial paper
   - 4-week, Treasury bills rate
   - Aaa Moody's Corporate bonds rate
   - Baa Moody's Corporate bonds rate

   Comment on which model provides a good representation for these interest rates.

2. Price the following bonds using appropriately chosen interest rates and corresponding calibrated model from Problem 1.

   Bond 1 : US Treasury T-Note, Annual Coupon Rate: 3.25% (paid semi-annually); Maturity: 6 Years; Rating: AAA

   Bond 2 : US Treasury T-Bond, Annual Coupon Rate: 7.25% (paid semi-annually); Maturity: 12 Years; Rating: AAA

   Bond 3 : Corporate Bond, Issuer: Johnson & Johnson; Coupon Rate: 5.55% (paid semi-annually); Maturity: 5 years; Rating: AAA

   Bond 4 : Corporate Bond, Issuer: Southwest Airlines; Coupon Rate: 7.375% (paid semi-annually); Maturity: 15 years; Rating: BBB

3. Define a bond option on the following bonds and determine the price of the bond option.

   (a) US Treasury T-BOND, Annual Coupon Rate: 7.25% (paid semi-annually); Maturity: 12 Years; Rating: AAA

   (b) Corporate Bond, Issuer: Southwest Airlines; Coupon Rate: 7.375% (paid semi-annually); Maturity: 15 years; Rating: BBB

4. Consider the following portfolio of US Treasury notes and bonds with weights for each bond given as number of bonds held in the portfolio.

- US Treasury T-Note, Annual Coupon Rate: 3.25% (paid semi-annually); Maturity: 6 Years; Rating: AAA, Weight: 10,000
- US Treasury T-Bond, Annual Coupon Rate: 6.25% (paid semi-annually); Maturity: 12 Years; Rating: AAA, Weight: 22,000
- US Treasury T-Note, Annual Coupon Rate: 2.25% (paid semi-annually); Maturity: 3 Years; Rating: AAA, Weight: 14,000
- US Treasury T-Bond, Annual Coupon Rate: 6.625% (paid semi-annually); Maturity: 15 Years; Rating: AAA, Weight: 18,000

(a) Compute the current value, duration, and convexity of the above bond portfolio.

(b) If the target liability being matched with the above portfolio remains the same, however the duration of the liability increases by 10% and convexity of the liability increases by 8%, by changing the weights as you see appropriate of the bond portfolio, immunize the bond portfolio for small changes in interest rates against the liability.

5. In Problem 3, a bond option was defined on the 12-year US Treasury bond. The portfolio of Problem 4 has dominant weight on this bond. Construct a hedge for the bond portfolio in Problem 4 using the bond option defined and priced in Problem 3. Define the objective of the hedge and determine the hedge ratio. How does the hedge change the duration and convexity of the portfolio?

6. Calibrate the following model for crude oil. Use publicly available data from International Monetary Fund (IMF) website; IMF Primary Commodity Prices available for 8 price indices and 49 actual price series from 1980 - current.

$$dY_t = \alpha(\bar{L} - \ln(Y_t))Y_t dt + \sigma Y_t dW_t. \tag{8.82}$$

7. Using crude oil futures prices from Chicago Mercantile Exchange, determine the cost of carry and convenience yield for crude oil.

(a) Determine the futures price for other crude oil futures contract you define.

(b) Define a European call option on crude oil, and determine the price of the call option.

8. A wheat producer anticipates producing 60,000 bushels of wheat at the upcoming harvest. The farmer cannot be definite about the amount of produce at harvest. The farmer expects to experience a variability of 8000 bushels. A lognormal distribution with the above mean and standard deviation would make a good representation of the wheat production of the farmer. If price per bushel at harvest is expected to be 85

cents per bushel, with a standard deviation of 7 cents per bushel, what futures contract hedge do you recommend to the farmer? What assumptions did you make to construct this hedge? Assume 5,000 bushels per futures contract.

9. Obtain exchange rate data for the following currencies from the International Monetary Fund (IMF) website, Exchange Rate Query Tool, to calibrate an appropriately chosen model for exchange rate evolution for each of the following currencies. Assume that the domestic currency is USD.

- Euro
- Chinese yuan
- Australian dollar
- Indian rupee

(a) Define a forward contract and a gap option for the exchange rate for any of the above currencies, and determine forward price and the price of the gap option.

(b) If you anticipate a need to pay 0.5 million Euros in 2 months, construct an appropriate hedge using i) the forward contract and ii) the gap option.

(c) If you expect to receive INR 20 million in 1 month, construct an appropriate hedge using i) forward contract and ii) the gap option.

10. Consider the following stocks and bonds from which different portfolios are being constructed.

- US Treasury T-Note, Annual Coupon Rate: 3.25% (paid semi-annually); Maturity: 6 Years; Rating: AAA
- US Treasury T-Bond, Annual Coupon Rate: 7.25% (paid semi-annually); Maturity: 12 Years; Rating: AAA
- Corporate Bond, Issuer: Johnson & Johnson; Coupon Rate: 5.55% (paid semi-annually); Maturity: 5 years; Rating: AAA
- Corporate Bond, Issuer: Southwest Airlines; Coupon Rate: 7.375% (paid semi-annually); Maturity: 15 years; Rating: BBB
- Consider three stocks evolving by continuous-time stock price evolution model of the form,

$$dS_{it} = \mu_i S_{it} dt + \sigma_i S_{it} dW_{it}, \tag{8.83}$$

for $i = 1, 2, 3$, where initial stock price is, $\vec{S_0} = [19; 53; 26]$,

$\vec{\mu} = [0.09; 0.05; 0.16]$ and $\vec{\sigma} = [0.10; 0.06; 0.25]$. The three correlated Wiener processes are described by the following correlation
matrix.

$$\rho = \begin{pmatrix} 1 & 0.3 & 0.1 \\ 0.3 & 1 & -0.05 \\ 0.1 & -0.05 & 1 \end{pmatrix} \tag{8.84}$$

Compute the Value-at-Risk at a desired confidence level and duration
of time for the following portfolios given by the number of bonds and
shares of above stocks held.

(a) $\vec{w} = [10{,}000;\ 22{,}000;\ 15{,}000;\ 20{,}000;\ 0;\ 0;\ 0]$

(b) $\vec{w} = [0;\ 0;\ 0;\ 0;\ 30{,}000;\ 8{,}000;\ 20{,}000]$

(c) $\vec{w} = [10{,}000;\ 22{,}000;\ 15{,}000;\ 20{,}000;\ 30{,}000;\ 8{,}000;\ 20{,}000]$

For each of the above portfolios, construct some hedges by using bond or
stock options with desired strikes towards your chosen risk management
objectives. Recompute the VaR for the hedged portfolio and compare
with the VaR of the unhedged portfolio.

11. Perform a detailed stress testing of the stock-bond portfolio in Problem
10.

Chapter 9

Credit Risk Management

Genesis of credit risk would not have trailed far behind development of trade and commerce in earliest human society. Therefore, it can easily be dated back at least three to four thousand years. Yet credit risk remains a risk type that still poses significant challenges for assessment and management. There are several reasons for this, the most important of which is that credit risk, at all times past and present, has reflected the needs, aspirations, and goals of individuals, business enterprise and financial intermediaries. It has also reflected the environment, both economic and social, the individuals and the business enterprise function in. As these aspirations and goals have evolved, and as the environment for these decisions has gotten increasingly more complex, the characteristics of credit risk have also continued to evolve. An evolving risk is expected to continue to pose challenges for accurately assessing and monitoring it, and to optimally manage it.

Credit risk would arise when individuals or enterprises acquire goods or services in return for a promise of future payment for them. Such arrangement enables the individual, household or business enterprise to create the possibility of generating value by using the goods and services, without initially having the resources to acquire the goods or services. Whatever the terms, duration or form for the payment for the goods or services, credit risk arises with the likelihood that the individual or the enterprise will breach the contract for the payment.

In discussing equity risk in Chapter 7, we viewed equity ownership as a mechanism for financing ventures of an enterprise. As an alternative to equity financing, historically and at the present time, enterprises have also resorted to borrowing funds for financing their ventures with a promise of repayment in the future in set form, terms, and duration. Individuals, households, and governments of different times, levels, and regions also utilize this mechanism for financing their projects. In each of these cases, financing of projects with such terms of future repayment also gives birth to credit risk.

Credit risk is also embedded in a variety of other contracts beyond those used for financing ventures or acquiring goods and services with a promise of repayment in the future. All the derivatives contracts we have studied thus far, and will continue to study in this and later chapters, carry credit risk arising from the fact that the counterparties engaged in the contract will not deliver their due, as per the contract. This is frequently labeled as counterparty risk.

From the reasons for the genesis of credit risk, a fundamental link between interest rate risk and credit risk is established. Repayment of funds or payment for goods and services is done over a duration of time. In order to compensate the creditor for the time value of money, as extensively discussed in Chapter 8, a term-structure of interest rates may be applicable to the payments. Moreover, the extent of likelihood of a counterparty not fulfilling the payment requirement as per the contract may also translate into an increase in the interest rate applied. This increase in interest rates applied to an account of heightened credit risk is often called credit spread.

To manage and mitigate credit risk, creditors would like to have as much information as possible regarding the debtors and the use they intend to put the funds or goods and services to. In this regard, the nature of credit risk differs significantly depending on the type of counterparty. Individuals and households as debtors have distinct characteristics when compared with business enterprise, while both of these differ in characteristics from governmental enterprise. Additionally, creditors would want to utilize every available and feasible mechanism to lower the adverse impact of credit risk. Subordination of funding sources utilized by the debtor and assurance using collateral are some such mechanisms, which we will study later in this chapter.

Assessing and managing credit risk not only entails determining and minimizing the likelihood of adverse events, it also involves determining what the repayment level may pan out to be should the counterparty hit financially turbulent terrain. This assessment is crucial for the assessment and monitoring of the loss from credit risk to the creditor. The percentage of original amount of funds due that are recovered is termed the recovery rate. Recovery rate will, in its own turn, depend on various factors, including the collateral and subordination characteristics of the credit. All the issues discussed above chart the platform on which topics of this chapter are developed. We begin with considering credit risk pertaining to individuals and households.

9.1   Retail Credit Risk

Retail banking is a segment of banking that serves individuals' and households' banking needs, but also often includes services for small business enterprises. Financial needs and credit risk characteristics of small businesses resemble those of individuals and households, as small businesses are often structured as sole proprietorships and size of funds are comparable. Therefore, retail banking or consumer banking offers deposit services, checking, savings, etc., as well as providing credit in the form of a variety of loans to small businesses and consumers.

Once perceived to be non-glamorous, retail banking has transformed over the last decade due to widespread access to banking using Internet technologies, deregulation facilitating innovations in products and marketing, customized to consumer needs, and greater emphasis and need for risk management. Some call the retail credit market a sleeping giant of the modern economy [87], at least up until the trigger of financial crises in 2007.

The 'giant' part of the sleeping giant title is justified by the following observations. In the US, retail banking and consumer lending is almost double the size of corporate lending. Household debt in the US was in excess of $13 trillion in 2010, where US corporate debt stood higher than $6.0 trillion in 2010. Home mortgages and home equity loans account for almost $11 trillion of this amount, in excess of 80% of the total retail credit market. The next largest category is consumer credit-card debt.

Retail credit has various different forms to meet the needs and demands for credit of households and small businesses. Given home mortgages make such a sizeable fraction of consumer loans, these can be further classified into two major groups, fixed-rate mortgages (FRMs) and adjustable-rate mortgages (ARMs), each with a variety of loan terms. In home mortgages, the concept of loan-to-value ratio (LTV) is an important one that defines the proportion of property value that is debt-financed versus equity-financed. It is a key risk variable. Home equity loans are a hybrid between consumer loans and mortgage loans, created by utilizing home equity as collateral. Other consumer loan types include installment loans, revolving loans, such as personal lines of credit that may be used repeatedly up to a specified limit (credit cards), automobile loans, and similar loans secured for education, personal property, and financial assets.

A similar variety of loans may be taken out by small businesses to meet their investment and operations needs, which are secured by the assets of the business. Usually loans of up to $200,000 would qualify for being considered within retail banking. Two new types of consumer loans that are attracting significant interest at present are micro-credit and payday loans. Micro-credit loans are very small loans given out to those in poverty in order to help them launch a business. Giving this opportunity to these individuals, which helps them sustain themselves and their families, can become a path out of poverty. Payday loans, on the other hand, are small emergency need loans with very short-term durations. They carry extremely high interest rates and effectively serve as an advance on an individual's next paycheck.

Differentiation between retail and commercial banking is justified based on certain fundamental differences in their characteristics both from a business as well as a risk management perspective. These differences will guide us in the rest of the chapter in developing risk management methods for retail and commercial credit risk. So, in what ways are retail credit risk characteristics different from those of commercial credit risk?

1. Each individual unit of retail credit exposure is in bite-sized pieces when compared to the overall assets of a bank. Therefore, default of a single customer is not costly enough to threaten the bank's solvency. This may not be true in case of commercial credit, since single exposure in commercial credit could be a significant fraction of a bank's total assets.

2. Individuals and households function as independent units, therefore retail customers tend to act financially independent of each other. This is
not the case for commercial credit risk. Firms function in an interdependent environment, manufacturers depending on suppliers and retailers,
and vice versa, firms are simultaneously affected by cost and availability of raw materials, market risks, and other macroeconomic factors.
Therefore, commercial credit risk is economically intertwined in particular geographical or industry sectors.

3. Due to the above two characteristics of bite-size individual exposure in a
large pool that are also independent of each other, retail banks can make
better predictions of percentage of portfolio expected to default and the
losses it may cause. While this more reliable estimate of expected loss
number in retail banking can be treated as operations cost, in commercial credit losses can be a threat to a bank's solvency and must be dealt
with using a multi-prong response of risk management.

4. In retail banking, since the expected loss rate dominates the bank's
credit risk exposure, it is possible to price it in the products to the customers. According to the riskiness of customers, retail credit products
can be designed to have higher price for higher expected loss customers.
In commercial credit, pricing of products cannot be an effective mechanism to recoup losses, since it is the unexpected or the upper-tail losses
that dominate the credit risk exposure.

5. Consumer behavior is more easily observed, especially with the advent
of information technologies, making it possible for the well-run banks
to make it a priority to collect and mine consumer data for possible
signals for occurrence of default or failure to make some payments. This
ability to monitor gives the bank time to take preventive actions to
reduce credit risk. The counterparties in the case of commercial credit
are complex entities, not amenable to close monitoring or easy prediction
of future financial health, therefore the luxury of a preemptive action is
not afforded in the case of commercial credit.

With the above distinctions in sight between retail and commercial credit
risk, we need to define the goals and objectives of credit risk management,
and develop methods for measuring and tools to effect the management of it.
Since banking, both consumer and commercial, plays a fundamental role in
supporting the economic activity in an economy, several of these goals are also
driven by the regulatory environment in place for the proper functioning of
the banking sector. Regulators recognize the differences in the characteristics
of retail and commercial credit, and appropriately respond to the needs for
risk management. In Chapter 3, the evolution of the Bank for International
Settlements (BIS) supported Basel Accord was briefly reviewed. The Accord's
central focus on credit risk will repeatedly emerge in this chapter.

Measures for credit risk are developed based on three important statistical quantities. The three statistical measures summarize credit risk both at individual exposure level, as well as at the portfolio level, and can be combined to create measures for the overall impact of credit risk on a credit portfolio or a bank. The statistical quantities are as follows.

Probability of default (PD): This measure is an estimate of the likelihood of a default to occur for an underlying contract in a given period of time. *Default* refers to a debtor not meeting his or her legal obligations according to the debt contract, for instance, being unable or unwilling to make a scheduled payment, or violating a condition as per the debt contract. When the underlying is a credit portfolio, the probability of default would further need to be specified as probability of first default, second default, and so on. Probability of nth default refers to the likelihood of nth default to occur in a portfolio in a period of time, no matter which specific units cause the n defaults.

Exposure at default (EAD): When a counterparty defaults, it can put a specific amount at risk for the creditor, depending on the terms of the debt contract. It could be either a specific installment, or a specific component of a specific installment, or the entire principal. Once the contract is in place, the creditor may not have additional control for modulating the EAD, but must estimate the EAD to determine the possible extent of loss due to default. However, the creditor can incorporate the EAD in the design of the credit product in response to the riskiness of the counterparties. EAD for single exposures can be combined to create the EAD for a credit portfolio.

Loss given default (LGD): This is the actual loss the creditor will suffer when a counterparty defaults. Once probability of default and exposure at default are determined, estimation of loss given default is the third crucial quantity that completes the assessment of impact of credit risk. The covenants included in a debt contract are designed to maximize the likelihood and extent of recovery from a default event. The collateral and other assurances are put in place for a high recovery rate.

For regulatory monitoring, banks must also provide estimates of the three statistical quantities for clearly differentiated segments of their retail credit portfolios. Segmentation of credit portfolio is performed by an individual exposure's riskiness in the portfolio, riskiness measured in terms of credit score or an equivalent measure, and vintage of exposure. With the above three crucial quantities defined for credit risk, we move on to developing methods for assessing them at the individual exposure level.

9.1.1 Measuring Retail Credit Risk

Information is power. In the information age, banks have continually attempted to develop better models for predicting credit risk of individuals in terms of the three measures described above. These models, that go as far back as the 1950s, are a constant effort to elicit more reliable and better usable knowledge from the information. Information about individuals or households is available in the form of 'Characteristics,' where each characteristic has several 'Attributes.' The goal of any model attempting to convert information into knowledge is to assign the attributes of the set of characteristics of an individual into an indication of their probability of default and the loss given default of the individual under given settings.

Let C_i for $i = 1, \ldots, N$ be N characteristics identified as important for determining credit risk of an individual or household. Let's say each characteristic has $A_i = \{a_{i1}, a_{i2}, \ldots, a_{iM_i}\}$ attributes, with M_i different levels for i^{th} characteristic. For example, household income is a characteristic, and its attributes can be identified as eight different brackets of annual income levels. If Y is an indicator of whether an individual will default in a given period of time, and L is the amount of loss suffered due to default, then a model sought gives a relation between C_i's and (Y, L), as follows.

$$Y = \sum_{i=1}^{N} w_i C_i + \epsilon_Y, \tag{9.1}$$

$$L = \sum_{i=1}^{N} \lambda_i C_i + \epsilon_L, \tag{9.2}$$

where ϵ_Y and ϵ_L are the error terms of the two prediction models.

Eqns. (9.1) and (9.2) hold stark resemblance with linear regression models, where C's are the independent variables and Y and L are dependent variables in their respective equation. Regression based prediction determines the predictive contribution or weight, w_i and λ_i, of each variable, C_i, to determine the outcome, Y or L. These methods are generally called statistical classification methods, set to classify 'good' customers from 'bad' customers, so that credit may be extended suitably towards a bank's risk appetite.

One of the earliest methods applied to credit risk prediction is discriminant analysis (DA), which focuses on Eqn. (9.1). Note that the dependent variable Y in Eqn. (9.1) is a binary variable, taking value '1' when there is a default, and '0' when there is none. If there are K observations of past default experiences and the corresponding attributes for the N characteristics, each row of the $[C]_{K x N}$ matrix is a point in an N-dimensional space. The goal of discriminant analysis is to draw a hyperplane, $Cw = b$, that separates maximum number of '1' responses from the '0' responses of each row of matrix, $[C]$. This problem has been successfully cast and solved as a variety of linear programming problems [31].

Even though Eqns. (9.1) and (9.2) predict the exact default outcome in a period of time and loss given default anticipated, most classification methods are designed to assign a score to each individual based on his or her attribute for each characteristic. Therefore, standard classification methods result in a scorecard, a mapping of characteristics to a score for the individual. A score is an ordinal indicator, higher the score better the creditworthiness, lower the credit risk. The user of the credit score must identify a cut-off score, so that individuals with scores above the cut-off are considered 'good,' and would be offered more favorable products, credit limit, attractive interest rates, while those below the cut-off score are classified as 'bad,' and would either be denied service or given more stringent terms. In order to keep the scorecards current, the process of building them and determining a cut-off should be repeated frequently.

Credit scoring is extended to behavioral scoring, on similar statistical principles of classification, to assess the credit risk of existing customers, as well as new applicants. Behavioral scoring is not only intended to make the decision of whether or not to grant credit to a specific applicant, but is geared towards life-cycle management of new and current customers. Therefore, the dependent or target variable is either whether the borrower would default in a given period of time, or what revenue he/she would generate, etc., in a given period of time, based on the available information on the borrower's recent repayment and purchase history. Such scores may also be used for marketing new products to the customers, adjust the terms of existing contracts, such as more/less attractive interest rates, or more/less attractive credit limits.

Growth in the automated receiving and processing of applications through the Internet, and through phone before the popularity of the Internet, meant applications were essentially private and so the products could be customized for the applicants' characteristics. Customization in terms of pricing strengthens the prospects for profitability and more effective risk management for a bank. Therefore, banks continue to use and improve their credit scoring methodologies to support their business objectives of profitability and market share, by optimizing all their decisions about their customers, from whether or not to offer the borrower a standard loan product to how to use credit scoring methodologies to help make variable pricing decisions and to determine the long-term profitability of a customer.

As stated earlier, profitability is as much about designing the 'right' product for a customer and marketing to the 'right' customer-base as it is about risk assessment and management. The credit scoring methodologies can be used by both marketers and risk management teams of a bank to make more concerted, robust decisions for a bank's credit risk management. As it turns out the models used by the marketers for customer segmentation by propensity to purchase different product types are very similar to the ones used by risk managers to build scorecards and for risk management.

Whatever the scope of scoring models being utilized in a bank, it is imperative that the models are frequently tested and adjusted for currency. It is known that economic conditions result in behavioral changes of individuals and households towards credit, moreover models suffer 'wear-and-tear' of becoming outdated due to individuals 'learning' to beat the model. This was, in fact, a major criticism meted out to Fair Isaac Corporation (FICO), a pioneer credit scoring company, during the 2007-2008 credit crises [26]. An alternative, albeit much more challenging and ambitious, would be to build a dynamic model for customer creditworthiness, which would allow forecasting the future dynamic behavior of the customer.

Regulatory requirements are also important in this context. As described earlier, credit scoring is focused on describing the ordinality of customer creditworthiness, i.e., whether one customer has higher creditworthiness relative to another. The precise implication of these scores in terms of actual probability of default and loss given default experience is an important validation for accurate measurement of credit risk. Prediction of default risk should be extendible to longer periods of time, rather than being focused on just the next couple of years, which has been an emphasis of credit scoring models. Clearly, since many retail products, such as home mortgages, extend into decades, and given the high fraction home mortgages make of the total pool of outstanding retail credit, longer-run prediction of probability of default would be quite valuable. Finally, even the fanciest of models cannot be blindly relied upon. Stress testing the models should be an integral part of decision making using the credit scoring models.

9.1.1.1 Credit Scoring Methods

The US Federal Trade Commission's Equal Credit Opportunity Act prohibits credit discrimination on the basis of race, color, religion, national origin, sex, marital status, or age. Creditors may not use this information when deciding whether to grant credit or when setting the terms of the credit. This makes it unlawful to deny credit on any other basis than financial, using factors like income, expenses, debts, and credit history for determining creditworthiness. Therefore, statistically based approaches are not just desirable, but are essential to justify consumer credit decisions.

Credit scoring is a principal tool for consumer credit risk management. In fact, every time you apply for a credit card, open an account with a telephone company, a utility company, submit a medical claim, apply for auto or home insurance, you are subject to a credit risk scoring model, sometimes without your explicit knowledge. This is because in almost all these cases, implicitly the service provider is exposed to credit risk based on your creditworthiness. Therefore, credit scoring is a widespread technique used beyond banking, by telecommunication companies, utility companies, and insurance providers, etc. to examine the likelihood that a potential customer is unable to make good on the due payments for the services.

Several statistical methodologies are utilized for developing credit scores. We provide a brief overview here of some of the methodologies adopted for credit scoring models; for more details, the reader should refer to a dedicated resource on this topic, such as [86]. Credit scoring is a statistical procedure that converts information about a credit applicant into a score, where the score is an indicator of the individual's creditworthiness, that is probability of repayment or probability of default (PD). As stated earlier, this is an ordinal number with a higher score indicating lesser likelihood of a customer to default.

Credit scores are created using credit scorecards, each containing several characteristics C_i, and each characteristic contains several attributes A_j. Each attribute is associated with a weighting number, which is assigned to the individual if the attribute describes the individual. A weighting is an indicator of odds of repayment given the attribute based on past performance. A range of statistical techniques get employed to generate the weights based on attribute information.

From Default Rates to Credit Scores

As for any statistical procedure, data for past default and non-default instances are required along with corresponding characteristics and attributes for developing credit scoring models. Based on this information, a score, $S(\mathbf{C})$, is then constructed as a function of the characteristics $\mathbf{C} = \{C_i, 1 \leq i \leq N\}$ of a potential borrower. This score is translated into the probability estimate that the borrower will be 'good.' The strength of prediction of 'good' (no default) from 'bad' (default) in test data is indication of quality and accuracy of a scoring model. Once a scoring model is built, the critical assumption thereafter in the use of the model is that the score is all that is required for predicting the probability of an applicant being 'good.'

At the start of this section, we examined one method for building a scorecard based on discriminant analysis (DA). We now develop logistic regression based credit scoring model, which is one of the most common approaches in use. Assuming again K past observations of default and corresponding information of attributes for all characteristics, we seek the following prediction,

$$P(Y = 1|\mathbf{C}) = \pi(\mathbf{C}) = \frac{e^{S(\mathbf{C})}}{1 + e^{S(\mathbf{C})}}, \tag{9.3}$$

therefore,

$$P(Y = 0|\mathbf{C}) = 1 - \pi(\mathbf{C}) = \frac{1}{1 + e^{S(\mathbf{C})}}, \tag{9.4}$$

Applying Eqns. (9.3) and (9.4) for probability of Bernoulli distribution of Y, we obtain the likelihood function as follows.

$$\mathcal{L}(S(\mathbf{C})) = \prod_{i=1}^{K} \pi(\mathbf{C}_i)^{Y_i}(1 - \pi(\mathbf{C}))^{1-Y_i}, \tag{9.5}$$

where $\{(Y_i, \mathbf{C}_i), i = 1, \ldots, K\}$ are the observations being used to construct the scoring model. The likelihood function in Eqn. (9.5) is maximized to obtain the scores, $S(\mathbf{C})$.

There are strong contenders to linear or logistic regression approaches to developing credit scoring models. One popular alternative to logistic regression is the use of classification trees. Classification and regression tree

(CART) analysis finds its origins both in statistics and machine learning. CART analysis identifies rules based on characteristics' values to get the best split in observations based on the dependent variable, Y or L, using notation of Eqns. (9.1) and (9.2). Once a rule is selected to split a node into two, the process is recursively applied to the 'child' node, until no further gains can be made by splitting or some pre-set stopping rules are met. Each branch of the tree ends in a terminal node, where each terminal node can be described by the sequence of rules used to arrive at the node from the top-most node. Therefore, each terminal node is uniquely defined by a set of rules and each observation falls into one and exactly one terminal node.

CART analysis, therefore, does not end up with a scorecard, instead a group of customers are described by different combinations of their characteristics, where each group is classified as either 'good' or 'bad,' or any other desired label. To apply a CART-based model to make decisions, information regarding attributes for different characteristics of an individual are used to determine which group the individual belongs to, and whether that group has been labeled 'good' or 'bad.'

Beyond logistic regression and CART analysis, neural nets, support vector machines, genetic algorithms, nearest neighbor methods and ant colony optimization are also used for retail credit risk models. If one is expected to explain why an applicant's application was rejected for credit, a 'black box' method, such as neural nets and support vector machines, would become unsuitable. One can still attempt to devise a classification tree that mimics the performance of the 'black box' neural nets or support vector machine based model, and hence provide reasons for concluding an applicant as 'bad' or 'good.' An additional way for improving credit scoring models is to combine methods. For example, classification trees can be developed where one of the characteristics is in its own turn a 'score' obtained using a logistic regression method. Similarly, one may develop a regression model where one characteristic is chosen as the different nodes of a classification tree.

From Credit Scores to Default Rates

Obtaining credit scores, or similar indicators for probability of default based on classification trees, neural nets, etc., for a customer-base is a crucial first step in credit risk management. The next step is to determine a cut-off score for a specific product, under the knowledge that no predictive tool is 100% accurate. In order to determine a cut-off, besides the credit scores, default and no-default data is also used. This is done to obtain insight on the extension of impact of predictive inaccuracy of the model.

False 'bads' and false 'goods' based on any cut-off score provide the lost profits and undesirable credit risk the bank is taking on, respectively. False bads and goods are similar to Type I and Type II error of statistical inference, when you reject a hypothesis when in fact it is true and when you fail to reject a hypothesis when in reality it is false. Any conclusion made for the population

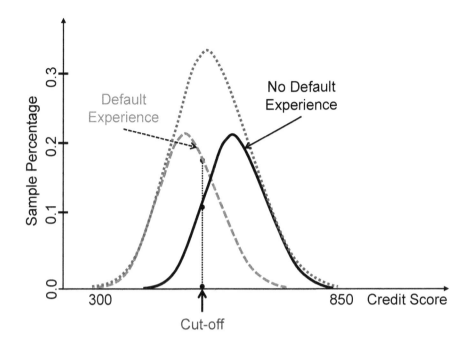

FIGURE 9.1: Distribution of the population by their credit score, as well as distribution of individuals who have defaulted and who have not defaulted on their loans by their credit score. The figure also indicates a selected cut-off score, with its implication on false 'bads' and false 'goods.'

based on limited data is bound to have this condition. Figure 9.1 shows the false good and false bad based on the choice of a cut-off score.

The cut-off score is the first risk management tool or knob for effecting risk management at the individual exposure level. Based on a choice of cut-off, a bank can first judge the quality and accuracy of the scoring model. For a model with satisfactory accuracy, the choice of cut-off determines the loss rate and profitability of the retail product it is being applied to. Over time, still utilizing the same scoring model, one can fine-tune the knob to continue to maximize profit margin. As stated earlier, a running business has opportunity to collect plenty of real-time data and update key knobs frequently, perhaps monthly or quarterly.

Estimating Loss Given Default in Retail credit

Estimating loss given default based on historical observation is a more tricky task, even though there is regulatory emphasis on this component of credit risk. In some cases, such as home mortgages and auto loans, the prediction task is split into two steps. In the first, one determines if the house or car must be repossessed, and then in the second step a forecast for resale price for the property is estimated. In other unsecured consumer credit cases, where an obvious collateral is not in place to secure the recovery rate, the recovery rate depends on the decisions of the creditor and the uncertainty around the borrower's ability and intention to repay. A decision-tree based strategic response may be developed to decide whether to collect the debt in-house, use an agent or sell off the debt.

In general, modeling the recovery rate or loss given default in terms of the characteristics of the debtor and the debt product is challenging. Some approaches based on linear regression, logistic regression, non-linear transformation so as to fit beta or log-log distributions, mixture models, and quantile regression have been developed, however the performance of these models leaves more to be desired.

One of the advantages of a robust credit scoring framework adopted by a bank is that it allows for automation of the product marketing, underwriting and management, and brings consistency and reduced load for the bank with millions of customers. As discussed earlier, if done right, credit scoring can be used for more than just to decide whether to extend credit to a specific customer. It can also be used to customize the product to a customer's creditworthiness and needs. Moreover, credit scoring can be used to ensure that the products offered by a bank remain profitable, where profit margin is what remains once operating and default expenses are subtracted from gross revenues.

Banks don't necessarily have to develop the entire credit scoring framework in-house. There are three ways a bank can approach this issue. First, the bank can entirely depend on what is offered by standard credit bureaus. Credit bureau scores are created by methodologies developed by the Fair Isaac Corp (FICO), and are maintained and supplied by companies like Equifax and TransUnion. These are generic scores, therefore are lower cost, but are not tailored for any specific products. The FICO scores range from 300-850, where below 660 defines sub-prime lending. The second approach would be to use a pooled model. Pooled models are also built by outside vendors, like FICO, but they are based on data from a range of lenders with similar credit portfolios. These pooled models are tailored to a specific industry, such as credit card, home mortgages, etc. The third option is to develop a custom model, which is an in-house model built for a specific product and specific lines of business. The third approach is indeed the most costly approach, but if done well, it can offer a very significant competitive edge in selecting the best customers and offering the best risk-adjusted prices.

9.1.2　Retail Credit Risk Management

The credit scoring framework primarily focuses on the avoid versus keep actions of risk management. The goal of a good credit scoring system is to reliably differentiate 'good' from 'bad' customers, and attract as many 'good' customers as possible to construct a large credit portfolio. A large portfolio of credits that are each bite-size relative to the overall notional value of the portfolio, as well as their independence of each other with regards to credit risk, implies that the expected loss of the portfolio is a good estimate of the actual loss given defaults in the portfolio. Therefore, mitigation of risk is naturally achieved if the retail credit portfolio is constructed with these properties. As discussed earlier, transfer of credit risk is possible through risk-based pricing, which also requires accurate models for predicting credit risk of customers. Additional transfers of credit risk is pursued using securitization, which we will discuss in Section 9.3.2.1.

It is fair to conclude that the most critical component for robust retail credit risk management is a reliable credit scoring framework. A credit scoring framework is built around methodologies that convert information about the customers, or customer characteristics, to indicate whether they would be 'good' customers or 'bad' ones, based on scorecards or similar constructs. For maintaining this robustness, it is important to measure and monitor the performance of scorecards and other constructs. The goal of a scorecard, by the type of questions it asks and interpretation it assigns to the response in terms of repayment versus default, is to minimize the overlapping area of the distribution of the 'good' and 'bad' credits in Figure 9.1.

How do we assess when to adjust or rebuild scorecards? The most popular validation technique employed for this purpose is a cumulative accuracy profile (CAP). Numerically, this accuracy is summarized in the related summary statistic, called the accuracy ratio (AR). The concept behind the CAP curve and the AR statistic is quite straightforward. It compares the performance of the existing model in predicting observed defaults relative to two extreme models. First is the perfect model, which attributes all the observed defaults to the lowest scored customers in the pool, and the second is a random model, that spreads the observed default evenly through the entire range of credit scores of the customer pool. The actual model is represented by a curve indicating how the defaults actually spread along the range of credit scores of the customer pool.

Figure 9.2 shows on the horizontal axis the credit scores of the pool in increasing order, while the vertical axis shows the observed fraction of defaults in the pool. In the figure, the perfect model is depicted by the line quickly rising in the left from zero to all the observed defaults attributed to the lowest credit score customers. The area under the corresponding piece-wise linear curve is labeled A_p, for the area under the perfect model. Similarly the curve that gradually rises from zero to all the observed defaults counted as per the actual credit scores of the defaulting customers is the actual model curve. The area under this curve is labeled, A_r. The straight line spreading the observed defaults evenly through all customers is for the random model. Since the area under this line is common to both the area under the perfect model and the actual model, it can be subtracted from both, thus making A_p the area under the perfect model above the random model and A_r, the area under the actual

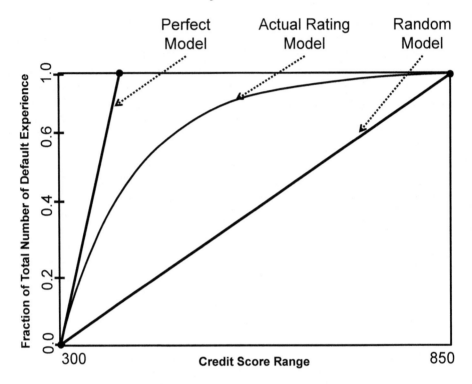

FIGURE 9.2: Level of accuracy in a specific credit scoring model relative to a perfect and a random model. This is summarized in the accuracy ratio, which is the area under the curve below the actual model profile relative to the perfect model profile.

model above the random model. The CAP statistic, AR, is constructed as,

$$\texttt{Accuracy Ratio}(AR) = \frac{A_r}{A_p}. \tag{9.6}$$

The closer the value of this statistic is to 1, the better the performance of the credit scoring methodology. The performance of a credit scoring model or other constructs, such as classification trees or neural nets, can thus be monitored at desired frequency through a CAP curve. When the accuracy ratio (AR) deteriorates to a critical level, the model must be reviewed for improvement or replacement.

Changes in credit risk of individuals may come about for a variety of reasons. Over time, customer mentality changes, or at specific financially stressed times, tighter circumstances may change customer behavior. Changes in bank's products or business model can also trigger changes in how customers respond, thus changing their credit risk characteristics. There is a general movement towards improving credit scoring and assessment of customer pools, including a higher emphasis on risk-based pricing of products.

Dependence of consumer credit risk on economic and market conditions points towards considering the possibility that while creditworthiness of individuals is time independent, how that translates to probability of default and loss given default may depend on economic or market factors. This suggests that consistent with empirical observations, the probability of default of customers varies with economic conditions, even if the creditworthiness of the individuals is not changing with time. Therefore, point in time credit scoring based estimates of probability of default must be converted into through-the-cycle probability of default estimates. There are efforts underway to include economic factors either directly into regression scorecard or utilizing survival analysis for this inclusion of economic factors.

Survival analysis estimates the default hazard rate, $h(t)$, where $h(t)\Delta t$ is the conditional probability of default in the time interval, $(t, t + \Delta t]$, given there has been no default in the time interval, $(0, t]$. For instance, in the proportional hazard model, the hazard function for default in time period $(t, t + \Delta t]$ for a customer with characteristics, $\mathbf{C}$, is constructed as a product of the baseline hazard function, $h_0(t)$, multiplied by an enhanced credit risk due to customer's characteristics. Namely,

$$h(t, \mathbf{C}) = e^{-S(\mathbf{C})} h_0(t), \tag{9.7}$$

where $S(\mathbf{C})$ is as before a credit score, such that the higher the score the less likely the borrower is to default. Both specific parametric or semi-parametric baseline hazard functions can be constructed to define the model. The survival analysis approach is extensively applied to commercial credit risk, which we will study in the next section.

Credit scoring methodology has many advantages for effective retail credit risk management, however as discussed above, there are still challenges ahead for continuing to improve these models. The drawbacks and challenges to the credit scoring framework involve reliance on historic data, which implies for new products where little or no data is available, the models may have limited effectiveness. Incorporating the impact of changing economic factors or societal norms regarding acceptance for default remains a challenge. Finally, complete automation or even semi-automation based on scoring models implies judgement of a human decision maker is taken out of the process. Automation can also exacerbate incentive misalignments in loan issuance. From a customer perspective, this can become an issue when customers 'learn' the system and begin to 'game' it. Therefore, beyond the efforts to improve credit scoring frameworks, there is a need to complement the credit risk management objectives by implementing a thorough stress testing routine.

9.2 Commercial Credit Risk

Commercial credit is when the obligor or debtor is not an individual or a household, but instead is a medium to large business, corporation, financial or non-financial institution. When governments of different levels of a country issue bonds to raise funds, these instruments also fall under commercial credit. Therefore, the obligors of commercial credit are complex entities, and a very broad range of instruments qualify as commercial credit. Traditional corporate loans, commercial loans, corporate bonds of different maturities, commitments and revolving lines of credit, and commercial contracts such as trade credits and receivables are all examples of commercial debt. While consumer credit is primarily meant to support consumption of goods and services by consumers, commercial credit is used to fund operations of a firm or government, or is used to finance new projects and investments.

Unlike retail credit, commercial credit exposure is not bite-sized, therefore default of a single obligor can threaten a bank's solvency. Commercial obligors may be directly or indirectly financially interdependent. They are often economically intertwined in particular geographical or industry sectors. When defaults occur in commercial credit, the losses can be quite large, making the unexpected loss dominate the overall credit risk exposure of a commercial bank. This also implies pricing the risk into debt contracts is not a feasible mechanism to manage this risk. Finally, due to the complexity of the obligors and their lack of transparency, signals are not received in advance for occurrence of default or change in creditworthiness. Therefore, commercial banks need to make an extra effort to assess their credit risk exposure.

Assessing commercial risk is a complicated task. There are many uncertain elements determining probability of default and loss given default. In order to address the challenge of measuring and monitoring commercial credit risk, many different approaches exist. Some of them utilize regression models based on information available for firms, while others make use of equity market or bond market data available for publicly traded firms to track likelihood of default. Others work only at the portfolio level to assess losses due to credit risk using mathematical and statistical techniques, such as those used in insurance.

Similar to credit scoring methodology in retail credit, the traditional approach for commercial risk is based on a credit rating system (CRS). Two rating agencies that have led the credit rating services in the US from early part of the 20th century are Moody's (from 1906) and Standard and Poor's (S&P)(from 1916). Moody's and S&P's have access to corporations' internal information, which they use for generating the ratings. In this section, we will present several approaches for modeling commercial credit risk, starting with the credit rating methodology.

9.2.1 Credit Rating System

A credit rating system (CRS), like the credit scoring methodology in retail credit, creates a credit assessment of a firm based on complex attributes of the firm. Unlike in retail credit, where only financial information about the individual or household may be used to develop the scorecards, in credit rating systems, financial, managerial, quantitative as well as qualitative and legal information is used to construct the ratings.

Ratings must ascertain the financial health of a firm, whether earnings and cash flows are sufficient to cover the firm's debt obligations, analyze quality of assets of the firm, examine its liquidity position, as well as nature of the industry and clients, and status of new clients. Ratings should also account for the potential effect of macroeconomic factors on the firm, political risk, currency risk, etc. A firm's competitiveness in its industry, expected growth of the industry, anticipated technological changes, regulatory changes, and labor relations can also impact a firm's creditworthiness.

Therefore, a credit rating system needs to have a way to organize and systematize all the information about firms and develop procedures so that credit analysts can rationally, consistently, coherently and comparably generate ratings for across the firms and time. The rating agencies develop public credit ratings of small and large corporations and government issued debt, while banks devote significant resources to develop their in-house (internal) credit ratings for small and large firms, especially those that lack public rating.

These ratings, public or internal, are used for a range of credit risk management decisions, such as decisions regarding loan origination, loan pricing, and loan trading. They are used for monitoring the credit risk at the portfolio level, where consistency in the rating system matters a lot. Capital allocation and capital reserve determination, profitability analysis, and management reporting also utilize credit ratings. Based on a CRS the probability of default (PD) and loss given default (LGD) statistics are estimated, which serve as key inputs for regulatory capital calculation. We will examine this method in the next section.

As stated earlier, Moody's and Standard and Poor's (S&P) are the oldest, most well-known and reputable credit rating agencies that provide publicly available ratings for a very large set of debt instruments. A third important rating agency is Fitch Ratings, and together the three rating agencies control more than 55% of the ratings market. There are other rating agencies active in the US, and there are several rating agencies in different countries, focused on their domestic credit markets. Rating agencies emphasize that the credit rating service they provide is not meant to be an investment recommendation for a security. The focus of the rating assessments is on the potential downside loss, rather than an outlook on the potential upside gain.

Ratings are issued for the issuers of bonds or for the obligor; these are called issuer credit ratings. These ratings assess the overall capacity of the obligor to meet its financial obligations. Additionally ratings may be issued for specific issues of debt, which are the issue-specific credit ratings. These issue-specific ratings for long- and short-term debt account for creditworthiness attributes of the issuer as well as the terms of the issue, such as subordination, collateral, and creditworthiness of any guarantor.

Unlike credit scoring, where a numeric score from 300-850 is assigned to an individual, rating agencies assign an ordinal symbol to obligors or issues. S&P rating categories start at the highest creditworthiness of 'AAA,' followed by the lower levels of, 'AA,' 'A,' 'BBB,' 'BB,' 'B,' 'CCC,' 'CC,' 'C,' and 'D.' 'D' indicates that the obligor has already defaulted on its debt. Modifiers of '+' or '-' are used to further refine these rating levels. On the other hand, Moody's rating categories start at 'Aaa,' dropping to lower levels of, 'Aa,' 'A,' 'Baa,' 'Ba,' 'B,' 'Caa,' 'Ca,' and 'C.' S&P's 'AAA' to 'BBB' rating levels and Moody's 'Aaa' to 'Baa' are generally considered investment grade, while ratings below that are considered to have significant speculative characteristics.

9.2.1.1 Risk Assessment by Credit Rating Migration

Default is an event, when an obligor is unable or unwilling to make a scheduled payment, or violates a condition as per the debt contract. The focus of credit rating agencies cannot just be the occurrence of default. Their objective is to create a perspective on the creditworthiness of obligors at all times. Therefore, at discrete points in time, they assess and consider revising their credit ratings of obligors, depending on how their creditworthiness has evolved. This evolution of credit quality is very important for banks, insurance companies, investors, and other financial institutions holding a portfolio of commercial loans or corporate bonds.

The change in credit rating over time is called credit migration or debt migration. If substantial amounts of historical observation of these migrations are available, they can be used to summarize average transition rates for all rating levels. These transition rates are summarized in a debt migration matrix. One can consider the rating level of a bond, R_t, at time t to be a Markov chain. We had studied Markov chains in Chapter 5 as discrete-space stochastic processes. Discrete-time Markov chains are stochastic processes whose value in the immediate future time period depends only on the state of the process at the present time, and is independent of the values of the process in the past. This key temporal property of stochastic processes is known as the **Markov property**. The debt migration matrix would serve as a transition matrix of the Markov chain, R_t.

Figure 9.3 shows an example of a debt migration matrix, built based on credit migration observations of S&P rating histories for the period 1981-2004. Estimates of migration based on data from a period of 23 years indicates that these estimates are aggregates through business cycles and economic conditions. Each row of the matrix is a probability distribution of the change in rating of a bond in one year, given its current rating. For instance, $\{0.03\%, 0.26\%, 4.05\%, 89.70\%, 5.05\%, 0.76\%, 0.07\%, 0.083\%\}$, is the probabil-

From/To	AAA	AA	A	BBB	BB	B	CCC/C	D
AAA	93	6.18	0.66	0.07	0.08	0.01	0	0
AA	0.61	91.03	7.53	0.64	0.09	0.08	0.01	0.005
A	0.08	1.99	91.69	5.55	0.49	0.18	0.01	0.008
BBB	0.03	0.26	4.05	89.7	5.05	0.76	0.07	0.083
BB	0.04	0.11	0.56	5.26	83.8	8.95	0.73	0.548
B	0	0.07	0.23	0.5	4.67	84.36	5.71	4.448
CCC/C	0.06	0.01	0.34	0.56	1.1	7.99	47.02	42.896
D	0	0	0	0	0	0	0	100

FIGURE 9.3: An example of debt migration shown in a year, a bond which started with a given rating has range of probability of migrating to all the other rating levels. This debt migration matrix is a one-year constructed using S&P's rating histories, from 1981-2004. Estimation method is cohort method and all values are in percentage points.

ity distribution of the rating of a bond that has a rating of 'BBB' at the start of a year. The corresponding states are, {'AAA,' 'AA,' 'A,' 'BBB,' 'BB,' 'B,' 'CCC'/'C,' 'D' }, where three ratings are merged into a single category for the purposes of credit migration assessment.

Along the diagonal, we have probabilities of a bond retaining its current rating. We notice that as the rating falls below 'BBB', the rating has increasing tendency to migrate away from its initial rating. The probability of default is also steadily increasing as the rating drops, with a peak of 42.89% for the 'CCC/C' rating group. Once a bond defaults, it is assumed to stay in that state. This is shown in the last row of the debt migration matrix, thus making default an absorbing state in Markov chain terminology.

If we can assume the Markov chain to be a stationary process, which means its migration matrix remains the same for all years, we can construct a multiple year migration matrix using the single-year migration matrix. If P is a single year credit migration matrix, the two-year migration matrix is obtained by the matrix product of the one-year matrix with itself, $P^2 = P \times P$. Similarly, n-year credit migration matrix can be constructed as, P^n, each row of this matrix is interpreted as probability distribution of rating of a bond in n years, given it started with a certain rating initially. In a study it was shown that when

Seniority	Mean	Median	Standard Deviation
Senior Secured	56.4	55.0	28.0
Senior Unsecured	36.5	26.0	29.0
Senior Subordinated	30.5	23.0	25.8
Subordinated	32.2	29.0	22.7
Junior Subordinated	27.1	15.3	25.8
Preferred Stock	10.1	4.2	21.1

FIGURE 9.4: Descriptive statistics for recovery rates by debt seniority. Calculations are based on Moody's data for period 1970-2008, taken from Mora (2012).

an additional year of data is used to create the transition matrix estimate, it only has a modest impact on the estimates of the migration matrix [63]. Therefore, even while not entirely accurate, a reasonable approximation for multi-year impact of credit risk can be assessed using a multi-year transition matrix constructed assuming stationarity.

Transition matrices play a major role in credit risk evaluation systems, such as in the pioneer credit risk assessment system released by J.P. Morgan in 1997, called *CreditMetrics*^{TM}. If the current value of a bond with rating, R_0, is V_0, in a year its rating could be any of the other levels with the probability distribution given by the applicable row of the debt migration matrix. Corresponding to each potential future rating, R_1, of the bond, its value can be assessed as V_1. For this valuation, forward rates would be required and the appropriate bond pricing approach developed in Chapter 8 will need to be applied. Additionally, for the case of default, an estimate of recovery rate will be needed to determine the value of the bond in case of default. As such, predicting recovery rate is tricky. The table in Figure 9.4 shows statistics for recovery rates by the seniority of the bond. More customized models for recovery rates may be considered in specific cases [65].

The change in bond value in one year is, $\Delta V = V_1 - V_0$. In order to create a credit risk measure, we construct the Credit Value-at-Risk (Credit VaR) by reporting how the 'worst'-case scenario at a chosen confidence level, c, is relative to average change in value of the bond. The average change in value of the bond is, $E[\Delta V] = E[V_1 - V_0] = E[V_1] - V_0$. We denote, $P(c)$, to be the change in value of the bond in the worst-case scenario, at the $(1 - c)$ percent confidence level (therefore for a 99% confidence level, we have $c = 1\%$). Credit VaR at $(1-c)\%$ confidence level is, $(E[V_1] - V_0) - P(c)$. Figure 9.5 displays the Credit VaR quantity for the change in bond value distribution due to credit migration.

Considering measures for credit risk of a single bond is not as useful as doing the same for a portfolio of bonds. In the case of a portfolio, V_0 is the current marked-to-market value of the portfolio of N bonds, expressed as a linear combination of portfolio weights, $\{w_i, i = 1 \ldots N\}$, and current value of each bond in the portfolio. For determining the future value of each bond, a credit migration matrix will need to be applied to each bond, picking rows from the matrix as relevant for the bond's initial rating. Even when the credit migration of the bonds is considered independent of each other, creating all combinations of the bonds' simultaneous migration in a portfolio of more than two bonds becomes quite cumbersome. Simulation analysis can be applied to create a large sample of possible joint migrations of the bonds.

For j=1:SampleSize
    For $t = 1 : N$
        Generate R_{1i} by using the 8-outcome probability distribution
            for the current rating, R_{0i}, of ith bond;
        Determine the future value of the bond, given its future
            rating, R_{1i};
    End
    $V_{1j} = \sum_{i=1}^{N} w_i$ future value of the ith bond;
End

As before, $P(c)$ is the change in value of the portfolio in the worst-case scenario at the $(1-c)\%$ confidence level. If the bonds in the portfolio do not migrate in the year, the portfolio's forward value (FV) is obtained by future price of the constituent bonds under their current rating. This allows computing of the promised rate in absence of credit risk, PR, by the following relation, $FV = V_0(1 + PR)$. Similarly, using the realizations of V_1, the expected value of the portfolio, $E[V_1]$, is obtained, which allows computing of the expected return of the portfolio in presence of credit risk, $E[V_1] = V_0(1 + ER)$, where ER is the expected return. This allows computing the expected loss due to credit risk as, $EL = FV E[V_1] = V_0(PR - ER)$. Credit VaR at $(1 - c)\%$ confidence level is again obtained as, $(E[V_1] - V_0) - P(c)$.

We have presented the credit VaR computation and analysis in the simplest setting. There are many improvements worth considering in response to empirical features of credit risk. For instance, we have stated summary statistics for recovery rates, however in practice, recovery rate or loss given default may need to be more accurately estimated to complete this analysis. Regression-based models have been developed that seek to relate recovery rates with economic factors, characteristics of the loan and the obligor in the corporate setting [3].

Both default rates and recovery rates vary with time; moreover they display a negative correlation in their movement, which is non-trivial at the level of -0.40 [65]. When aggregate default rates go up, recovery rates become lower. Recovery rates display procyclicality, with the aggregate recovery rate closely

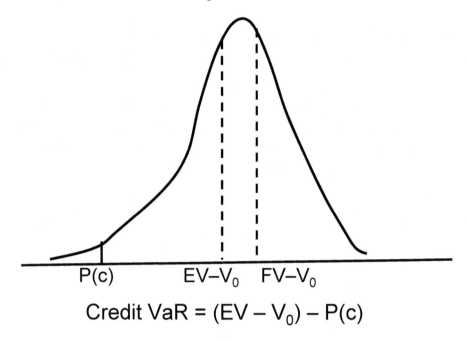

$$\text{Credit VaR} = (EV - V_0) - P(c)$$

FIGURE 9.5: Display of change in portfolio value due to credit risk.

tracking the business cycle. Aggregate recovery rate is seen to have a positive correlation with real GDP growth, with a correlation coefficient equal to 0.45 for the period 1978-2010 [65]. Defaults also tend to cluster around periods of economic slowdown, with the frequency of defaults increasing substantially during recessions. However, all recession periods don't seem to have the same impact on default frequencies, for instance the 2001-2002 period was much worse than the 1990-1991 period in terms of default experience.

In assessing credit risk, the first most important factor is the credit standing of specific obligors. If a bank focuses on investment grade ('AAA'-'BBB') obligors, this will result in low probability of default of any individual obligor in its portfolio. However, if a bank's strategy is to create a market share in more risky speculative-grade obligors, who pay a much higher coupon rate on their debt, the pricing of debt is a significant issue. The institution should charge the appropriate interest rate spread to each obligor to compensate for the excess risk exposure. Pricing alone does not safeguard the bank against credit risk. The bank should also set aside the appropriate amount of risk capital, in order to limit the chance of itself defaulting due to significant losses from obligors' defaults.

In the case of consumer credit, obligors are expected to be independent in terms of their default risk. In commercial credit, risk can easily get concentrated if attention was not paid, since obligors are not independent in terms of their credit risk. Attempting diversification of obligors in terms of number of obligors, their geography and industry is a valuable exercise. A bank with only a few big-ticket corporate clients, most of which are, say, in commercial real estate, is rightly considered more risky than a bank with many corporate loans to obligors distributed over many industries. Banks catering to narrow geographical regions would suffer from a slowdown in economic activity of that particular region. Extension to the credit migration methodology to incorporate correlations can be developed, for instance by Bernoulli random variables based coupling to create degree of dependency by rating classes and the sectors of the debtors [44].

As discussed above, state of the economy has a non-trivial impact both on default frequency and recovery rates. During the times of economic growth, the frequency of default falls sharply, whereas the reverse phenomenon is observed during periods of recession. Periods of high default rates are also characterized by a low rate of recovery on defaulted loans, i.e., banks tend to find that the various assurances and collateral that they use to secure the loans are less valuable during a recession. Therefore, a bank's credit risk management strategy should account for this.

The factors discussed above, namely individual obligor's credit risk, its correlation with other obligors, and the effect of economic conditions, can also interact with one another. Interaction of concentration and economic downturn further enhance their individual impact. Economic downturns uncover hidden tendencies of obligors to default together, which is the clustering of default phenomena. Finally, the maturities of loans in a portfolio is also important. Longer-term loans are generally considered more risky than short-term loans. Therefore, if banks can build a portfolio that has time diversification, i.e., it is not concentrated in particular maturities, it can reduce portfolio maturity risk, which also helps in managing liquidity risk.

We next develop other models for credit risk, which may be used as alternatives to credit migration methodology.

9.2.2 Models for Credit Risk

The traditional way in which rating agencies and large banks rate the credit risk of bonds and corporate loans is a combination of qualitative, judgmental and quantitative assessment. Our continued goal is to put objective and absolute numbers on the risk of default of single obligors and measure the credit risk in a bank's entire credit portfolio using statistical and economic tools. Much of this effort is an attempt to apply successes in modeling market risks to modeling credit risk. This involves utilizing continuous-time stochastic processes to model key quantities so that credit risk can be measured uniquely for a single obligor, instead of applying an aggregated credit migration framework. Secondly, observing evolution of credit risk in any time granularity is important, rather than just annual migration through potentially different ratings.

The mathematical sophistication that made such a difference to management of market risks and derivative pricing and trading for market risk, has penetrated credit risk with varying degrees of impact. We begin with first considering a classical firm-specific regression-based model for predicting creditworthiness of a firm.

Altman's Z-Score Method

In 1981, Edward Altman and James La Fleur developed a credit scoring model for firms using a combination of traditional financial ratio analysis and discriminant analysis [4]. An objective overall measure of corporate health, called the Z-score, was arrived at by combining five measures based on reported accounting and stock market variables. Therefore, the score is constructed for a publicly traded company based on publicly available information. The Altman's Z-Score model is the following linear model, in which the five measures are objectively weighted and summed to arrive at an overall score, Z. The overall Z-score is used to classify the firm into a predetermined grouping of creditworthiness.

$$Z = 1.2X_1 + 1.4X_2 + 3.3X_3 + 0.6X_4 + 0.999X_5, \qquad (9.8)$$

where X_1 is the ratio of Working Capital to Total Assets of the firm, hence measuring liquidity of the firm, X_2 is the ratio of Retained Earnings to Total Assets, X_3 is the Earnings Before Interest and Taxes (EBIT) to Total Assets ratio, which is a measure of Return on Assets (RoA). The fourth variable, X_4, is the ratio of Market Value of Equity to Book Value of Total Liabilities of the firm, which measures leverage of the firm, and the final variable, X_5, is the Sales to Total Assets ratio, which measures efficiency of the firm.

The Z-score is interpreted as the higher the Z-score, the stronger a firm's financial health is. In particular, if $Z > 3.0$, the company is considered unlikely to default. For Z-score lying in the range $3.0 > Z > 2.7$, we should be 'on alert', in the case of $2.7 > Z > 1.8$, there is a good chance of default. Finally, in the case of a low enough Z-score, $1.8 > Z$, the probability of financial embarrassment is very high. As in any statistically motivated prediction scheme, there is room for Type I error, where a firm that was predicted to not go bankrupt according to the Z-score, actually did go bankrupt. Similarly the reverse error, Type II error, also occurs where according to the Z-score the firm should have very high probability of default, and it actually ends up surviving the financial turbulence.

9.2.2.1 Structural Model of Credit Risk

Altman's Z-score model is a fundamental model, which means it digs into the financial statements of a firm to detect the firm's financial health and the likelihood of the firm hitting financial distress. Structural models of credit risk take a similar theme. They develop models for the evolution of the total assets of the firm or the firm value, V_t, and utilize this model to determine the probability of default and loss given default.

The contingent claims approach (CCA) utilizes option pricing concepts for assessing credit risk. Given a model for the evolution of value of the firm, V_t, it considers the equity of the firm, E_t as an option on the firm's asset, V_t. Hence the name, contingent claims approach. Therefore, if V_0 is the value of the assets of a firm initially, we choose a future time T, and identify V_T to be the value of assets of the firm at time T. If E_0 is the value of equity of the firm initially, and E_T is the future value of equity at time T, then $E_T = \max(V_T - D_T, 0)$. Here D_T is the amount of the firm's debt maturing at T.

The form of the function, $\max(V_T - D_T, 0)$, matches that of the pay-off at maturity of a call option on value of assets with the strike price set at D_T. If the firm value falls below the debt maturing at T, D_T, then the firm defaults, and equity value of the firm is zero. Assuming the Black-Scholes option pricing formula applies, where the volatility of the value of the firm is σ_v, the probability that firm will default on its debt at time, T, is $P(V_T < D_T) = \Phi(-d_2)$, where

$$d_2 = \frac{\ln(\frac{V_0}{D_T}) + (r - \frac{\sigma_v^2}{2})T}{\sigma_v\sqrt{T}}, \tag{9.9}$$

and $\Phi(.)$ is the cumulative distribution of the standard normal random variable. In practice, the volatility of value of the firm, σ_v may not be easily obtainable. In this case, volatility of the equity of the firm, σ_e, is used as a proxy.

Therefore, the contingent claims approach is able to estimate probability of default in a rather simplistic way, customized to the risk of a specific firm. Its limitation, however, is that it considers debt maturing only at one specific time, and is not able to give an estimate of loss given default. The Moody's KMV (Kealhofer, McQuown, and Vasicek) approach uses a concept called 'distance to default' and translates this into an estimate of probability of default and recovery rate. It doesn't lock itself to a specific time of maturity of debt, and treats debt level as a combination of short-term and long-term debt of the firm. Figure 9.6 displays the evolution of firm value and the 'distance' of the firm value from the debt level of the firm, summarized as the 'Distance to Default' quantity.

If the firm value evolves by the model,

$$dV_t = (\alpha - \delta)V_t dt + \sigma_v V_t dW_t, \tag{9.10}$$

where α is the return on asset for the firm and δ is the cash payout made to claimholders, then the distance to default, $E[V_T - D_T]$, for some future time T, normalized by the standard deviation, is estimated as,

$$E[\frac{\ln(V_T) - \ln(D_T)}{\sigma_v\sqrt{T}}] = \frac{\ln(V_0) + (\alpha - \delta - \sigma_v^2/2)T - \ln(D_T)}{\sigma_v\sqrt{T}}. \tag{9.11}$$

The expected recovery rate, conditional on default, is estimated as, $E[V_T|V_T <$

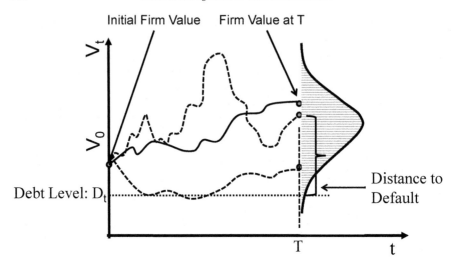

FIGURE 9.6: Display of evolution in firm value over time relative to debt level of the firm for measuring distance to default.

$D_T]$. The expected recovery rate can be derived as,

$$E[V_T|V_T < D_T] = V_0 e^{(\alpha-\delta)T} \left(\frac{\Phi[-\frac{\ln(\frac{V_0}{D_T})+(\alpha-\delta+\sigma_v^2/2)T}{\sigma_v\sqrt{T}}]}{\Phi[-\frac{\ln(\frac{V_0}{D_T})+(\alpha-\delta-\sigma_v^2/2)T}{\sigma_v\sqrt{T}}]} \right). \qquad (9.12)$$

Although several enhancements of the above structural models for estimating commercial credit risk have been developed, the challenge of utilization of a structural model remains their need for firm-specific data in order to build the model for the firm, which may not always be readily available. Moreover, however complex these models may be made, they must still make significant simplifying assumptions regarding the firm's debt and asset characteristics.

9.2.2.2 Reduced-Form Model of Credit Risk

Instead of using the total firm value, debt and equity level of a firm for predicting default and loss given default, a reduced-form model of credit risk uses prices of traded debt instruments of the firm to elicit estimates of credit risk of the firm. Specifically, the difference in prices of risk-free bonds and risky-bonds issued by firms are used to assess the likelihood of the firms to default. Therefore, the reduced-form models are also customized to the risk of a specific firm. Assumption behind the approach is that the credit spread is solely due to increased default probability.

As suggested in consumer credit modeling using survival analysis, we define a default intensity or a hazard rate, h_t, for a firm. The probability of default for the firm in Δt period of time is, $h_t\Delta t$. As an extension of the bond pricing derivation of Section 8.1.2, it can be shown that the price of a risky bond satisfies the following differential equation.

$$\frac{\partial P}{\partial t} + \frac{1}{2}\sigma^2\frac{\partial^2 P}{\partial r^2} + (\mu - \lambda\sigma)\frac{\partial P}{\partial r} - (r+h)P = 0, \tag{9.13}$$

along with the end condition, $P(r_T, T) = F$, where T is the bond's maturity. Price of a risky bond, as a solution of the above equation and end-condition is

$$P(t, r_t) \quad = \quad E[e^{-\int_t^T r(s)+h(s)ds} F], \tag{9.14}$$

where the stochastic short-term interest rate, $r(t)$ evolves driven by a Wiener process, W_t, in the risk-neutral measure by the following equation,

$$dr_t = (\mu(r_t, t) - \sigma(r_t, t)\lambda(r_t, t))dt + \sigma(r_t, t)dW_t. \tag{9.15}$$

As a further generalization, the default intensity, $h(t)$, is modeled to be time-dependent and stochastic, governed by the following model.

$$dh_t = \gamma(t, r_t, h_t)dt + \delta(t, r_t, h_t)dW_{2t}. \tag{9.16}$$

This is the doubly stochastic default intensity reduced-form model. Default intensity is estimated using historical data on risky bonds prices, as well as various credit derivatives, such as credit default swaps. Since the reduced-form models are calibrated using price data, the calibration is in the risk-neutral measure, not the real-world measure. Therefore, the model is not used in estimating real-world probability of default or loss given default. Reduced-form models are extensively utilized for pricing credit derivatives.

9.3 Credit Risk Hedging Instruments

Innovation in credit risk transfer is obviously important for banks, but actually it has had a significant effect in the much wider world. Different types of credit derivatives ended up playing a prominent role in the occurrence and evolution of the financial crises of 2007-2008, which triggered the regulatory changes for broadly defined 'swaps.' Figure 9.7 shows the volume of net credit protection sought over a period of two years by protection seekers and by the nature of underlying debt the protection was sought for. Data from the period after the crises also shows the recovery of volumes in this market.

Credit derivatives and securitization are not only useful for the banking industry, they are also relevant to the management of credit risks borne by leasing companies and large non-financial corporations, in the form of account receivables. This is seen by the participation of non-financial corporations in

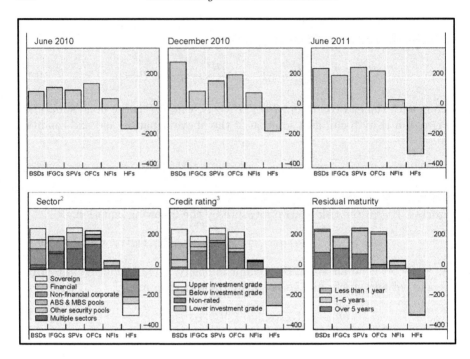

FIGURE 9.7: Volume of net credit risk protection bought by reporting dealers from different counterparty groups using credit default swaps in 2010-2011 period taken as a snapshot of time after the 2007-2008 credit crises. Acronyms used are: BSDs: banks and securities dealers; IFGCs: insurance and financial guaranty companies; SPVs: special purpose vehicles; OFCs: other financial companies; NFIs: non-financial institutions; HFs: hedge funds. (Courtesy *BIS Quarterly Review*, December 2011)

Figure 9.7. For instance, for producers of capital goods, which are often provided to the customers for long-term credit or at long-term leases. Credit derivatives also allow non-banking institutions to participate in credit risk and get a piece of the risk-return benefit of credit risk. Private and institutional investors, hedge funds, and insurance companies also participate in this market, as seen in Figure 9.7.

The traditional way of mitigating credit risk, beyond diversification, was by the use of bond insurance, collateral, guarantees, early termination, or selling off a portion of the loan in the secondary loan market. These mechanisms have drawbacks of being too close to the customer, hence customer relations and reputation can sour should things go wrong. Since these traditional mechanisms are a mutual agreement between the transacting parties of the underlying debt, they don't allow separating credit risk from the underlying positions, hence don't allow redistribution of credit risk among a broader class of financial institutions and investors. The essential advantage of credit derivative instruments is to allow spreading the risk so that its effect on a single entity is not devastating.

Credit derivatives have so far been off-balance-sheet arrangements that allow one party, the beneficiary or protection seeker, to transfer the credit risk of a reference asset to another party, the guarantor or protection seller, without actually selling the underlying reference asset. This mechanism has made it possible to strip credit risk from the market risk of the underlying asset, and transferring credit risk independent of funding and relationship management concerns.

Securitization is the other arm in managing credit risk, where a pool or portfolio of credit-risk exposures are segmented and a variety of securities with different risk profiles are extracted from it. We mentioned securitization as one mechanism for credit risk transfer in the context of retail-credit, mortgages and credit-card receivables. Similar credit restructuring techniques are also applied to the corporate credit sector, to enable lenders to repackage corporate loans to notes, securities or credit derivatives with a variety of credit-risk features. A side-benefit from credit derivatives and securitization is they help in 'price discovery' of credit risk, i.e., they make clear how much economic value the market attaches to a particular type of credit risk. This can in turn lead to more liquidity, more efficient market pricing, and more rational credit spreads for all credit-related instruments.

In the still evolving credit markets, credit risk is not simply the risk of potential default. It is also the risk that credit spreads will change affecting the relative market value of the underlying corporate bonds, loans and other derivative instruments. In effect, the 'credit risk' of traditional banking is evolving into the 'market risk of credit risk.' In more efficient capital markets, large investment-grade firms can borrow directly from investors via issuing bonds, rather than going to banks. As a result, the environment has changed for banks, with a greater exposure to less credit-worthy obligors, their portfolios of loans and other credit assets have become more concentrated in risk. As discussed earlier, lower creditworthy obligors have more severe problems with defaults and decreased recovery rates in economic downturns.

Banks have had to develop response strategies, which has helped and has been helped by the growth in the credit derivatives market. Banks don't have to simply make loans and hold them until they mature. Instead they have adopted the 'underwrite and distribute' business model. The broad strategy for credit risk management of bank credit portfolio includes distribution of large loans to other banks through secondary loans market or by participation in large loans through bank syndication. Utilization of credit derivatives can help reduce loan exposure by hedging concentrated loan positions or by selling down concentrated loan positions. And finally, focusing on high-risk obligors, while selling or hedging low-risk, low-return loan assets to free-up some bank capital.

As is necessary for any contract, in the context of credit derivatives, counterparties must make sure they understand the amount and nature of risk that is transferred by the derivative contract and how much is retained. It is also important to understand the impact of correlations between underlying assets, economic factors, and counterparty risk of credit derivatives, which relates to enforceability of the contracts. Through the credit crises of 2007-2008, some tough lessons were learned in this regard.

We present an overview of the most important credit derivative instruments used for hedging credit risk in the following sections.

9.3.1　Single-Name Credit Derivatives

Both the pace of innovation and volume of activity in the credit derivatives markets have soared in the last decade, leading up to the financial crises of 2007-2008. After more than a decade of rapid growth, the volume of outstanding credit default swaps peaked at roughly $60 trillion in 2007. After that it nearly halved, even though the turnover has continued to rise. The Bank for International Settlements (BIS) has reported that in December 2010 the notional amount of credit derivatives outstanding stood at a total of $29.89 trillion, with a gross market value of $1.35 trillion. $29.89 trillion is a nontrivial figure in its own right, and is a sizeable fraction of $158 trillion of outstanding global stock of debt as of 2010.

With the strengthening credit derivatives market, the role of traditional players, such as bond insurers, declined, while hedge funds and other institutions have started playing an increasingly prominent role, as also seen in Figure 9.7. Credit derivatives offer a cheaper access to high-yield markets since these instruments allow unsophisticated institutions to piggyback on the massive investments in back-office and administrative operations made by banks.

Among different credit derivatives, single-name ones are the simplest. They are called single-name because the underlying asset is issued by a single obligor, for which protection is being sought. For example, consider two banks, one bank has special expertise in lending to the airline industry and has a $10 million worth, A-rated loan exposure to an airline company. The other bank is in an oil-producing region and has an outstanding loan of $10 million worth to a A-rated energy company. If the two banks have a natural concentration in airline and energy industry, respectively, they are not diversified, and therefore, are vulnerable to a downturn in their favored industry.

Since an airline company is generally better off with declining energy prices, and an energy company is better off with rising energy prices, there is a negative correlation between credit risk arising from these two sources at the two banks. Since it is less likely that airline and energy industries will run into difficulties at the same time, the two banks could sell their loans to each other. However, this runs the risk of upsetting their customers, since the loan will now be serviced by an unfamiliar bank. Alternatively, both banks can swap just some of the credit risk underlying the two loans. Say, they decide to swap 50% of the principal of the two loans, i.e., $5 million of each other's loans. After this swap, should either loan suffer losses due to default, the other bank will be responsible to make up for 50% of the losses. After swapping their risks, both banks can be reassured, and can continue to exploit their proprietary information, economies of scale and existing business relations with corporate customers by extending more loans to their natural customer base.

This is a simple example of a credit swap, where both banks are acting simultaneously as protection buyer and protection seeker. The key point, however, is that they are seeking or selling protection for a single underlying loan, hence participating in a single-name credit derivative. We now look closer at perhaps the most popular single-name credit derivative, the credit default swap or CDS.

9.3.1.1 Credit Default Swaps

Credit default swaps (CDS) are the most popular credit derivatives, with single-name CDS making up for a total notional amount of $18.14 trillion in 2010, and a market value of $884 billion. Credit derivatives are mostly structured or embedded in swaps or options, and are normally of a shorter maturity than the underlying loans or bonds, which is the case for credit default swaps.

CDS can be thought of as an insurance against default of some underlying instrument or as a put option on the underlying instrument. For instance, a credit default swap may specify that a payment be made if a 10-year corporate bond defaults at any time in the next 2 years. The protection buyer, or seller of credit risk, makes periodic premium payments to the protection seller of a negotiated basis points times the notional amount of the underlying bond or loan. The party buying the credit risk makes no payment unless the issuer of the underlying bond or loan defaults. In the event of a default, the protection seller pays the protection buyer a default payment equal to the notional amount times a pre-specified recovery factor.

Figure 9.8 shows an example of cash flow implication of a credit default swap. An appropriately defined credit event relative to the underlying being a single credit risk bearing instrument, a loan or a bond, triggers the protection response. The protection is in terms of recovering from the default event. The recovery factor captures what the bank may recover of the notional value after default from collateral, etc. The default should be clearly defined in the contract. There is often a materiality clause requiring that the change in credit status be validated by third-party evidence.

CDS is a par-value product, therefore it does not hedge the loss on the bond from its market value, which may be an issue if the bond is trading far from its face value. The CDS will end up under-hedging if the bond is selling at a premium, and over-hedging if the bond is selling at a discount. Variations on the plain CDS are designed, such as an amortizing default swap, where the face value amortizes as the maturity is approached, or a binary credit default swap, in which a fixed amount is paid in the event of a default.

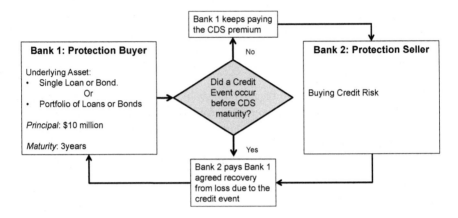

FIGURE 9.8: Display of cash flow and structure of single- or multi-name credit default swap.

The periodic premium paid by the protection buyer until a default event is called the CDS spread, which is determined so that the expected net present value of cashflows between the protection buyer and seller is zero. We utilize the reduced-form approach of credit risk modeling to give an overview of pricing a CDS. Let s be the CDS spread for a CDS contract maturing in T_1 years for a bond maturing in $T_2 (> T_1)$. Assume that the protection buyer pays the periodic premium at times, $\{t_0, t_1, \ldots, t_N\}$, where we assume $t_N = T_1$ for simplicity. Then the CDS protection buyer leg has the following expected present value,

$$\sum_{i=0}^{N} P(0, t_i) E[s \mathcal{I}_{\tau > t_i}], \tag{9.17}$$

where the expectation is in risk-neutral measure, τ is random time of default, $\mathcal{I}$ is the indicator function used to indicate occurrence of default, and $P(0, t_i)$ is the price of a zero-coupon bond with a face value of \$1 maturing at t_i. If the recovery rate of the bond is R, then the payment of the protection seller per dollar of nominal value of the CDS contract is $1 - R$. Therefore, the CDS protection seller leg has the following expected present value:

$$E[P(0, \tau)(1 - R)\mathcal{I}_{\tau < T_1}]. \tag{9.18}$$

By matching the two legs, we obtain the CDS spread to be the following.

$$s = \frac{E[P(0, \tau)(1 - R)\mathcal{I}_{\tau < T_1}]}{\sum_{i=0}^{N} P(0, t_i) E[\mathcal{I}_{\tau > t_i}]}, \tag{9.19}$$

The quantities, $E[\mathcal{I}_{\tau > t_i}]$, are estimated using the hazard rate $h(t)$ calibrated under the risk-neutral measure.

9.3.1.2 Spread Options

We have seen several times in the earlier discussions of credit risk that credit risk is about more than just default risk. Credit risk is also in the deterioration or improvement in creditworthiness of an obligor. These changes result in significant loss in value of debt instruments, therefore is another crucial credit risk to consider hedging. A spread option is a single-name credit derivative, with its pay-off depending on the credit spread of a single underlying credit-risk sensitive asset or a bond. More specifically, the option has its underlying as the credit spread of the underlying bond, which is the yield spread between a specified corporate bond and a government bond of the same maturity.

The strike price of the spread option is set as the forward spread at maturity of the option, and the pay-off is the greater of 0 or the difference between the market-observed spread at maturity and the strike price, times the multiplier. The multiplier is meant to convert the spread into a dollar value, and is usually taken as, duration of the underlying bond times notional amount. Therefore, the pay-off of a spread option is given as,

$$\max(s_T - f_s, 0)FD_B, \tag{9.20}$$

where f_s is the forward spread, F is the face value of the bond and D_B is the duration of the bond. Given that a bond is itself a derivative of the underlying interest rate and default intensity process, the risky bond pricing framework can be applied to determine the price of a spread option. Therefore, the price of a spread option is obtained as,

$$s(0, r_0, h_0) = E[e^{-\int_t^T r(s)+h(s)ds} \max(s_T - f_s, 0)FD_B], \tag{9.21}$$

where the risk-free interest rate, r_t, and the default intensity, h_t, evolve by the following equations.

$$
\begin{align}
dr_t &= (\mu(r_t, t) - \sigma(r_t, t)\lambda(r_t, t))dt + \sigma(r_t, t)dW_{1t}, \tag{9.22} \\
dh_t &= \gamma(t, r_t, h_t)dt + \delta(t, r_t, h_t)dW_{2t}. \tag{9.23}
\end{align}
$$

We next look at some examples of multi-name credit derivatives.

9.3.2 Multi-Name Credit Derivatives

The multi-name credit derivatives are defined on a portfolio of credit-risk sensitive instruments. There are a variety of multi-name credit derivatives, those that are defined on real assets and others that are synthetically constructed. It is in the case of multi-name credit derivatives that the default correlation between obligors takes on great significance. As discussed earlier, default correlation is an important risk factor, since in retail credit risk, while we assume it to be low, in strained economic conditions, default behavior of obligors can change and default correlation can emerge. In commercial credit risk, default correlation is always an important factor. As a result, a less diversified portfolio of credit-risk sensitive instruments can have a significantly higher risk due to default correlation.

We first consider some multi-name credit default swaps; later in this section, we will focus on the large segment of securitization products, as examples of multi-name credit derivatives. In a multi-name credit default swap, the protection buyer pays a premium up to the specific default the swap is meant to provide protection for. At the event of the specific default alone the protection seller pays the lost payment due to the default. Refer to Figure 9.8 again, where in the case of first-to-default swap the swap is defined on a portfolio of loans or bonds as the underlying, and the first credit event from the inception of the swap will trigger the protection seller to make up for the loss of the protection buyer. Multi-name CDSs make up a comparable volume as the single-name CDSs, for a total notional amount of $11.73 trillion in 2010, and a market value of $467 billion.

The first-to-default spread will lie between the spread of the worst individual credit in the portfolio and the sum of the spreads of all the credits in the portfolio, therefore these are cheaper for the protection buyer than buying a CDS for each credit in the portfolio. First-to-default swap can be extended to n^{th}-to-default swap, where a bank will seek help with recovering losses related with n^{th} credit risk event from the inception of the swap. First-to-default swap is clearly the most valuable protection, however a bank may want protection up to a certain depth of losses in its loan portfolio. These swaps give the flexibility of credit risk management strategy a bank may want to construct.

9.3.2.1 Collateralized Debt Obligations

Collateralized debt obligations (CDOs) is a specific application of securitization to a portfolio of debt instruments. Given the variety of debt instruments, CDOs take specific form depending on the specifics of the debt instruments in the collateral pool. Accordingly, we have mortgage-backed securities, which have home mortgages in the pool of debt instruments, asset-backed securities have other kinds of retail credit instruments, such as leasing receivables, automobile loans, personal loans, and revolving loans, such as credit card receivables. Similarly, in the commercial credit space, collateralized loan obligations (CLOs) or collateralized bond obligations (CBOs) are also examples of collateralized debt obligations (CDOs), where these securities are collateralized by means of high-yield bank loans or corporate bonds.

Securitization is the process by which a set of cash flows from a retail credit portfolio, such as mortgage payments on a mortgage portfolio, or commercial credit portfolio, such as a portfolio of commercial loans or bonds, is transformed into payouts of securities through various legal and financial engineering procedures. Terms like mortgage-backed securities (MBSs), asset-backed securities (ABSs), collateralized mortgage obligations (CMOs) have existed for decades, and have been made particularly well-known through the

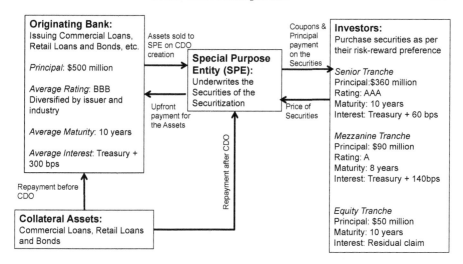

FIGURE 9.9: Display of cash flow and structure of securitization to create collateralized debt obligation securities.

financial crisis of 2007. This process of securitization started in the mortgage segment of credit instruments in the 1970s, and got steadily stronger and more broadly applied over the years. In an increasingly competitive environment, banks found it advantageous to not issue loans and hold them until maturity. Freeing up capital to find new profitable avenues for investment created a much desired liquidity in banks' assets. Moreover, banks often also saw securitization as a means to free up some regulatory capital.

Beyond the retail segment, CMBSs or commercial mortgage-backed securities, applied the same principle to commercial mortgages, as did the CMOs, which primarily differ in the temporal structure of expected payments. CMO introduced the notion of tranches, where payments for the securities issued are divided into tranches, the first tranche receiving the first set of payments, other tranches taking their turn. A tranched bond follows the seniority structure as in classes of senior and subordinated bonds. Figure 9.9 displays the cash flow underlying the creation of a CDO structure. The CDO structure, of which a CMO is a special case, is shown to have tranches that result in what is known as a 'waterfall' of cash flows.

Advancing this to create CLO and CBO was an efficient securitization structure that allowed cash flows from a pool of below investment grade loans or bonds to be pooled together and prioritized, so that some of the resulting securities could achieve investment-grade rating. The main difference between CLOs and CBOs is of course the type of collateral, but this implies that one benefits from a better recovery rate given default compared to the other. The recovery rate given default is much lower (30-40%) for unsecured bonds versus the 70% for secured loans. This is because loans are periodically paid off and loans have shorter duration than high-yield bonds. It is therefore easier to produce securities with investment-grade ratings from CLOs than from CBOs. Beyond these, there are synthetic CDOs, which have a pool of CDSs as the collateral. There are also CDO^2 and CDO^3, which are CDOs of CDOs.

A CDO is issued through a special purpose vehicle (SPV), special purpose entity (SPE), or a trust, which issues the securities. The proceeds are used to buy high-yield notes that constitute as collateral. The collateral is the bank-originated credit market assets, for instance, home mortgages, credit card receivables, or auto loans, or a portfolio of high-yield corporate bonds. The SPV creates a set of securities based on pooling together many similar assets serving as collateral, whose aggregate income provides returns on the issued security. Therefore, the pool of collateral is the SPV's assets, while the securities issued, which form the CDO, are the SPV's liabilities. Finally, once the securities are sold to outside investors, the corresponding asset-liabilities are taken off the bank's books. However, if the securities are not fully subscribed/sold, the bank must buy them back.

The objectives and benefits of securitization are plenty. By securitization, both commercial and retail banks can generate liquidity, since banks obtain a principal payment up front. This can be lent out again to other customers, thus putting the bank's assets to work more efficiently. If the CDO issue is well-subscribed, banks can substantially shift the credit risk of their loan portfolio to the investors, and therefore through this process reduce the economic risk and the economic capital. However, in cases where only a small portion of the economic risk of the portfolio is transferred to reduce the amount of risk capital regulators require banks to set aside, this is referred to as regulatory arbitrage.

Financial engineering supporting securitization must develop an estimation of default risk. As stated earlier, in multi-name credit derivatives, default correlation becomes very important. Therefore, the approach to modeling single-entity credit risk, such as by the contingent claims approach or reduced-form approach, must be extended to incorporate the interdependence in credit risk of obligors. Correlation of defaults and loss given default or recovery rate for the pooled assets must be estimated. Based on these models for portfolio-level assessment of credit risk and its impact on portfolio loss in an extended period of time supports the design of CDO tranches.

A definition of each tranche must define a lower and an upper attachment point for each tranche. An attachment point is an indicator of the amount of losses in the collateral pool before it hits the tranche; and detachment point is the level of losses experienced by the collateral pool which entirely erases or wipes-out the tranche. Therefore, attachment and detachment points of each tranche must be determined based on the default risk, both in terms of probability of default (PD) and loss given default (LGD), of each instrument in the collateral pool. Moreover, the interdependence of these credit risk characteristics, as well as the impact of economic or business cycles on the collateral, must be considered in the portfolio-level credit risk assessment.

The attachment and detachment point of each tranche determines the risk underlying the securities of that tranche. Accordingly, the risk premium in terms of yields or spread must be determined for the tranche, corresponding to the loss probabilities. The spreads are often stated with a reference interest rate, such as LIBOR or Treasury rates. Rating agencies must then grant ratings to each tranche depending on the quality of collateral, guarantees, tranche definitions and yield spread.

Securitization has several advantages, and for the successful and effective use of this mechanism for credit risk management attention must be paid to some issues. Securitization provides a significant motivation for efforts in improving consumer and commercial credit risk modeling, in order for a more robust risk transfer by banks using securitization. The upside of increasing liquidity of credit risk is washed away if credit is transferred to a broader segment of the economy based on poor understanding of the underlying risk, poor design of instruments or incentives, or lack of transparency in the issuance of securities through securitization.

Portfolio credit risk is in itself a complex entity, on top of that designing increasingly complex securitization based products poses an added danger. It is important that a bank recognizes the true economic value of any credit-risk cash flow that it retains from securitization. The residual assets can be sizable in nominal value but very concentrated in their riskiness, which can result in insolvency. Moreover, investors participating in this market should recognize the nature and riskiness of securities they expose themselves to, and use the information available to make sound investment decisions.

9.4 Portfolio Credit Risk Management

In this chapter so far, we have considered several aspects of credit risk and credit risk management. In this section, we summarize the considerations for credit risk management at the portfolio level. The first distinction we had made based on fundamental differences in risk characteristics was between retail and commercial credit risk. The risk management strategy must be developed for a credit risk portfolio based on this distinction, since the individual obligor risk characteristic lend the portfolio definite and distinct risk characteristics. As discussed in Section 9.1, retail credit risk is bite-sized, with limited financial interdependency between obligors. However, we also noted that consumer behavior changes with time, and also depends on the economic conditions. In the case of commercial credit risk, single exposure is not bite-sized, therefore default of a single obligor can threaten a bank's solvency. There is a significantly high degree of direct or indirect financial interdependence between commercial obligors, by business relationships, particular geography or industry, and sectors. Moreover, economic conditions exasperate the default frequency in commercial credit risk, as well as seem to induce stronger correlation.

The mitigate response of risk management is an effective response for both retail and commercial credit. Given the fundamental role of banking, both commercial and retail, banks must stay in the business of offering credit to individuals and firms. However, improvements towards better and updated models, based on better information about their counterparties, are a key to risk mitigation decisions in retail and commercial credit risk. Based on these models, banks can construct their strategy of preemptive efforts for selecting customers and determining their market share. The goal of developing business is to diversify obligors by industry, geography, and time.

Each of the credit risk modeling approaches developed in previous sections - credit migration, structural models, reduced-form models, among others - can be utilized for developing a risk mitigation assessment. However, once risk characteristics of a portfolio are captured in models, the models may be utilized for guiding risk mitigation, transfer, as well as for the keep response of portfolio credit risk management. The portfolio-level view of default risk, probability and number of defaults in a window of time, loss experience due to defaults, beyond similar assessment of individual obligors, requires models for correlation between credit risk of obligors. In a portfolio of N debt instruments, where N could be quite large, assessing every pair-wise correlation could be prohibitively large. This challenge is often surmounted by utilizing factor models, where a relevant, few significant factors, and their mutual correlation, is used to describe the default correlation between N obligors in the portfolio.

The credit migration, reduced-form or structural models are advanced for portfolio credit risk assessment by adopting this fundamental principle for incorporating default correlation between obligors, which may then be utilized for various risk management considerations. In credit migration methodology, this correlation in credit migration, and hence occurrence of default, is captured via asset return of the obligor. For instance, equity return of the N obligors may be modeled using correlated N-dimensional normal distribution, and equity return distribution of each obligor is mapped onto migration probability distribution for each obligor. Thereafter, correlated credit migration is obtained by simulating the N-dimensional equity returns of the obligors. Equity returns for N obligors, in turn, may be modeled in terms of K factors capturing systematic risk, industry risk, sector risk, country risk, and global economic risk affecting the obligor.

In the structural model or reduced-form model, correlation can be incorporated by making the driving stochastic processes for firm-value (in Eqn. (9.10)) and default-intensity (in Eqn. (9.16)), respectively, to be correlated. Here again, to maintain tractability the pair-wise correlation between driving stochastic processes may be described in terms of a few key factors. Finally, the copula approach for capturing correlation described in Section 5.4.2 can also be applied here. This approach would be specifically applied to default time, τ_i, and corresponding survival function $p_i(.)$, of each obligor. For more details on modeling correlation in credit risk models, the reader may refer to Duffie and Singleton [21], and Ammann [5].

There are multiple mechanisms for transfer of credit risk, appropriate for different types and different levels of credit risk exposures. In the retail credit segment, there is an increasing effort to develop risk-based pricing, which prices products in response to the risk the customer exposes the bank to. This is feasible in the case of retail credit, since risk is bite-sized in this domain. When defaults occur in commercial credit, the losses can be quite large, making the unexpected loss dominate the overall credit risk exposure. This implies that transferring the risk by pricing it into debt contracts is not a feasible mechanism to manage the default risk in a commercial setting.

In Section 9.3, we discussed many single-name and multi-name credit derivatives, including securitization, which are effective tools for transferring portfolio credit risk. In a lumpy commercial credit portfolio, if there are a few key large exposures, these can be managed by traditional credit risk management approaches, or can be managed by hedge instruments like credit default swaps. If the portfolio has several large but similar sized exposures, with no single one standing out in its risk profile, a multi-name credit default swap may be a better suited risk transfer. The portfolio credit risk models discussed above provide the necessary information regarding the default risk characteristics of the portfolio. Depending on the depth of protection sought, first-to-default through n-th-to-default swaps can be utilized, up to the right level of $n(\leq N)$.

In the case of retail or commercial credit risk, when possible, securitization is an effective mechanism for risk transfer. This is provided the design of the securitization is done based on responsive and robust models and analysis. Again, the portfolio credit risk models are crucial to this design, as discussed in Section 9.3.2.1, for identifying the attachment-detachment points of tranches, as well as pricing each tranche for its credit risk. Unsubscribed securities of securitization must be accurately acknowledged as retained risk, with risk characteristics of the tranches the unsubscribed securities belong to. In general, the risk left-over after mitigation and transfer responses of credit risk management is retained risk. Any portfolio credit risk management strategy must conduct detailed assessment of characteristics of retained risk. The credit VaR framework developed in Section 9.2.1.1 is one such assessment. Credit VaR, as in the case of market risk, must be complemented with stress testing.

We discussed the different challenges to commercial credit risk, and its modeling, in Section 9.2.1.1. Given default rates and recovery rates vary with time, as well as display marked dependence on business cycles, stress testing should design stress shocks and scenarios to elicit the impact of these non-stationarities in credit risk. Defaults also tend to cluster around periods of economic slowdown, with the frequency of defaults increasing substantially during recessions. Moreover, all recession periods don't seem to have the same impact on default frequencies. The factors regarding an individual obligor's credit risk characteristics, their correlation with other obligors, and the effect of economic conditions, can also interact with one another. Interaction of concentration and economic downturn further enhances their individual impact. Economic downturns uncover hidden tendencies of obligors to default together, which is the clustering of default phenomena. All these features of retained portfolio credit risk must be evaluated through stress testing. Banks need to safeguard their solvency in light of retained risk through sufficient availability and management of economic capital, while also satisfying the requirements of regulatory capital. Finally, the portfolio credit risk that cannot be satisfactorily managed by the above responses should be avoided.

9.5   MATLAB Tools for Credit Risk

MATLAB mathematical software has a vast array of functions for working with financial variables and methodologies in its Financial Toolbox. We list a few of these functions here relevant for credit risk assessment. The reader is advised to look up the extensive help documentation available with MATLAB to see the details of these and other related functions. At the bottom of each function description in MATLAB help documentation, look for 'See Also' to explore other related functions. Resources such as MATLAB Primer [20] are also useful.

Credit migration: `transprob, transprobbytotal, portvrisk`

Credit scoring: `classify, classregtree`

9.6   Summary

Credit risk is one of the oldest of financial risks. It is an important risk for everyone, but of particular importance to some firms whose business model requires extending credit in some form or another to its customers, clients, suppliers, etc. Depending on the nature of the counter-party exposing a firm to credit risk, we separated the discussion of credit risk management by retail and commercial credit risk. In the initial part of the chapter, we developed modeling techniques for retail credit risk, where as for market risk, the decisions for risk management include selection, followed by management. From retail, we moved to commercial credit risk, and presented a discussion of currently used and newly developing models for commercial credit risk management. We followed the presentation of the credit risk models by their use in the valuation of a variety of credit risk derivatives, both single-name and multi-name. Finally, we looked at ways to manage the credit risk of a portfolio of credit risk-sensitive instruments utilizing the risk management framework developed in Chapter 2. After the credit crunch and the financial crisis of 2007-2008, credit risk and its management have taken a front-seat. We will discuss the current advances for portfolio credit risk management.

9.7 Questions and Exercises

Review Questions

1. What is credit risk? Despite being one of the oldest risks, why does credit risk continue to pose significant challenges for assessment and management?

2. What is counterparty risk?

3. What is the link between interest rate risk and credit risk? What is credit spread?

4. Besides charging a credit spread, how do creditors safeguard their interest regarding credit risk?

5. What is retail banking? Who does it serve, and how?

6. What is the variety of consumer credit products? What are micro-credit loans and payday loans?

7. How are retail credit risk characteristics different from commercial credit risk characteristics?

8. What are the three important statistical measures used to summarize credit risk?

9. What are characteristics and attributes of individuals or households used to model retail credit risk? Give 3 examples of each.

10. What is discriminant analysis? How is it useful in retail credit risk modeling?

11. What is credit scoring? How are the scores interpreted?

12. What is behavioral scoring? How are these scores different from credit scores, and how are they used?

13. What opportunities does improving credit scoring methodologies offer banks? Why is it imperative that the models are frequently tested and adjusted for currency?

14. How widespread is the usage of credit scores?

15. What is logistic regression, and how is it used to construct credit scores?

16. What are the contending approaches to developing credit scoring or retail credit prediction models?

17. What are false bads and false goods in applying credit scoring to accepting or rejecting a loan application? How does it compare to Type I and Type II error of statistical inference?

18. How does a cut-off score determine the loss rate and profitability of a retail credit product?

19. What are the methods and challenges in estimating loss given default in retail credit?

20. What are the three ways a bank can approach the effort of building credit scoring models? What are the advantages and disadvantages of each?

21. What is Cumulative Accuracy Profile? What is it used for?

22. What is the Accuracy Ratio (AR) statistic? How does it help monitor the accuracy of a credit scoring model?

23. What are the reasons for the credit risk of individuals to change with time?

24. How can economic factors affect individual credit risk? How can this relationship be incorporated in retail credit risk models?

25. What challenges do credit scoring methodologies face going forward?

26. What is commercial credit? Who are the typical obligors in commercial credit? Give examples of commercial credit instruments.

27. What is a credit rating system? What are rating agencies? Give some examples.

28. What information is incorporated in assessing creditworthiness in a credit rating system?

29. How are credit ratings used for credit risk management?

30. Why are credit ratings not meant to be an investment recommendation?

31. What is the difference between issuer versus issue-specific credit rating?

32. What are the rating levels in S&P and Moody's rating systems? What do these levels mean?

33. What are investment grade and speculative grade ratings?

34. What is a discrete-time Markov chain? What is Markov property?

35. What is an absorbing state of a Markov chain?

36. When is a Markov chain stationary? How can this property of debt migration be utilized for multi-year credit risk assessment?

37. What is credit VaR? How is credit VaR of a portfolio of loans computed?

38. Empirically how are default rates and recovery rates related through economic cycles?

39. What are the important factors in assessing credit risk using credit migration methodology?

40. What is Altman's Z-score? How does this model predict default risk?

41. What is a structural model of credit risk? How is Altman's Z-score model an example, albeit a simple one, of a structural model?

42. What is a reduced-form model for credit risk? How are these models calibrated and used?

43. What is doubly stochastic default intensity reduced-form model?

44. What is a credit default swap? What is CDS spread?

45. What is amortizing default swap and a binary credit default swap?

46. Describe how the CDS spread is computed.

47. What is a spread option? How is the pay-off of the option stated?

48. What is first-to-default swap? How does the first-to-default spread compare with the spread of the worst individual credit in the portfolio? Explain.

49. What is n^{th}-to-default swap? How are these swaps useful for credit risk management of a portfolio of credits?

50. What is securitization? What are MBSs and ABSs?

51. How does an MBS differ from a CMO? What are CDOs?

52. What is a synthetic CDO?

53. What are the benefits of securitization? What are the cautions or disadvantages of securitization?

54. What are the financial engineering tasks behind securitization?

55. What are the mitigation responses for portfolio credit risk?

56. How is portfolio credit risk managed by transfer of risk?

57. What should be the emphasis of stress testing for a portfolio of credit risk?

58. What is the purpose of Credit VaR framework for portfolio credit risk?

Exercises

1. The 100,000 customers in a retail credit portfolio have credit scores given by the following distribution, $270 * Beta(1.2, 1.5) + 580$. Therefore the cutoff score used for this product is 580. The portfolio experiences a total of 5% default. The default experience by credit score in this portfolio is described by the following distribution, $\chi^2(5) + 580$. What is the distribution of false goods? Extrapolate by an appropriate distribution to determine the profile of false bads.

2. Construct the cumulative accuracy profile curve and Accuracy Ratio (AR) statistic for the retail credit portfolio in Problem 1. Is this admissible by your chosen accuracy requirement?

3. Consider the following portfolio of US Treasury notes and bonds with weights for each bond given as number of bonds held in the portfolio.

   - Corporate Bond, Issuer: Johnson & Johnson; Coupon Rate: 5.55% (paid semi-annually); Maturity: 5 years; Rating: AAA
   - Corporate Bond, Issuer: Southwest Airlines; Coupon Rate: 7.375% (paid semi-annually); Maturity: 15 years; Rating: BBB
   - Corporate Bond, Issuer: Spring Nextel Corp; Coupon Rate: 9.25% (paid semi-annually); Maturity: 10 years; Rating: B
   - Corporate Bond, Issuer: Royal Caribbean Cruises Ltd; Coupon Rate: 7.25% (paid semi-annually); Maturity: 6 years; Rating: BB

   Compute the annual Credit Value-at-Risk in MATLAB at a desired confidence level for the following portfolios given by the number of bonds held.

   (a) $\vec{w} = [10,000;\ 0;\ 0;\ 0]$
   (b) $\vec{w} = [30,000;\ 8,000;\ 0;\ 0]$
   (c) $\vec{w} = [10,000;\ 22,000;\ 15,000;\ 20,000]$

   Assuming stationarity of transition matrix, compute 2-year and 3-year Credit VaR for all the above portfolios.

4. Compute the Z-score of a firm picked from the following four sectors, and comment on the (relative) default likelihood of these firms.

   (a) A utility company

   (b) A pharmaceutical company

   (c) Consumer goods

   (d) Airline company

5. For a company with a total debt, D_T, of $10 million, all of which matures in 1.5 years, what is the probability of default, if the volatility of firm value, $\sigma_v = 17\%$ and current firm value, $V_0 = \$25$ million. Consider the short-term risk-free interest rate to be 2.3%.

6. For the firms you selected for Problem 4, under what assumptions are you able to apply the structural approach to estimate the probability of default in a chosen period of time?

7. In Problem 5, if the firm has an annual return on asset of 18% and an annual claimholder payout of 10%, what is the distance to default for a time-period of 1.5 years and expected recovery rate, given default?

Chapter 10

Strategic, Business, and Operational Risk Management

The specific risk types we have studied thus far most prominently figure in financial markets and institutions. In this chapter, we turn our attention to risks that appear more broadly and are relevant for a broader variety of firms. Firms engaged in producing a variety of products and services must evaluate forward-looking prospects at different timescales, and assess the impact of different risk exposures on the firm's financial health and profitability. Based on the extent of forward view the firm adopts for this risk assessment, we label the risk management effort as strategic, business, and operational. Figure 10.1 summarizes the relative temporal scope of strategic, business, and operational risk management.

In this chapter, we address several issues of strategic, business, and operational risk management from the broader corporate risk management perspective. The risk management objectives are developed on specific themes for these risk types in order to provide an overview of an otherwise rather vast topic. For non-financial firms, a framework developed for integrally, comprehensively and consistently managing all relevant risks of the firm is called enterprise risk management. We also utilize the temporal range of strategic, business, and operational risks as a context to present asset-liability management objectives of financial services firms in this chapter.

10.1 Strategic Risk Management

Even if risk management may not be an explicitly stated activity in a non-financial corporation, management of risks happens in various garbs. In general, non-financial firms tend to focus more on risk mitigation, as opposed to a financial institution's approach of risk optimization, i.e., constructing optimal risk-return trade-off. Focusing on its core business, a typical non-financial firm has greatest incentives to conduct its business with the least negative impact of risks. This has also to do with these firms' inability to swiftly shift the markets they function in due to potentially high capital investment required for their products and services.

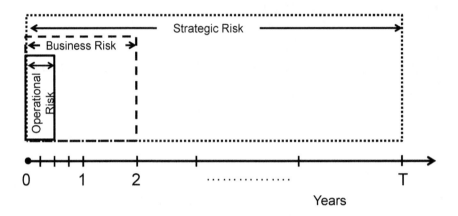

FIGURE 10.1: Overview of strategic, business, and operational risk for their temporal scope.

Theoretical justification for explicit risk management effort of firms goes back to the results developed by Nobel laureates Franco Modigliani and Merton Miller in the 1950s. Irrelevance of corporate financing and risk management decisions arises from the basic notion that in an efficient market, which offers investors a variety of investment opportunities, the investors can perform risk management on their own rather than relying on a firm's managers. The assumptions behind irrelevance of risk management extend the results to several other types of irrelevance - irrelevance of choice of capital structure, debt and leverage, security holder indifference, and hedging and insurance irrelevance. In reality, the stringent assumptions behind these results don't hold. Therefore, while these results provide valuable theoretical insight to these issues, in practice managers actively make these financial management decisions for a firm.

Firms devote significant attention to increasing expected cash flows from their projects, or reduce the variability of these cash flows, with an eye towards reducing the cost of capital by managing risk. Managing risk helps reduce the conflict of interest between security holders and managers, as it aligns the agent's (manager's) incentives with those of principal (owners) of the firm. Risk management of the firm's risks also reduces conflict among security holders by minimizing the possibility of risk shifting between debt-holders and shareholders. Another side benefit of active risk management is the opportunity it offers for controlling and exploiting informational asymmetries. Risk management undertaken by the managers can function as a substitute for monitoring, for instance by increasing the signal to noise ratio behind managerial risk management decisions in presence of informational asymmetry. While active risk management may be a valuable activity for a firm to devote resources to, half-baked efforts can back-fire. There are numerous instances of strategic risk management failures one can cite.

10.1.1 Objective of Strategic Risk Management

Capital allocation and budgeting is a fundamental activity undertaken in the financial management of a firm, where strategic decisions about a firm's investments are made. Risk management must be an integral part of this activity, whether this is explicitly acknowledged or not. The question one must address is what objective(s) should the manager serve while making these risk management decisions, which indeed are a subset of broader financial management decisions. Managers make these decisions on behalf of the security holders of the firm, therefore normatively speaking, managers should maximize the security-holder welfare through risk management decisions.

If E_t is the market value of outstanding shares of equity at a time, t, and D_t is the market value of debt outstanding at the time, then the value of the firm at time t is $V_t = E_t + D_t$. If the firm pays δ_t dividend and i_t interest to shareholders and debtholders, respectively, earns a net cash flow of C_t and makes investment of I_t in new projects in the next period, then the firm's cash flow in the period is, $\delta_t + i_t = C_t - I_t$. This is assuming that the number of securities outstanding remains constant for this period, where appropriate adjustments can be made when there is a change in number of securities outstanding. Combining the two relations, we obtain the total wealth of the security holders at the end of the period in terms of value of the firm, net cash flow and new investments as follows.

$$V_t + C_t - I_t = (E_t + \delta_t) + (D_t + i_t). \tag{10.1}$$

Therefore, the security holder welfare is maximized by maximizing $V_t + C_t - I_t$. However, net cash flow, C_t, is obtained based on prior investment decisions, therefore the firm should maximize, $V_t - I_t$, which is the excess of the firm's value over the investment expenditures needed in the period to generate that value. One may further note that value of the firm, V_t, can also be expressed as,

$$V_t = \sum_{i=1}^{\infty} \frac{E_t[C_{t+i} - I_{t+i}]}{(1 + \rho)^i}, \tag{10.2}$$

where ρ is the cost of capital of the firm. Therefore, maximization of the value of the firm is equivalent to maximizing the expected discounted value of all future net cash flows and new investments of the firm. For increasing firm value, the manager can either increase the expected future net cash flow or decrease the cost of capital of the firm through risk management decisions.

Increasing the firm's expected future net cash flow can be pursued strategically for the entire firm, or for each future project of the firm. The latter forms the core of capital allocation and budgeting decisions. One may assume that projects refer to cost reduction efforts, launch of new product or services, expansion to new markets, however from the risk management perspective, the project could also be how to change the impact of corporate tax-structure on the net cash flow of the firm.

10.1.2   Approaches for Strategic Risk Management

In capital budgeting, the 'avoid' or 'keep' responses from the avoid-mitigate-transfer-keep decisions of risk management are well-represented. When a project's risks are seen as unfavorable, it is avoided, while when it seems to fit within the manager's risk appetite for its risk-reward offering, the project may be adopted. The usual mitigate decision of portfolio optimization, by taking advantage of correlation between asset returns, can also be applied here, however additional constraints and limitations must be accounted for here. Negative correlation between future cash flows of two or more projects can help increase the expected future net cash flow of the firm. Such opportunities must be cost-effectively exploited whenever possible.

One of the major sources of destruction of a firm's value is when the firm falls on financial distress. Pure risk, by definition occurrence of which results in losses without any gains, can result in significant financial distress to a firm, pushing the firm in many cases to the brink of bankruptcy. Direct and indirect costs of bankruptcy, or even being at the brink of bankruptcy, can eat away a firm's value. In the case of pure risk, mitigation must be taken on the risk-reduction interpretation, and a firm must strategically identify its pure risks to construct a risk reduction strategy. Risk reduction or control lowers the loss due to pure risk events, and in doing so, improves the expected future net cash flow from a firm's projects. Risk management of pure risk is discussed at greater length in Chapter 11, specifically in Section 11.4.

Insurance contracts play a significant role for the transfer of pure risk. An insurance contract, much like an option contract utilized for risk transfer in case of speculative risk, is a state-contingent claim which pays off beyond an agreed-upon level, called the deductible, in certain states of the world. As for option contracts, in an insurance contract the firm seeking risk transfer must pay a premium to the protection seller, or insurer. Despite these similarities, the terminology used in insurance and options differ significantly, and most books on one topic do not discuss the other at all.

As will be discussed at length in Chapter 11, not all pure risks are insurable, or affordably insurable, by a firm. Additionally, a firm can explore alternative risk transfer (ART) tools as a range of non-traditional transfer of risks. These alternatives include comprehensive risk policies or 'total risk' policies, multi-trigger policies, self-insured retentions, risk retention groups, liquidity provision facilities in presence of pure risk events, securitization, and weather derivatives, to name a few. These non-traditional risk transfer products have become possible and popular due to a closer integration of insurance companies and capital markets. For more details on ARTs, the reader should refer to Culp [19] and Lam [54].

Risk mitigation or control for a firm's future projects also has a temporal or a sequential component. This must be a crucial aspect of project evaluation and implementation, and in fact of corporate strategic planning and capital budgeting. The real options framework is an important valuation framework that responds to the sequential opportunities in projects, where decisions are made regarding an investment strategy of a firm depending on how project risks unfold. The framework can incorporate sequential responses to the risks, including decisions of timing of investment, timely scaling up or scaling down of a project, and exiting a project.

Real options can essentially be seen as bundling of standard options on a project's or firm's assets, however these options are not tradable. Therefore, valuation of real options must be done using cost of capital in the net present value framework. We discuss several types of real options embedded in sequential assessment of projects.

Waiting to Invest: When the risk about the project is not so much about the future cash flows the investment on the project will generate, but instead about the risk in interest rate that will be used for discounting future cash flows. In light of this risk, this option considers delaying investment in capital-intensive projects.

Option to Defer: This option is a generalization of the 'waiting-to-invest' option, where riskiness about many more aspects of a project make it worthwhile to consider delaying implementation of the project. Investment cost of the project, time until the opportunity remains, improves or disappears, or other relevant market and credit risk factors affecting the project need to be included in this analysis. The option of acquiring or undertaking the project at a future time resembles exercising a call option. Such options often arise in the context of extracting natural resources, or real estate development.

Option to Abandon: This is the reverse of the 'option to defer' real option, where a decision regarding termination of all production and operations must be taken, along with selling the current assets utilized for production. Option to abandon is common in capital-intensive industries like transportation, telecommunication, and resembles a put option on the value of continuing operations.

Time to Build: This is the option of staged investment coupled with an option to abandon the investment project as new information is revealed. This staged structure resembles a compound option. This option is particularly valuable when information is released gradually, such as in R&D for pharmaceutical drug development, venture capital financing of young enterprises, or large-scale construction projects.

Option to Alter Operating Scale: This is a less extreme response to changing demand than abandoning the project, by deciding to contract, expand or temporarily shut down and restart production. This should be seen as a relevant response in fashion and fad sensitive industries, such as food, entertainment, and fashion apparel.

Switching Option: This is the option of switching inputs or outputs in the production process depending on cost, price or other risk factors. For instance, fuels for production may be switched, or fuels to generate electricity may be switched. On the output, sales contracts can be constructed to allow switching a model from a class of models, as is interestingly done in the airline industry.

Option for Interactive Growth: Staged investment in a project can result in opening opportunities for growth in other areas and projects. Merger and acquisition is a typical example of this kind of option, since an M&A can result in many other projects benefitting from consolidation and growth.

In evaluation of these real options, and in general for project evaluation, simulation analysis can play a significant role. Discrete-time, discrete-space to continuous-time, continuous-space models developed in Chapter 5 can be applied as seen appropriate for the cash flows of a project to determine the value of the project, or to develop the real options strategy.

Firms may seek transfer of risk even for their speculative risk exposures, particularly in presence of deviations from the assumptions underlying Modigliani and Miller propositions. Firms most significantly have exposure to speculative risks from fluctuations in the price of their raw materials or finished products. Depending on the firm's financing structure, as well as the firm's scope of activities, there might also be interest rate, exchange rates and equity risk exposures. In Sections 7.3.2, 8.2.3, and 8.3.2, we had developed an optimal hedging framework and its application to various market risk types, which can be applied here. In these sections, we had also discussed various types of derivative contracts available for risk management.

For the raw materials or finished goods, firms may utilize spot market, futures market or maintain high levels of inventories to support their production and sales activities. Firms often favor storage of physical commodities and maintaining inventory of raw material or finished goods. This can be explained by some important motivations. First, this may be driven by the need to make sure there is no disruption in the firm's production process or fulfilling demand for finished goods. Some firms may engage in precautionary storing to protect from unanticipated shocks in demand or supply, or solely for speculative reasons in anticipation of price increase in the future. In some cases, firms may store commodities as a passive response to avoid transactional cost of getting rid of it.

There is a crucial relationship between spot prices and futures prices of commodities, captured by *convenience yield* discussed in Section 8.3.2.1. When the demand for immediacy is high, inventories are high, therefore the relative premium a commodity can command in the future relative to the present is reasonably small. In this case the convenience yield is small and the term-structure of futures prices is upward sloping. We call this a contango market. However, when inventories start to deplete, the spot price starts to rise as the demand for the commodity increases. Spot price being higher relative to the futures price of the commodity is an indication that inventories are tight today relative to the future. A resulting downward sloping term structure of futures prices is called a backwardation, depicted by a high convenience yield. In some cases, the term structure of future prices can show both contango and backwardation, indicating that after a time-point in the future, the market expects the inventories to run out. Therefore, convenience yield and its volatility is a strong indicator of inventory levels of a commodity in the market, and relative levels of spot and futures prices of the commodity. This insight also helps us understand how demand or supply shocks to a commodity will move the term structure. A firm's hedging strategy should take into account these issue in light of the firm's specific operations needs for the commodities.

Earlier in this discussion, we had concluded that risk management should attempt to increase firm value by either increasing the expected future net cash flow or decreasing the cost of capital of the firm. Moreover, active risk management can give managers opportunities for controlling and exploiting informational asymmetries, and in the process better manage the conflict of interest between security holders. One such strategic risk control available to managers is by utilizing structured liabilities or debt. Structured debt instruments are defined as debt instruments whose cash flow is linked to the value of some underlying asset, reference rate, or index. Structured liabilities can help the issuer manage its credit risk, decrease cost of capital, as well as increase its debt capacity. Credit rating sensitive notes, putable bonds or convertibles are examples of such structured liabilities. When the manager believes the firm's prospects are going to significantly improve in the future due to its latent or intangible assets, such as R&D, intellectual property, patents, but the capital market has not caught up to this, structured liabilities can be used for exploiting this informational asymmetry.

The avoid-mitigate-transfer-keep responses form the pillars of management of projects', and therefore the firm's, strategic risks. Unfortunately, many firms have forgotten that an effective overall corporate strategy combines a set of activities a firm plans to undertake with an adequate assessment of the risks included in those activities. In other words, there can be no real strategic management without risk management, and risk management needs to be interwoven into all aspects of the firm's business. Strategic decisions about what activities to undertake should not be made unless senior management understands the risks involved. In fact, failures of strategic management arise from organizational gaps in understanding, developing and implementing strategic risk management. Firms can, in fact, select to approach strategic risk management as a total or selective risk management effort depending on how much resources they can afford for this activity.

In summary, while strategic risk management points the direction, business risk management must translate it into guidelines for conducting the business of the firm. Organizational structure must be developed and the culture inculcated that facilitates this translation in a seamless way, where the vision, limits and boundaries set forth by strategic risk management decisions guide the business to be run profitably.

10.2   Business Risk Management

Key revenue generating activities of a medium to large business enterprise are usually organized by strategic business units by geography, customer group, product, or some combination of all these in a matrix structure. These business or line units account for the majority of assets and employees, and also are the primary source of business, market, credit and operational risk. The strategic risk management guidelines developed in the previous section must be translated to each business unit. In the end, the business or line units have the closest contacts with customers and suppliers, therefore can provide unique perspectives on what is working and what is not and generate feedback on whether the strategy is effective and complete. The business units must also develop tactical risk management responses for any gaps in the strategic risk management guidelines. Therefore, failing to connect strategic risk management to proper running of the business can cause significant loss in firm value, numerous lost opportunities, and in some cases, a failed business.

A few different business models are utilized for translating strategic risk management guidelines for running a business. The first is that of a classical risk controller approach, where resources are allocated for risk management exclusively to avoid losses in excess of strategic risk tolerance. In this approach, risk management is not seen as a source of opportunity. Therefore for assuring compliance, significant effort is put into risk measures and procedures reviews, evaluation of effectiveness of hedging strategies, credit risk management and monitoring, and cash management. The second business model is that of firms who have evolved beyond the defensive view of risk management, and look to risk management as an opportunity for overall efficiency enhancement in the firm. These firms are still internally bound, creating their own solutions for proactively connecting strategic and business risk management. The third most evolved case is of firms that don't only see risk management as an opportunity for improving efficiency and profitability of the firm, but also provide products and solutions that make risk management standard for their industry. Whichever business model a firm may operate by for its risk management function, incentive alignment for eliminating any conflicts between a business unit's focus on revenue maximization and risk limits and policies provided for strategic risk management must be in place for a lasting positive impact.

As stated earlier, the business risk management strategy is obtained by translating strategic risk management responses as relevant for different business units of a firm, and developing them beyond what is relevant for the business unit. The fundamental risk of running a business is demand risk, the risk of how low or high demand for the products and services of the business unit will end up being in the relatively near future, say in the next quarters or year. Demand being much higher or lower than expected, both create significant challenges for the business unit in attempting to adjust its production and resource management to cater to the demand. Significant development in data analytics is driven by the need to better predict demand. Demand for different products and services may show cyclicality, daily, weather or seasonal dependency, or change structurally over time. Diversification in product and service offering can help mitigate some demand risk. Pricing is an effective tool for responding to demand risk, where increasingly the emphasis of business intelligence and analytics is on dynamic pricing based on customers' characteristics, behavior and propensity to pay. Finally, businesses can develop contracts for their products and services that are loyalty enhancing.

Integrally connected with demand risk is the risk of competition taking away part or all of demand for the business unit's products and services. Part of business risk management, including at the strategic level, must address how a firm will manage competition risk. Moreover, a firm that manages its demand risk well, along with effectively translating its strategic risk management guidelines to how business units are run, stands a good chance of managing competition risk well. Strategic guidelines should render a good mechanism to manage risk of quality of products and services delivered to the customer. Systems must be in place for robust customer relations management, and customer feedback should be systematically translated in improving business unit operations as well as in strategic planning.

A business unit is also exposed to the typical market risks, such as commodity risk and exchange rate risk, depending on the type of raw materials and finished products the firm and the business unit is engaged in, as well as in the global markets it is active in. Strategic guidelines for managing these risks must be applied while filling in the specifics where needed. The usual approaches of applying risk management for these market risk components discussed in Chapters 7 and 8 would be applicable here. In cases where a business unit offers its products on credit, interest rate risk and credit risk also become integrally relevant to the business unit's risks. The market risk analytics for interest rates, as developed in Chapter 8, and credit risk analytics for business risk decisions, as relevant for retail and commercial clients developed in Chapter 9, should be applied.

The pure risk exposures of a business unit is best understood at the business unit, since here is where the ground reality is visible to the management. The strategic guideline obtained for management of pure risk must be evaluated and supplemented at the business unit level as seen appropriate. Insurance contracts may have to be evaluated for their relevance with changing environments and characteristics of the business unit. The risk reduction methods will need to be tested and reviewed for their efficacy. Programs that promote risk monitoring, reduction, and control must be kept current. Beyond the traditional pure risk transfer mechanism discussed thus far, for business risk management, as in the case of strategic risk management, application of alternative risk transfer (ART) tools can also be utilized by a business unit. Finally, risk capital corresponding to the retained pure risk of the business unit will need to be periodically determined and reported.

Firms are constantly impacted by changing regulations, government policies, as well as industry trends. Technology risk or other innovation risk also can't be ignored in business risk management. At the strategic level, real options framework is a powerful method for evaluating the impact of these risks on the firm's interests. At the business unit level also real options framework can be applied with higher granularity of detail relevant to the risk for the business unit to guide in generating fine-tuned risk management responses.

The purely financial component of business risk lies in managing cash flows of the firm or business unit so that the liabilities of the business unit are well supported. Identifying, measuring, and monitoring liquidity risk are important components of business risk management. In the next section, we will develop an asset-liability management framework, which integrally addresses liquidity risk management issues primarily from a financial institution perspective. However, this framework may also be applied in non-financial corporations to develop a strategic assessment of liquidity risk, which provides a sound basis for liquidity risk management at the business unit level. As discussed in the previous section, sound strategic risk management can help enhance the borrowing capacity of a firm, which in turn helps manage liquidity risk. Along with improved debt capacity, tactical risk control for liquidity shocks can be done through actual and synthetic asset divestiture, such as securitization.

Strategic uncertainties such as business plan assumptions, competitor responses, and technology changes should be measured and managed through robust business risk management. Even a company with a well-thought-out strategy must establish feedback mechanisms and contingency plans to ensure that the company's strategy is sound over time. Companies with unbending strategies can face extinction. Therefore, it is fair to say that a nimble business is a successful business. Finally, business risk management needs to connect management of business risks with the business unit's and, at a comprehensive level, the firm's operational risk; these are the disruptions to 'business as usual' for the firm or business unit. Before we delve into operational risk management, we next present a framework for interlacing strategic-business-operational risk into asset-liability management for a financial institution.

10.3 Asset-Liability Management

Asset-liability management (ALM) is balance-sheet risk management. Described as such, it constitutes stitching together strategic, business and operational risk management to insure short-term viability and long-term profitability of an enterprise. Asset-liability management is most practically and integrally important for firms whose core role is financial intermediation, such as financial services firms, thrifts, banks, financial subsidiaries, insurance companies, leasing companies, and pension funds. Some non-financial firms, such as nimble manufacturing firms and trading firms, that must actively manage their balance sheet for the purpose of running their business, can also benefit from the principles of asset-liability management.

In the late 1970s, the focus of monetary policy switched from stabilizing interest rates to controlling monetary aggregates, thus interest rates became more volatile. The sudden sharp rise in interest rates between 1980 and 1982 precipitated many banks and thrifts to become insolvent, and they ultimately had to be liquidated. This triggered an enhanced attention to asset-liability management. In this backdrop, ALM was considered an important risk management activity because the traditional business model of financial institutions, such as commercial banks, was to finance long-term loans with relatively short-term liabilities. This position of significance for ALM faded away in the more stable 1990s, when the practical importance of ALM and liquidity management took a back-seat yielding to other short-term priorities of the firms.

The financial crisis of 2008 triggered a renewed interest and greater emphasis on asset-liability management. Through these crisis years, even the largest of multinational financial institutions were significantly stressed by liquidity risk and needed to seek external help to ensure survival. Management of large institutions, regulators, and even the general public witnessed how well-reputed and trusted institutions folded up and were not able to adequately respond to the deep liquidity crisis. Thereafter regulators have attached a higher importance to management of liquidity risk, which is a subset of ALM, by expressly bringing it into overall risk management frameworks and developing new measures to ensure sound liquidity management.

As will be evident to the reader, asset-liability management is essential for the seamless and profitable growth of a firm. Although the primary focus of ALM is managing balance sheet risks, it must also focus on balancing profitability with managing risks. In the process, objectives of asset-liability management can proactively seek to guard the firm's bottom line as well as maximize profitability. As stated above, liquidity risk management has emerged as an important function of asset-liability management. Liquidity risk, more specifically funding liquidity risk, is defined as the risk of not meeting the expected and unexpected immediate and future cash flows needs effectively. A well managed liquidity function must include a liquidity contingency plan, buffers of liquid assets, and liquidity policies and limits set at acceptable and manageable levels.

As an integration of strategic, business and operational risk to ascertain smooth functioning in the short-run and profitability in the long-run, asset-liability management constitutes activities of comprehensive and challenging scope. In Figure 10.2, we display the essential components of asset-liability

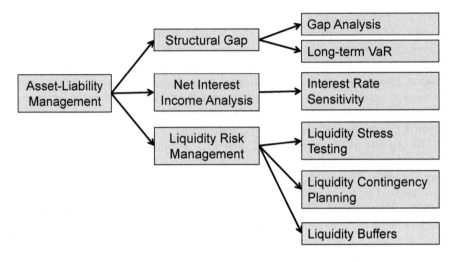

FIGURE 10.2: Elements of asset-liability management and its relation with liquidity risk management.

management, including liquidity risk management, structural gap risk and net interest income risk or net earnings risk. The figure also indicates some of the tools available to address each of the ALM tasks and objectives. We next examine these components and tools of ALM in detail.

10.3.1 Components of Asset-Liability Management

Financial institutions perform the essential intermediary function of creating and absorbing liquidity in the financial system. In other words, financial institutions facilitate the fundamental transformation of financial resources into maturity structures needed in the economy at a time. In the process, financial institutions undertake significant maturity mismatch risk, interest rate risk, credit risk and foreign exchange risk. The role of ALM, along with its components of structural gap risk, net income risk and liquidity risk management, is to help manage and mitigate the adverse impact of these risks, while maintaining them within accepted levels. In its comprehensive scope, while managing these risks ALM should help achieve profitability and an optimal allocation of capital.

The following are the components of asset-liability management and an overview of tools designed to achieve the above goals. The components clearly span the strategic, business and tactical range of financial institution's risk management considerations.

Managing Structural Gap: This component most squarely tackles the strategic issues in a firm's balance sheet. A mature ALM function of a financial institution actively and continuously monitors all the assets

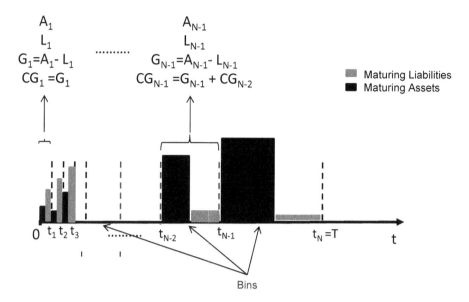

FIGURE 10.3: Gap structure organized as gap and cumulative gap in each bin through the planning horizon, T.

and liabilities on a firm's balance sheet. More specifically, it organizes this information in the form of the gap between asset and liabilities over different ranges of time horizons. The emphasis in this assessment is first on creating a summary view of the relative status of asset against liabilities, and then on balancing maturities and cash flows from both sides of the balance sheet.

The goal here is not to eliminate any maturity or cash flow mismatches, for these mismatches may well be the source of profitability of the firm. The objective is to identify the mismatches and manage them to keep them within the risk tolerance of the firm. Therefore, the tools for managing structural gap respond by strategically balancing the gaps by issuing timely guidelines to business units to focus on appropriate products, investments and strategies for modifying and managing the gap. Liquidity management also deals with cash flow mismatches emerging from assets and liabilities, but it is focused on the short term of less than a year.

Following are the main tools for managing structural gap.

Static Gap: This is an as-it-stands analysis that summarizes assets and liabilities maturing in short, medium and long term after segmenting the planning horizon into bins. The analysis seeks the difference in cash flows from asset and liabilities in each bin as per their maturity structure. The gap in each period points to the expected funding gap and excess funds available at different time-points. A gap structure is constructed in Figure 10.3

Cumulative Gap: As assets and liabilities mature through the planning horizon, the funding gaps and excess funds accumulate and cancel each other out. Cumulative gap, which is the accumulation of all prior periods' gaps, constructs an as-it-stands temporal view of these accumulations and cancelations. Therefore, cumulative gap shows the time-points when assets fully support liabilities, or brings out any weaknesses in this regard. Cumulative gap is also a static analysis.

Duration Gap: It is true that neither assets nor liabilities remain static, therefore static gap provides a snapshot, but fails to provide how the gaps may change with time. Therefore, as gaps dynamically evolve, they need continuous monitoring as the balance sheet changes. Modified duration, defined in Chapter 8, as a measure of interest rate sensitivity is used to assess the impact of interest rate risk on the gap. Moreover Macaulay's duration is a measure of effective maturity or center of gravity of discounted cash-flows of an asset and a liability. Therefore, duration gap measured as difference in Macaulay's duration of asset and liability is also informative. Beyond duration, for the second-order sensitivity, gap convexity can also be assessed as part of this analysis.

Long-Term Asset/Liability Ratio: A common financial ratio of comparing long-term assets, those maturing beyond a year, against liabilities of similarly long maturity serves as an indication of long-term balance sheet performance. As a risk management response, acceptable limits may be set to this ratio. Implementation of this acceptable limit combined with liquidity gap controls can help to eliminate any imbalances and help maintain a structurally sound balance sheet.

Dynamic Gap: Besides interest rate risk, there are other risks listed above that are relevant in their effect on assets and/or liabilities. Dynamic gap simulates the future gap positions as a consequence of these broader sets of risk factors, and in terms of more advanced, broader risk measures than interest rate sensitivities. In this dynamic analysis of impact of risks, risks like business volume risk, early repayment risk, deposit roll-overs risk, etc., can also be incorporated.

Managing Net Interest Risk: A financial institution relies on the above measures, namely gap, cumulative gap, duration gap, gap convexity, and other measures developed in dynamic gap analysis, to evaluate the impact of interest rate risk. This is the most significant risk for ALM. Based on the levels of these measures, appropriate strategic or tactical responses must be developed for modifying the undesirable effect of the risk. As stated earlier, other risk types can also be included in preparing a response towards the firm's asset-liability management goals. Financial institutions must also assess the impact of interest rate changes, new business models, change in product-types and mix, and roll-over of deposits on net interest income. Gradual changes or significant shocks to the net interest income must both be noted, and their impact on balance sheet of the firm should be determined.

Managing Liquidity Risk: Daily or monthly gaps on short-term asset-liabilities ladders ensures that cumulative gaps will operate within pre-set limits. Liquidity gaps must be adequately managed by optimizing borrowing capacity in the money-market, liquidity contingency plan, liquid asset buffers, and setting liquidity policies and limits in tune with level of risk considered acceptable and manageable in the firm, as well as per regulatory guidelines.

In the next section we develop the risk management framework applied to asset-liability management further by developing each of the above measures and responses for achieving the ALM objectives.

10.3.2   Risk Management in ALM

The scope of ALM is broad, and therefore to perform risk management within asset-liability management, a multi-tier strategy may be necessary. Each tier is designed to give a level of insight on the implication of assets and liabilities, their maturities, cash flow and risk exposures for the firm. We first consider static and stochastic gap analysis.

10.3.2.1   Gap Analysis

Gap is defined as the difference between the amounts of interest rate-sensitive (and/or other risks sensitive) assets (A) and interest rate-sensitive (and/or other risks sensitive) liabilities (L) maturing or repricing within a specific time period or bin (shown in Figure 10.3). Therefore, the primary objective of static or dynamic gap analysis is to measure interest rate risk (or other risks) as it affects assets and liabilities. For the rest of this presentation, we will focus on interest rate risk, although the framework and terminology can be extended to other relevant risks.

The measure of static gap risk is rather rudimentary in just considering the nominal value of assets and liabilities maturing during a time period. The assessment horizon $[0, T]$ is broken down into specific periods of time, $\{t_1, \ldots, t_N\}$, and gap is defined for each period as,

$$G(t_i) = A(t_i, t_{i+1}) - L(t_i, t_{i+1}), \tag{10.3}$$

where $G(t_i)$ (or G_i in Figure 10.3) is the gap for period (t_i, t_{i+1}), $A(t_i, t_{i+1})$

are the nominal value of assets maturing in that time period, while $L(t_i, t_{i+1})$ is the nominal value of liabilities maturing in the period. The gap of each period relates to interest rate risk by the fact that it is affected by the same segment of the term structure of interest rates. This is assuming asset and liabilities in a bin derive value from the same yield curve. When this is not true, addition factors, such as credit spread, must be incorporated to assess the gap risk for each bin.

Positive-gap in a specific period of time results from $A(t_i, t_{i+1}) > L(t_i, t_{i+1})$, and when there is positive gap in short-term bins, we say that assets reprice before liabilities. This implies that short-term assets are funded by long-term liabilities. An increase in short-term interest rates will therefore lead to an increase in net interest income (NII), while decrease in interest rates will lead to a decrease in NII. Similarly, negative-gap in a specific period of time arises from $A(t_i, t_{i+1}) < L(t_i, t_{i+1})$, and when there is negative gap in near-term bins, it is liabilities repricing before assets. This implies that long-term assets are funded by short-term liabilities, which is the case in Figure 10.3. Here an increase in interest rates leads to decrease in NII.

Issues to consider while developing a gap analysis include the time intervals, or bin widths, to consider for determining the gaps. Time slots defined of varying duration allows focusing the gap in the near-term more closely than that at distant-future. Therefore, we may construct the gap for up to one month, (1-3] months, (3-6] months, (6-12] months, (1-3] years, (3-5] years, and above 5 years, etc. This non-uniform bin-width is the approach taken in Figure 10.3.

Once gaps are defined, management of risk can be achieved by setting a gap-limit for specific time-slots. This is the maximum allowable difference between assets and liabilities within a specific bucket. For example, if short-term interest rates go up and assets reprice before liabilities in the gap structure, in the presence of non-parallel shift of the term structure interest rates, the asset values will fall and gap will narrow down. For the gap dropping below a gap limit, a response with an appropriate gap management strategy must be developed. Use of derivatives, such as interest-rate futures, swaps or swaptions, can help modify the interest rate sensitivities of assets or liabilities, thus changing characteristics of the gap risk without changing positions in the assets or liabilities.

The time intervals or bin-widths for gap analysis may appear arbitrary, and moreover there could be timing mismatch within a bin. Although a bin is idealized as one time point, within a bin, in reality an asset may reprice at the end-time of the bin, whereas a liability may reprice toward the beginning of the bin. Moreover, when we compare assets with liabilities in their maturity times in gap analysis, limitations may also appear from the fact that there may be lag in the timing of interest rate changes as they affect the assets versus liabilities maturing in a bin. These lags could be due to internal lag in making decision for increasing loan rates. Finally, many assets, for instance loans, may have embedded options, such as early repayment of mortgage, or liabilities, as early withdrawal from savings account. For including these complexities, the next tiers of ALM must be implemented. Gap analysis remains attractive for its simplicity as the first tier of ALM analysis.

10.3.2.2 Cumulative Gap Analysis

While the time intervals or bin-widths of gap analysis may appear arbitrary, this can be partially addressed by considering cumulative gaps. Cumulative gap aggregates the gap over several bins to consider the cumulative impact of asset-liability mismatches. Accumulation of gap over several bins makes each bin less relevant, while emphasizing how over time assets balance out with liabilities. Therefore, cumulative gap is defined as,

$$CG(t_i) = \sum_{k=0}^{i} (A(t_k, t_{k+1}) - L(t_k, t_{k+1})), \qquad (10.4)$$

for $i = 0$ to $i = N$, where preferably for the ALM horizon $CG(t_N) = 0$, since by the planning horizon, all positive and negative gaps should have canceled each other out. If, however, the cumulative gap remains negative in the long-run, this would not bode well for the firm's long-term viability. Appropriate response must be constructed in this case towards the firm's strategy for managing the cumulative gap.

Cumulative gap complements gap analysis, however it still misses capturing sensitivity of asset and liabilities to different risk factors discussed above.

10.3.2.3 Duration Gap Analysis and Gap Convexity

Gap and cumulative gap analysis are static as far as quantifiable impact of interest-rate risks or other risks go. It can give some qualitative idea of how the asset-liability mismatch will change if interest rates change, but falls short of becoming quantitative regarding this impact. As seen in Chapter 8, duration is a measure of interest rate sensitivity of any cash flow series. It summarizes cash flow characteristics taking into account both the size and timing of these cash flows. It does not hide cash flow timing mismatches within the maturity bins as gap or cumulative gap analysis can.

We find the market value and duration of the rate-sensitive assets, market value and duration of the rate-sensitive liabilities. The difference in duration of assets and liabilities in each bin of gap analysis clarifies what the actual timing mismatch exists between cash flows of assets and liabilities, and the extent of interest rate sensitivity this gap carries. Therefore, the duration gap is defined as,

$$DG(t_i) = D_A(t_i, t_{i+1}) - D_L(t_i, t_{i+1}), \qquad (10.5)$$

where $D_A(t_i, t_{i+1})$ is the duration of assets maturing in the period (t_i, t_{i+1}), whereas $D_L(t_i, t_{i+1})$ is the duration of liabilities maturing in the same period.

In Section 8.2.3 of Chapter 8, the duration risk measure was used to develop a hedging or risk transfer strategy for obtaining an immunized bond portfolio. Similarly here, duration and convexity can be used to develop a hedging strategy for gap risk in bins where the gap risk exceeds risk limits. Hedging could be done at the macro level, where one type of instrument is used for a basket of assets or liabilities, or both, or it can be at the micro level, where a specifically tailored hedge instrument is utilized for hedging the risk of each asset or liability. Clearly, the former is a coarse approach, while the latter is painstaking but allows significant customization to the actual hedging need for each type of risk. One can utilize bond futures, interest rate swaps, options, swaptions, or interest rate caps and floors, each with its own advantages, disadvantages, and hedge ratios, as hedge instruments.

Duration as a measure of risk has its shortcomings, the most important of which is that it only measures first-order impact of interest rate risk, and only from parallel shifts of the yield curve. Shortcomings of duration as a measure for immunizing against interest-rate risk carry over to duration gap analysis. Moreover, only when interest rate changes are small does it provide an accurate measure of risk. This latter issue can be addressed by similarly defining gap convexity, which summarizes second-order impact of parallel shift of the term structure on the gap. Our assumption throughout of applying the same rate change on assets and liabilities does not incorporate basis risk, i.e., the interest rates relevant to assets and liabilities may not be perfectly correlated. The shortcomings of the duration gap and gap convexity analysis can be eliminated in the third tier of dynamic gap analysis.

10.3.2.4 Dynamic Gap and Long-Term Value at Risk Analysis

ALM related forecasting can be pursued with two themes of analysis, deterministic versus stochastic analysis. The deterministic approach would clearly be a simpler setup designed to derive some basic intuition of how the balance sheet may look for the planning period. The deterministic forecast would be based on the risk manager making explicit assumptions about the interest rates and other risk factors' movements, and forecasts for different scenarios at different time-points in the planning period. These were the tiers of ALM discussed thus far.

In the stochastic modeling approach of dynamic gap analysis, various forecasting models would need to be developed, calibrated and solved, either analytically or by simulation methodologies, to assess the balance sheet characteristics at desired confidence levels over the planning horizon. The modeling framework additionally allows for simulating the impact of strategic decisions for asset or liability choices, of hedging strategies, and in determining what the gap at future time-points might look like. Simulation analysis for a firm's balance sheet also allows an assessment of future earnings risk, and the impact of risks on the firm's earnings can be summarized as earnings-at-risk at a certain confidence level.

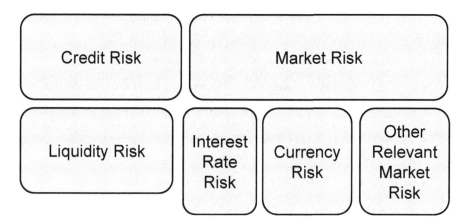

FIGURE 10.4: Risk types included in balance sheet risk management and asset-liability management.

In our progression towards dynamic gap approach, duration gap and gap convexity analysis offered improvements over static gap analysis, at least with regards to interest-rate sensitivity. However, both frameworks are somewhat limited, since they do not take into account the fact that the balance sheet evolves with time, where many more risk factors and issues beyond interest rate risk determine its health (see Figure 10.4). Inclusion of stochastic models for risk factors related with other market risks, liquidity risk, and credit risk shown in Figure 10.4 may also not be enough in some cases. Relevant risk factors underlying strategic, business, operational, regulatory, and legal risks may also need to be included for a comprehensive analysis. For instance, a financial institution may originate new retail products, adopt a new technology to deliver its products, and maturing assets and/or liabilities may be rolled over as they approach maturity, not necessarily into instruments with similar characteristics, or securitization may be adopted in some segments of assets of the firm. All these aspects should be incorporated in a comprehensive stochastic analysis in ALM.

Long-term Value at Risk (LT-VaR) is an extension of Value-at-Risk (VaR), where the time horizon is chosen to be much longer, at least one to several years depending on the planning horizon for ALM, much on the lines of Credit Value-at-Risk (Credit VaR). The objective of LT-VaR is to generate the statistical distribution of net worth of a firm at different time horizons, say next quarter, end of one or two years, in order to produce the worst-case net worth at a given confidence level, say 99%. As indicated in Figure 10.1, combining different horizons into a single analysis is why ALM is a combined strategic-business-and-operational risk management. Moreover, it is quite comprehensive in its risk types consideration.

When simulation is used for the purpose of estimating LT-VaR, the following information must be developed for the implementation.

- Correlated term structures of all relevant interest rates.

- Implied volatilities of various instruments in order to price derivatives.

- Stochastic models for evolution of all other relevant risk factors.

- Interest-rate sensitive prepayments of mortgages and other loans, as well as changes in deposits and savings balances.

- Loan default frequencies and loss given default distributions.

- Renewal/retention or new origination for retail products.

At each step of simulation, pricing models must be used to assess the value of assets and liabilities at that point of time. The computation then proceeds as in the case of VaR. That is, we assess the value of assets and liabilities at present time, as well as after a given amount of time elapses, and construct the change in value distribution of the stochastic, dynamic gap. Based on this constructed change in value distribution, the LT-VaR can be determined at the desired confidence level.

10.3.2.5   Scenario Analysis and Stress Testing

Gap, cumulative gap, and dynamic gap analysis provide insight in qualitative and quantitative terms of cash flow weaknesses in the balance sheet leading to liquidity risk, based on which responses for liquidity risk management can be developed. However, liquidity risk can emerge or intensify due to certain stressed market conditions or sudden exogenous shocks. These situation can only be anticipated, evaluated, and responded to via a detailed scenario analysis and stress testing. In today's highly interconnected financial institutions, liquidity issues arising in one institution can easily affect other firms.

Liquidity risk scenarios must be considered from short-term and long-term perspective. After the financial crisis of 2008, ensuring liquidity buffers for prolonged and sustained stress scenarios is considered a wise choice. Scenario analysis and stress testing for liquidity risk management should pave the way for shocking the balance sheet under various scenarios and assumptions. This should enable validating liquidity contingency plans, and fine tune where weaknesses emerge. In order to keep stress testing realistic, it is important to ensure that assumptions behind stress scenarios are in anticipation of changing realities, and capture the impact on liquidity buffers under stressed conditions. Scenarios for behavioral changes that impact cash flow assumptions of a firm must also be included in stress scenarios, early withdrawals, bank-runs, deposit roll-overs, and prepayment events.

Usually an ALCO, or asset-liability committee, serves as the reviewing and approving authority for several key ALM decisions, including balance sheet structure, gap analysis, capital adequacy ratios, proactive management of balance sheet, liquidity risk scenarios, and stress testing. Reliability and accuracy of ALM reports, as well as their dependability for the purposes of gap forecasts, projected cash-flows, and balance sheet planning are paramount for achieving the goals of ALM.

Match or mismatch of assets and liabilities and the cash-flows from the assets and liabilities can be critical for the solvency and creditworthiness of a firm or subdivisions of a firm. For certain firms, such as banks and insurance companies, asset-liability management makes the core of the firm's risk management objectives. We began this section with introducing the risks underlying asset-liability management, which go squarely beyond market risk, to include credit risk, funding risk, liquidity risk, trading risk, etc. We then developed tools to assess asset-liability risk and methods to manage them, and how simulation can be utilized for implementing these tools, especially when the scope of the asset-liability management is complex. Scenario analysis and stress testing also significantly aid in asset-liability management to determine and prepare for the impact of specific conditions on the firm.

10.4 Operational Risk Management

Operational risk is the oldest and yet nearly the newest risk! When compared to many of the risks studied thus far, operational risks are not speculative risks; they don't have an up-side, only downside. Hence, they are pure risks. So far the focus on managing these risks has been developing practical techniques for minimizing the chance of loss, whether this meant putting security guards or establishing independent internal audit teams, or building robust computer systems. Putting an economic number on the size of the operational risks faced, or managing them systematically as a risk class is a relatively new response, essentially due to regulatory impetuses.

Financial institutions are paying increasingly greater attention to wide-ranging frameworks for enterprise-wide operational risk management, including relating operational risk directly to risk capital. It seemed inappropriate to leave a whole class of operational risk out of the calculations of risk capital, which can potentially have very significant, detrimental impact. For overseeing risk management efforts in a firm, along with a chief credit and a chief market risk officer representation in the overarching risk committees, it seems logical to have a similar representation of operational risk control.

In the past decades, a series of catastrophic scandals have increasingly highlighted the importance of operational risk and the need for addressing it in a more comprehensive manner.

Nick Leeson and Barings Bank (1995): Founded in 1762, Barings Bank was Britain's oldest merchant bank and Queen Elizabeth's personal bank. Nick Leeson was employed by Barings to profit from low risk arbitrage opportunities between derivatives contracts on the Singapore Mercantile Exchange and Japan's Osaka Exchange. A scandal ensued when Leeson left a $1.4 billion hole in Barings' balance sheet due to his unauthorized derivatives speculation, causing the 233-year-old bank's demise.

Orange County, California (1995): The risky positions of the then Treasurer-Tax Collector of Orange County, Mr. Robert Citron, led to the bankruptcy of the county.

Jerome Kerviel, Societe Generale of France (2007): Bank officials claimed that throughout 2007, Kerviel had been trading profitably in anticipation of falling market prices; however, they accused him of exceeding his authority to engage in unauthorized trades totaling as much as 49.9 billion euros, a figure far higher than the bank's total market capitalization. Bank officials claimed that Kerviel tried to conceal the activity by creating losing trades intentionally so as to offset his early gains.

Kweku Adoboli, UBS trader (Sept 2011): A UBS European equities trader in London, Mr. Kweku Adoboli, caused the Swiss bank a loss of over $2 billion, as a result of unauthorized trades. The incident had raised serious questions regarding the bank's risk management policies at a time when it was trying to rebuild its operations and bolster its flagging client base.

JP Morgan's trading loss (May 2012): The bank announced a multi-billion-dollar loss on a soured trade, which is described as a major failure of the bank's risk management practices.

JP Morgan, Credit Suisse (Nov 2012): J.P. Morgan Chase & Co and Credit Suisse Group AG paid a combined $416.9 million to settle U.S. civil charges that they misled investors in the sale of risky mortgage bonds prior to the 2008 financial crisis. JP Morgan paid $296.9 million, while Credit Suisse paid $120 million in a separate case, with the money going to harmed investors.

The increasing complexity of financial instruments and information systems increases the potential for operational risk events. Mispricing, ineffective hedging, and unauthorized actions by employees can result in very large losses to a firm, not to mention the reputational damage that ensues. Large amounts of data is processed for running any firm's operations, and specifically in financial institutions. Any errors in data feeds can have very significant distortion in the firm's assessment of its risks.

10.4.1 Assessing Operational Risk

To accurately assess these risks, measure them and respond to them requires a precise definition of operational risk. The definition should be so constructed that it facilitates comparison of operational risk profile of different firms. We define operational risk as the risk of loss resulting from inadequate or failed internal processes, people and systems or from external events. Each of the components is further elaborated as follows.

People risk: Loss caused due to incompetence, fraud, errors of judgment by people involved in the firm.

Process risk: This is further subclassified as,

    Model risk: Model or methodology error, mark-to-model error, model implementation or usage error.

    Transaction risk: Execution error, product complexity, booking error, settlement error, documentation, or contract risk.

    Operational control risk: Exceeding limits, security risk, or volume risk.

Systems and Technology risk: System failure, computer breakdowns, programming error, information risk, telecommunications failure.

Legal risk: Exposure to fines, penalties, punitive damages from supervisory actions, private settlements.

Above is a long, yet non-exhaustive, list, where the classification serves the purpose of eliciting this all-pervasive risk type. The list does not include business and reputational risk, even though operational risk directly affects these risk types of a firm.

In order to create a clear demarcation between what qualifies as loss due to operational risk and what does not, losses should not include cost of controls, preventive actions, quality assurance, or investments in upgrades, new systems or processes. These are essentially risk management responses developed as per the firm's risk appetite for operational risk. A caution is needed to avoid double counting a risk as market, credit, business, as well as operational. For instance, if a loan officer extends a bad loan against bank guidelines, any loss arising from this must be treated as operational loss, and not credit loss.

Having defined operational risk and its major subtypes, we must next identify event types so that the extent of loss from these events can be estimated. A taxonomy of drivers of operational risk, as developed in the Basel capital accord, contains the following seven loss event types.

1. Internal fraud. This includes acts to defraud, misappropriate property, circumvent regulations, the law, company policy. For example, intentional misreporting, employee theft, insider trading.

2. External fraud. These are acts of a third party that defraud, misappropriate property, or circumvent the law. For example, robbery, forgery, damage from computer hacking.

3. Employment practices and workplace safety. Here acts inconsistent with employment, health and safety laws or agreements are included.

4. Clients, products, and business practices. Loss arising from unintentional or negligent failure to meet a professional obligation to specific client or nature or design of product. For example, misuse of customer information.

5. Damage to physical assets. These are events of natural disasters or other events, such as terrorism, vandalism, earthquakes, fire, floods.

6. Business disruption and system failures. Loss arising from business disruption or system failures. For example, hardware or software failures, telecommunication problems, utility outages.

7. Execution, delivery and process management. Failed transaction processing, process management, relations with trade counterparties or vendors. For example, data entry errors, collateral management failures, incomplete legal documentation.

Quantifying operational risk is a *key challenge* in implementing any operational risk management framework. Defining and classifying operational losses that may arise from the risk helps. Due to its all-pervasiveness, lack of rigor and process would make assigning numbers and developing a meaningful risk management response for operational risk impossible. Operational losses should be as specific as possible, which we can define in three key terms – cost to fix, write-downs, and resolution expense.

Cost-to-fix: This is best defined to include only external payments that are directly linked to the operational risk incident. For example, legal costs, consultancy costs, costs of hiring temporary staff.

Write-down: This is the loss or impairment in the value of any financial or nonfinancial assets owned by the firm due to the operational risk event.

Resolution expense: Finally, cost incurred for the process of correcting the individual event, including out-of-pocket costs and write-downs, and returning to a position or standard comparable to the firm's original state before the loss event, including restitution payments to third parties.

With the above assessment of components of loss due to specific operational risk events, the total cost of a specific operational risk event type can be estimated as:

$$\textbf{Operational Loss} = \textbf{payments to third parties} + \textbf{cost to fix} \\ + \textbf{write-downs} + \textbf{resolution expense.}$$

In an operational loss model, besides the extent of loss estimation from different types of operational risk events, for measuring the impact of operational risk, we also need to assess the frequency with which these events are expected to occur. The combined frequency of loss and level of loss from operational risk events produces a complete loss model for operational risk. Data from loss events, either internally available or acquired from external sources, is crucial to develop such models.

10.4.2 Managing Operational Risk

Managing operational risk is just as difficult, if not more, as measuring it. With larger size, broader spread and complex structure of organization of a firm, and its products and services, the problem becomes that much more challenging. Adopting a sound and rigorous approach to estimating and developing loss models for operational risk is the first essential step. However, this must be complemented with a detailed set of guidelines that aid the risk management process. The following eight key elements are useful in constructing this guideline.

1. Common language and agreed-upon terminology for risk identification - people risk, process risk, system and technology risk - must be identified for qualitative self-assessment or statistical assessment.

2. Business process maps make the business process associated with the firm's dealings with clients, suppliers and customers, so that this is transparent to management and auditors. The process map may be extended to create a full operational risk catalogue, describing people, process, systems, and technology risk arising from each organizational unit.

3. Choice of risk metrics for operational risk must be made, using quantitative methodology based on historical loss experience and scenario analysis, to derive loss frequency and loss severity distributions.

4. Establish clear guidelines for practices that monitor, control or reduce operational risk. For instance, from the point of view of an investment bank that runs several trading desks, this may translate to policy guidelines about the following. Establish policies on

   (a) traders and back-office segregation
   (b) out-of-hours trading
   (c) off-premises trading
   (d) legal document vetting
   (e) vetting of pricing models that underpin trading decisions, etc.

   Some of these are defined, encouraged or required by regulators, others will be created to define best practices of the firm.

5. Managing operational risk exposure with appropriate actions to reduce or hedge operational risk. This involves considering the cost-benefit trade-off for insuring for those operational risks that can be insured. Since far from all operational risks are insurable, risk reduction or mitigation is also important response to consider.

6. Identifying a risk management reporting scheme, that is which risks and responses to report more usefully and to whom - senior management, the board for firm-wide operational risk profile, to operations and administration committee, or capital and risk committees.

7. Tools for risk analysis and procedures for when these tools should be deployed. Appropriate measures, up-to-date databases of internal and industry-wide operational loss data, well-designed scenario analysis, and a deep understanding of key risk drivers in the firm's business lines. All these should feed into risk financing of retained risk in the calculation of operational Value-at-Risk (OpVaR).

8. Finally, appropriate attribution of operational risk capital to every business, which creates proactive incentives for operational risk management.

The risk management response to operational risk is exactly as we have applied to all other risk types considered so far: avoid-mitigate-transfer-keep. In order to create the response strategy, we need to evaluate efficacy of mitigate, transfer and avoid responses. Further for the monitoring and control of retained operational risk, a measure must be created that evaluates the joint impact of all operational risks in one metric. We explore operational Value-at-Risk, or OpVaR.

10.4.2.1   Risk Measures for Operational Risk

The Advanced Measurement Approach (AMA) is the proposed method for computing operational Value-at-Risk, or OpVaR, based on analytical techniques that are widely used in the insurance industry to measure the financial impact of an operational failure. The aim is the determination of OpVaR, for which we need to determine the following for the retained operational risk types and events.

1. The expected loss from operational failures

2. The worst-case loss at a desired confidence level

3. Required economic capital for operational risk

4. Concentration of operational risk

The firm's activities are divided into Lines of Business (LoB), and for each line of business an Exposure Indicator (EI) is assigned. The EI is determined from

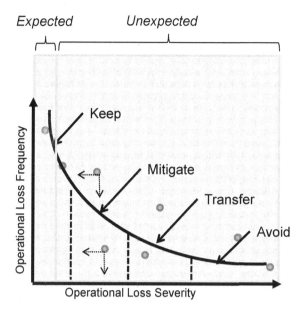

FIGURE 10.5: Classification of operational risk by severity and frequency to guide through developing a risk response strategy.

probability of operational risk events in that LoB, as discussed earlier. We term this probability of event, or PE. We also need the estimates of loss given events (LGE) for that LoB, which was also discussed in Section 10.4.1. The EI for all LoB's are combined to create the firm's operational risk distribution, and from that OpVaR is computed at the desired confidence level. Alternatively, OpVaR can be computed for each LoB separately in order to obtain insight regarding relative riskiness of each LoB.

10.4.2.2 Operational Risk Management Strategy

For generating the avoid-mitigate-transfer-keep strategy for operational risk, we need to classify losses from operational risk events into a) expected, b) unexpected level. This nomenclature is used to simply bisect the overall loss distribution into segments. The expected losses may be absorbed as cost of doing business. These failures may be explicitly or implicitly budgeted for in the annual business plan and covered by the pricing of the products and services. Therefore, there is no need to try to allocate risk capital against it. Therefore, the expected loss is kept and funded as regular cost of conducting business. The unexpected losses, however, are large enough that a response strategy must be developed for these, including assigning risk capital for kept unexpected losses. Figure 10.5 sketches the classification of operational risk by severity and frequency to guide through developing a risk response strategy.

For developing a risk management response to each operational risk type, the unexpected losses can be further subdivided into:

Severe loss: These are severe losses from operational risk events, but not catastrophic. Unexpected severe operational failures should be either mitigated to reduce their harmful impact or covered by an appropriate allocation of operational risk capital.

Catastrophic losses: These are the most severe, but also the rarest operational risk events. However, it is the kind that can destroy the firm entirely. Firms would want to tighten procedures to protect themselves against catastrophic events, which is the avoid response of a risk management strategy. Alternatively, they should consider using insurance contracts or alternative risk transfer mechanisms to transfer these catastrophic risks. Keeping these risks and safeguarding against them using risk capital may not be feasible due to the sheer magnitude of their impact. If transfer is not feasible, the catastrophic risks should be avoided.

The equi-expected-loss curve in Figure 10.5 gives an acceptable level of expected loss; management should take appropriate action to mitigate risks located above the curve. Operational risk is pure risk, therefore more operational risk does not provide higher reward, it only destroys value for all security holders of the firm. But attempting to mitigate all operational risks is prohibitively costly, therefore transfer or avoid responses must be invoked where needed.

Insurance has been historically used to transfer the effects of key operational risk events. For instance, a firm may purchase insurance to protect itself from large single losses arising from acts of employee dishonesty, robbery and theft, loans made against counterfeit securities, and various forms of computer crimes. These are low-probability, high-severity losses. Policies may also be available for lawsuits and major disasters, such as fire and earthquakes. However, insurance is a mechanism for pooling and transferring common loss exposures with the industry or across economies, which we will study in greater detail in the next chapter.

Availability of insurance for specific risks therefore depends on the ability of an insurer or group of insurers to generate sufficient premium volume and an adequate dispersion of risk to make a market and enable them to take on the risk of others. It also depends on the insurer's ability to avoid the problem of moral hazard, i.e., the insured should have a strong interest in preventing any costly events. Insurance deductibles are one way for insurance providers to protect themselves against moral hazard. If the policies are not implemented in honesty, they may get canceled or not be renewed. Problems may also arise for the insured in the form of counterparty risk if the insurance provider is not able to pay the benefits at the occurrence of insured risks. Therefore insurance is a mechanism of transfer for operational risk, but it has its limitations.

Finally, the operational risk retained by the firm must be supported by risk capital. Mechanism for attributing capital to operational risk should be risk-based, transparent, scalable and fair. Capital requirements and policies should vary directly with levels of verifiable risk, provide incentive to manage operational risk, improve operational decisions, and should be incorporated in the determination of risk-adjusted return on capital.

10.5 Summary

In this chapter, we covered the spectrum of strategic, business, and operational risks of a firm. Clearly, managing these risks is not only of paramount importance to firms in the financial sector, but is important to address for any firm of any size. We differentiated the three risk segments by their time-stamps. Operational risk is the risk of here-and-now, business risk being the risk of running the operations of a firm profitably for the next year or so, while strategic risk addresses risks that will impact the firm in the long-run. The boundaries between the three risk types are not very crisp and the time-stamps of here-and-now, one year, and the long-run can have a significant overlap. Moreover, not all risk management problems for these three risk types can be addressed in a quantitative setting. Therefore, in this chapter, we started with first distinguishing between the three risk types, along with highlighting where the boundaries between them blur. This was followed by identifying specific issues under each of the three risk types and picking specific example contexts where a quantitative analysis holds merit. We elucidated how simulation can be useful in addressing these quantitative problems of risk management for strategic, business, and operational risk. We used this context to develop the asset-liability management framework for financial institutions, which intertwines the three segments for balance sheet management of the firm.

10.6 Questions and Exercises

Review Questions

1. What is strategic risk management? How does it relate to business risk and operational risk management?

2. What are the Modigliani-Miller results for corporate risk management?

3. Why do firms engage in risk management?

4. What are the major components of strategic risk management for a firm?

5. What are the objectives of strategic risk management?

6. How are 'avoid' and 'keep' decisions made in capital budgeting?

7. What is a major source of destruction of a firm's value? How can this be avoided?

8. What are alternative risk transfer tools?

9. What are real options? How are they useful in strategic risk management?

10. How are futures prices related with spot prices? When do we see a contango? What is backwardation?

11. What is meant by the term structure of futures prices?

12. How can structured liabilities be useful for strategic risk management?

13. When do firms fail in strategic risk management?

14. What are the different components of business risk management?

15. How does business risk management respond to demand risk?

16. How can competition risk be managed?

17. How can real options methodology be helpful in business risk management?

18. How can business risk relate to market risk or credit risk? Give examples and discuss.

19. How is liquidity risk management related with business risk?

20. What is asset-liability management? What kind of firms can benefit from explicit efforts for ALM?

21. What are the major components of asset-liability management?

22. What is structural gap?

23. What are the net interest income and net worth of a bank?

24. Why is interest rate risk considered important for ALM?

25. What is gap analysis?

26. What is cumulative gap analysis? How is it an improvement over gap analysis?

27. What is duration gap analysis? How is this an improvement over gap analysis?

28. What is LT-VaR? How is this an improvement over all other methods of conducting ALM?

29. How can stress testing and scenario analysis be helpful in ALM?

30. What is operational risk? Investigate some historical instances of operational risk and discuss their impact.

31. What are the different sources of operational risk for a firm? Discuss the taxonomy of operational risks?

32. How is operational risk quantified?

33. What are the guidelines for managing operational risks?

34. What is Operational Value-at-Risk (OpVaR)? How is it computed for a line of business (LoB) of a firm and for the entire firm?

35. Why is operational risk segmented into expected and unexpected levels?

36. How are expected operational risks managed?

37. How are unexpected operational risks managed?

38. What role does insurance play in the management of operational risk?

39. What is OpRisk capital? What is it used for?

Exercises

1. The Dord Motors company is considering introducing a new sports car model, named *The Racer*. The management is trying to assess the prospects for this new model. While understanding the project's profitability is a difficult task, it is an important task before the project is taken up. For this purpose, they have put together estimates for fixed and variable costs, projected future sales and prices at which they intend to sell this model. Each of these project features are described in a model, as follows.

   - Fixed cost of developing *The Racer* is equally likely to be either $3 or $5 billion.

   - Variable cost per car manufactured for the first three years are: For year 1 it is equally likely to be $5000 or $8000, for year 2 it is going to be $1.05 *$ (year 1 variable cost), for year 3 it is going to be $1.05 *$ (year 2 variable cost).

   - Sales projections are determined as average sales in year 1 as 200,000, with a standard deviation of 50,000 cars. A normal distribution is chosen to describe year 1 sales. Average sales of year 2 and 3 are expected to be at the sales level of the previous year, with a standard deviation of 50,000 cars.

- Pricing for year 1 is set at $13,000. Years 2 and 3 prices will be determined based on the previous year's price and sales. Specifically, year 2 price = 1.05 * (year 1 price)+$30*(percentage by which year 1 sales exceed expected year 1 sales), year 3 price = 1.05 * (year 2 price)+$30*(percentage by which year 2 sales exceed expected year 2 sales).

Set up a simulation based assessment framework in MATLAB for the management's evaluation of the profitability of *The Racer*.

2. An oil company has a finite period lease to drill on an unexplored piece of land with potential reserves, and suppose the following investment costs must be incurred for various interrelated activities.

   (a) The land can be explored for cost I_e any time between t_0 to t_1.

   (b) Construction cost I_c will be incurred for processing facility commenced at time t_2, where construction will be done only if oil reserves are discovered at or before t_1.

   (c) Construction of the processing facility may be terminated at any time before its completion at t_3.

   (d) During construction, from t_2 to t_3, management can reduce the scale of extraction facility by α percent and recover a portion I_k of its latest outlay if demand is perceived to be weak.

   (e) After the facility is in production at t_4, management can expand scale of the facility by β percent for an additional investment I_β.

   (f) After the facility is in production at t_4, management can temporarily shutdown the plant for one period by paying its variable operating cost during that period, I_v.

   (g) After the facility is in production at t_4, management can abandon the plant and sell the assets or switch the assets to an alternative use for value V_a.

Construct and discuss all the real options embedded in the above prospects for the oil company.

3. Conduct an asset-liability analysis for the following balance sheet.

   Assets: These constitute US Treasury bonds.

   - US T-Bonds, Maturity 2-years, Yield 0.25%: $256,125,000
   - US T-Bonds, Maturity 5-years, Yield 0.76%: $95,625,000
   - US T-Bonds, Maturity 10-years, Yield 1.84%: $230,190,000

   Liabilities: These constitute short-term liabilities and equity.

- 3-Month LIBOR: $538,295,000
- Equity: $43,645,000

Conduct a gap, duration gap, gap convexity, and dynamic gap analysis for the above balance sheet.

4. Consider the following distributional fits for loss frequency and loss severity of 5 operational risks identified in a line of business (LoB) of a firm.

   (a) $N_{1t} \sim Po(\lambda)$, Poisson distribution with $\lambda = 15$ per year; $L_1 \sim$ Weibull(a, b) with scale parameter $a = 5$ and shape parameter $b = 0.8$

   (b) $N_{2t} \sim Po(\lambda)$, Poisson distribution with $\lambda = 20$ per year; $L_2 \sim \chi^2(\nu)$ Chi-square distribution with degrees of freedom $\nu = 25$

   (c) $N_{3t} \sim Bin(n, p)$, Binomial distribution with $n = 50$ and $p = 0.05$; $L_3 \sim$ Lognormal distribution with mean $\mu_L = 100$ and standard deviation $\sigma_L = 15$

   (d) $N_{4t} \sim NegBin(r, p)$, Negative Binomial distribution with $r = 20$ and $p = 0.13$; $L_4 \sim$ Gamma distribution with shape parameter $a = 5$ and scale parameter $b = 5$

   (e) $N_{5t} \sim Bin(n, p)$, Binomial distribution with $n = 5$ and $p = 0.2$; $L_5 \sim$ Weibull(a, b) with scale parameter $a = 500$ and shape parameter $b = 0.8$

   Construct a quantitative assessment in MATLAB of the total annual loss of each operational risk, as well as the grand total annual loss from all the operational risks of the LoB combined.

5. In the frequency-severity dimensions of operational loss events, if we coarsely consider two levels - high-low, give examples of operational risks due to people, process, system, and technology in each of the following cases.

   (a) Risk Profile: (low, low); Risk Management Response: Retain.
   (b) Risk Profile: (low, high); Risk Management Response: Insurance or Non-insurance transfer.
   (c) Risk Profile: (high, low); Risk Management Response: Loss prevention and control.
   (d) Risk Profile: (high, high); Risk Management Response: Avoid.

6. For a severity-frequency curve for operational risk given by the following relation

$$\frac{\bar{N}^2}{2300 * \bar{L}^{0.2}} \tag{10.6}$$

make the avoid-mitigate-transfer-keep decisions for the risks in Exercise 4. $\bar{N}$ is expressed as number of occurrences per year and $\bar{L}$ is the total dollar loss per occurrence.

Chapter 11

Risk Management Using Insurance

Insurance is the transfer of that risk which we mostly don't like to think about, but should the need arise, we desperately hope we had thought of and prepared for it sooner. The goal of risk management using insurance is that we are well prepared on such occasions to eliminate harsh surprises and facilitate easier recovery. If we think of insurance as a concept of spreading the risk, then insurance has been utilized from the advent of human societies. Any device put in place to recover from an adverse event of fire, flood, wild-cat attack etc., by extending support to the suffering unit to recover from the event in material and/or emotional terms is essentially the concept of insurance.

With time, as societies became more complex and enterprises more sophisticated, there are numerous examples of safety nets being put in place by the rulers of the time or certain trade groups to improve well-being of the subjects or to encourage better commerce. The concept of a formal insurance contract, and its widespread availability and demand, is a more modern phenomenon. According to the fact sheet of The Geneva Association, a leading international think tank of the insurance industry, the global financial assets under management of insurers in 2009 were valued at US$22.6 trillion, which was 12% of global financial assets in 2009. While the worldwide insurance premium volume in 2009 stood at US$ 4.34 trillion, which is equivalent to almost 7.0% of global GDP. From these figures, it is clear that insurance has come a long way from its humble beginnings to becoming a massive worldwide industry today.

In order to consider insurance for risk management, we will need to look back at the risk management process developed in Chapter 2. Even though we are considering insurance for risk management at this late stage in the book, the traditional view of risk management primarily revolved around insurance serving as a key mechanism for management of risks. This was before derivative markets and the plethora of derivative instruments for every kind of speculative risk took off, which required and supported expanding the scope of risk management to speculative risks. Nevertheless, individuals, households and firms continue to consider insurance as the fundamental tool for risk management. The utilization and innovation of insurance products remains unabated, which explains the volume of the industry today.

One difference is that the traditional view of risk management focused on what we call 'pure' risk, while modern risk management doesn't make that rigid a distinction, since the boundaries between 'pure' and 'speculative' risk aren't that obvious and clear. Commodity prices, interest rate risk or credit risk exposure of a firm may not be due to speculative decisions of some business unit of a firm, instead these exposures may be integral to the unit's operations. For example, a farmer's exposure to interest rates or commodity price risk, a bank's exposure to interest rates or credit risk, an airline company's exposures to interest rates and jet oil price risk are unavoidable to the enterprise's core business. Moreover, one may have argued that credit risk, at least as it pertains to occurrence of default or bankruptcy, is more akin to pure risk, since it has two states, 'neutral' and 'unfavorable.' Yet a variety of instruments, as discussed in Chapter 9, allow for speculation of credit risk. Later in this chapter, we will examine other pure risks that are edging into the speculative territory.

We also note that all pure risks are not operational in nature. In the last chapter, issues regarding operational risk were discussed, along with the development of methods for operational risk management. Operational risks were defined as risks that don't have an up-side, only downside that disrupts 'business as usual.' We make a distinction here that while all operational risks are of the pure kind, all pure risks are not operational. Pure risk can be strategic in nature, for instance there is a definite certainty that death will be experienced by all, when that happens may be in the distant future. Therefore, for risk management of mortality and longevity risk, one must adopt a strategic perspective. Similarly, pure risk can be a business risk, in that a large citrus farm must determine the annual damage to crop anticipated due to hurricanes and incorporate its management as business risk.

With the explosion in risk types to consider for risk management, the modern view of risk management has created the concept of enterprise-wide risk management, or simply enterprise risk management (ERM), where attempts are made to integrally consider and manage all the dimensions and sources of risks in a comprehensive and consistent way for a firm, as discussed in Chapter 10. While this is an ambitious goal, it is a worthy one, given the high interconnectedness of risk exposures in today's world with changing characteristics and evolving boundaries between them. We have so far seen issues related with management of several risk components of the ERM spectrum. We now bring back the traditional view of risk management within the fold of the modern view of risk management, by studying the role of insurance in risk management. Insurance, as a pure risk transfer mechanism, plays an important role in the ERM spectrum, for a variety of types of firms, as well as for individuals and households.

In this chapter, we will begin with a discussion of basic concepts in insurance. This will include developing an extended classification of pure risk, with a view to develop a basic understanding of the insurance industry. This will be followed by developing a formal analysis for the basis of design of insurance contracts. As stated earlier, the notion of spreading of risk pre-existed the creation of formal insurance contract. We will investigate the mathematical principle that makes insurance work. This will be followed by creating a framework for making risk management decisions that utilize insurance, along with presenting detailed examples for modeling risk in specific important contexts. Finally, we will shift our attention to considering risk management for an insurer. Specifically, we will consider the pricing, investment, asset-liability management, securitization, and reinsurance decisions of an insurance provider from a risk management perspective.

11.1   Basic Concepts of Insurance

In the world of insurance the focus is on loss, and to avoid losses. This is in contrast to the risk-reward trade-off perspective of modern risk management, where risk may be sought to improve prospect of higher reward. For reconciling the emphasis on losses in the case of insurance and our prior emphasis on risk-return trade-off for risk management, a redefinition of risk may be required, as the likelihood of suffering loss due to occurrence of certain events. In the broader view of risk management, however, we had defined risk as the variability that can be quantified in terms of probabilities. Since the earlier, broader definition makes no reference to gain or loss, it is in fact inclusive of this narrower definition.

We still need the distinction we made in Chapter 1 between pure risk versus speculative risk, even though in some cases the distinction between the two is somewhat blur. Pure risk pertains to situations in which there are only possibilities of loss or no loss, i.e., outcomes are either adverse (cause loss) or are neutral (no loss). Examples we include here are premature death, job-related accidents, catastrophic medical expense, damage to property due to fire, lightning, flood, or earthquake. While speculative risk is where both profit or loss can occur. We have seen many examples of speculative risks in all the chapters thus far, various market risks, equity price risk, interest rate risk, strategic, and business risk, etc.

For our study of insurance, we further classify pure risk into the following three categories.

Personal: Personal risks are risks to the well-being of an individual or members of a household. Risk of poor health (health insurance), premature death (life insurance), insufficient income during retirement (annuities, social security), and unemployment (government programs), are some examples of personal risks. For each risk we have listed (in parentheses) the plausible insurance product that may be used to alleviate the personal risk.

Property: Individuals, households, and firms own property and physical assets that are exposed to the risk that they will suffer damage or loss due to various causes, such as fire, lightning, hurricane, earthquake, flood, vandalism, break-in, etc. These events can result in direct loss due to damage to the property or physical asset. Moreover, they may result in additional indirect loss due to the property or asset not being available for use. This is the opportunity cost of the asset. A third category, which may not be as evident at first thought, but is at least as important as the previous two categories of pure risk is the liability risk due to pure loss events.

Liability: An individual or a firm is legally liable if something he or it does results in personal harm or property damage to someone else. This consideration in pure risk category is very important, since for liability risk there is no upper limit on the amount of loss it can result. As a result, a lien can be placed on an individual's income, or a firm's physical and/or financial assets, to satisfy the legal judgment. Legal defense costs, which can be enormous, are an additional cost burden.

Our perceptions of risk very significantly guide our decision making regarding them. With this in mind, it is important to make a distinction between objective versus subjective risks and probabilities. We had already made this distinction in Chapter 2, highlighting the fact that our beliefs and experience lends us some subjective views regarding likelihood of events, which can have a role to play in the decision making for risk management.

Objective risk: It is a risk that can be assessed based on quantifiable, past observations. These observations provide an undisputed judgement of degree of variation in the actual loss. Objective probabilities are the long-run relative frequency of an event based on the assumption that the number of observations seen are representative of the population of such events, and that there is no change in the underlying conditions for the occurrence of the event.

Subjective risk: A person's perspective of uncertainty about the future occurrence of loss events, which could be affected by the person's past experience, state of mind, and mental attitude. Subjective probabilities that quantify the subjective risk are an individual's personal estimate of the chance of loss.

The subjective view of risk was not deemed irrelevant in our earlier discussion in Chapter 2. In fact, respecting the subjective view was considered specifically important in the case of unavailability of relevant or insufficient data for undisputed judgement of the likelihood of risks, which is when the boundary between objective and subjective probabilities becomes unclear.

Pure risk events occur and cause loss to person or property, or result in a liability. In order to sharpen our ability to refer to the source of this loss, we distinguish between peril and hazard.

Peril: It is the cause of loss to person, property, or resulting in a liability related loss. For example, in the case of fire burning down a house, fire is the peril, the cause of loss. The fire spreading into the neighborhood can create liability loss. Common perils that cause property damage include fire, lightning, windstorm, hail, tornadoes, earthquakes, theft, and burglary.

Hazard: Conditions or features of the property or person that creates or increases the chance of loss are hazards. Hazards can be further classified into three categories.

> **Physical Hazard:** A physical condition that increases the chance of loss.
>
> **Moral Hazard:** A behavioral condition, such as dishonesty or character defect in an individual, that increases the chance of loss.
>
> **Morale Hazard:** A fine distinction with ' moral hazard,' carelessness or indifference to a loss because of the existence of insurance.

Finally, in order to start our examination of insurability of pure risk and the ability of an enterprise to provide this insurance, we make a distinction between fundamental versus particular risk. There are a wide variety of pure risks, not all of which are readily insurable. The following distinction is crucial to move forward with the identification of pure risks that are more favorably managed by insurance.

Fundamental Risk: The risk that affects the entire economy or a large number of persons or groups within an economy. Examples of fundamental risk include, rapid inflation, cyclical unemployment, war, major hurricanes, and devastating earthquakes.

Particular Risk: The risk that affects only individuals, not an entire community. For example, car theft, residential fire, and health risk.

Fundamental risks have a more severe, widespread impact. Therefore, they may require government assistance in management and recovery. In the above discussion, we have provided the most important terminology and classification of pure risk to support the development of risk management using insurance in the rest of this chapter. Additional details on the above classifications of pure risks can be found in Rejda [70].

11.2 Principle behind Insurance

The American Risk and Insurance Association (ARIA) defines insurance as the pooling of fortuitous losses by transfer of such risks to insurers, who agree to indemnify insureds for such losses, to provide other pecuniary benefits on their occurrence, or to render services connected with the risk. This is a very concise and complete definition, since it summarizes what insurance is and does with a carefully selected choice of words. In this section we explore what each of these words means, and using this as a context describe the key principle that makes insurance work.

11.2.1   Characteristics of Insurance and Insurable Risk

An insured is an individual, household or firm seeking protection from pure risk under an insurance contract. An insured seeks to use insurance for its ability to transfer pure risk to the insurer as per the insurance contract. The insured does this transfer of pure risk with the hope that the insurer is financially stronger than the insured to bear the loss, and will pay for the loss.

Paying for the loss is indemnification. Indemnification is when the insurer helps restore the insured to his/her approximate financial position prior to the occurrence of the loss. Clearly, this transfer of pure risk is effected for a price. The insured pays a periodic, or in some cases a lump-sum, premium for the purchase of this protection from pure risk.

The basic principle behind insurance provision is the pooling of losses due to pure risk events. Pooling of losses is the process of spreading of losses incurred due to pure risk events experienced by a few over the entire group. Therefore, the insurer makes a business case by having a large set of customers transfer their pure risk (of a very specific kind) over to the insurer, where at a time only a small fraction of the insured will experience a loss event. The accumulated premium collected from the larger group of customers can be used to indemnify the few who suffer the loss.

This spreading of loss can be done viably by the insurer only if the payment towards indemnification is for fortuitous losses. Insurance is meant to pay for fortuitous loss, ones that are unforeseen and unexpected, and occur as a result of chance. If losses are intentional, and not accidental, insurance is unlikely to be feasible. Losses being accidental is one of the prerequisites of the fundamental mathematical principle that supports insurance, namely the law of large numbers.

11.2.1.1   Law of Large Numbers

Let's now examine the spreading of losses by pooling a large number of exposure units more closely. If the loss of each insured in a given period of time, say annually, is a random variable L_i, then summing L_i's for many insureds, say N of them, results in the average annual loss experienced by the insurer to be, $\bar{L} = \sum_{i=1}^{N} \frac{L_i}{N}$. The *weak law of large numbers* states that,

$$\lim_{N \to \infty} P(|\bar{L} - \mu| > \epsilon) = 0, \qquad (11.1)$$

where $\mu = E[L_i]$ for each i. In other words, this states that the (sample) average loss experienced by the insurer converges in probability to the expected value of loss from each individual exposure unit. The *strong law of large numbers* makes a stronger claim by stating that the convergence is an almost sure convergence. In particular,

$$P(\lim_{N\to\infty} \bar{L} = \mu) = 1, \tag{11.2}$$

which implies that the sample average loss converges almost surely to the expected value of loss from each individual exposure unit. Therefore, the two versions of the law of large numbers give the insurer increasing degree of assurance that only having a large pool of exposure units is enough for predicting the average loss experienced. Stated slightly differently, the law of large numbers gives,

$$P(\lim_{N\to\infty} N\bar{L} = N\mu) = 1, \tag{11.3}$$

where $N\bar{L} = \sum_{i=1}^{N} L_i$ is the total loss experienced by the insurer. Therefore, the insurer has increasingly accurate estimate of total loss experienced as the size of the pool of insureds gets larger.

The weak or strong law of large numbers (LLN) have given asymptotic assurances, i.e., what will happen if there are 'infinite' customers in the pool. However, in reality it is feasible to only have finite, hopefully very large, number of exposure units in a pool. In this situation, what can be said about how much variability will there be in actual total loss experienced by the insurer. A related result to the law of large numbers provides some insight regarding this question.

Consider the central limit theorem, which we stated in Section 1.2.2.3 of Chapter 1. For applying the central limit theorem, we would need that the loss experience of individual exposure units, L_i, are independent and identically distributed. It can then be shown that the total loss, $TL = \sum_{i=1}^{N} L_i$ is approximately normally distributed, with $E[TL] = N\mu$ and standard deviation of $TL = \sqrt{N}\sigma$, where $\sigma = \sqrt{var(L_i)}$.

This implies that the total loss, TL, on average increases at the rate of number of insured, but with increasingly lowered variability. This means a more accurate estimate of future losses can be created as the pool of insureds is increased. In this process, average loss can be substituted for actual loss. Therefore, if losses from individual exposure units have a mean loss level (μ) that is not too high, this is good for the mean level of the pooled losses ($N\mu$). In Figure 11.1, we have displayed the impact of growing customer pool on the distribution of the average loss. Even while the individual loss distribution is far from symmetric and normal, as shown by the density plot in the left-most panel of the figure, the average loss looks increasingly normal for a larger pool.

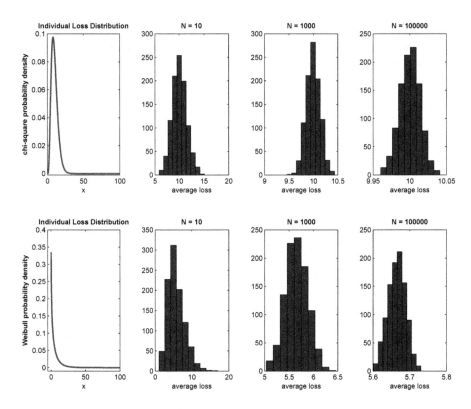

FIGURE 11.1: Display of the law of large numbers for two individual loss distributions.

11.2.1.2 Requirement of Insurable Risk

With an understanding of the fundamental principle that can support the design of an insurance product, we can now elaborate on the characteristics of a pure risk that has the potential of being insurable. Let's discuss each characteristic in detail to clarify when insurance works and why.

- The law of large numbers is a result for random variables. If losses from exposure units are intentional, the insurer would not be able to support the indemnification of all such losses, and the insurance contract will not be honored. Therefore, insurable losses must be accidental and unintentional.

- Before an insurer can design an insurance product that is viable, it must do extensive analysis of viability of the product. This requires access to data from past experience of loss due to the underlying pure risk. Once the contract is put in place, the loss must be measurable and quantifiable so that the indemnification can be performed. Therefore, losses must be determinable and measurable, and chance of loss must be calculable. Insurer must be able to calculate both average frequency and average severity of future losses with some accuracy. This will be necessary to assess the benefit of pooling, so that the necessary premium can be charged that is sufficient to pay all claims and expenses and yield a profit during policy period.

- The law of large numbers is an asymptotic result, therefore for its conclusions to hold, arbitrarily large number of exposure units are desirable. Therefore, large number of exposure units are necessary for insurability of a pure risk.

- For insurability, losses should ideally not be catastrophic. This means that the average loss of a single exposure unit (μ) should not be astronomical and a large number of exposure units should not incur losses at the same time. The correlation between losses from exposure units should be low, otherwise the LLN would not apply. If the mean loss of a single exposure unit is too high, in order to support indemnification of the incurred losses, it will be necessary to increase the insurance premium to prohibitively high levels. That is why in case of catastrophic losses, reinsurance becomes relevant and useful. Geographic diversification can also help introduce lower correlation between sub-pools. New financial instruments are becoming available for catastrophic losses, such as cat bonds, catastrophe insurance options, and weather derivatives, which we will examine later in this chapter.

- Finally, related to the previous point, the insurance premium must be economically feasible. For insurance to be an attractive mechanism for risk transfer, the premium should be substantially lower than the face value or the coverage of the insurance contract, and should be significantly lower than the amount of policy benefits in case of a loss. For this to be feasible the likelihood of loss should be relatively low.

Most of the above required characteristics for insurability of pure risk are necessary for the application of the law of large numbers. However, the law of large numbers only states that with an increasing pool, the total loss can be more accurately estimated. In order for insurance to work, for economic feasibility of insurance provision, the insurance premium should not be too high, which can make it unattractive for risk transfer. Reinsurance is the shifting of part or all of the insurance-related risk transfer originally underwritten by one insurer to another insurer. Reinsurance is a mechanism by which some catastrophic pure risk can also remain insurable. We will look at reinsurance again in Section 11.5.2. Due to the above necessary criteria for insurability of risk, insuring against most market risk, financial risk, production risks, and political risk are usually not feasible by private insurers.

11.3   Types of Insurance

The first order of classification of insurance is done in terms of life insurance versus non-life insurance. This has to do with certain unique characteristic of life insurance, most importantly the extremely long-term nature of the underlying risk and the eventual certainty of it. The second order of classification is on the lines we differentiated earlier between personal pure risk versus property, casualty and liability related pure risk. Private insurers play a significant role in providing life, health, property, casualty, and liability insurance products. We look at each closely for an overview of the types of insurance products available. This overview should help the reader appreciate the kinds of choice at hand when making risk management decisions using insurance.

Life insurance pays death benefits to designated beneficiaries when the insured individual dies. Life insurance products can be broadly classified into term life or cash-value life insurance. Term life provides insurance for a defined period of time, and requires renewal at the end of this time period. The terms can range from one year to as long as 20 years, or run up to a certain age. Whole life insurance is a cash-value life insurance that provides lifetime protection, where in one of its sub-types, ordinary life insurance, the face value is paid if the insured survives the high age limit of the policy, and another subtype, universal life insurance, requires flexible premium payments that provide lifetime protection that unbundles into protection and saving components.

Variable life insurance, on the other hand, requires a fixed premium, however death benefit and cash surrender value vary according to the investment experience of a separate account maintained by the insurer. Annuities are a related insurance product that serves the opposite role. It provides income should the insured outlive her savings, therefore it is a protection against longevity risk. One would think that living long is a blessing, however, if one outlives one's savings and the ability to participate in the labor market, the blessing can turn into a bane. Annuity products hedge this risk. Therefore, there are numerous life insurance products, with new ones getting offered every so often.

Health insurers sell a wide variety of individual and group health insurance products, where the latter kind are offered through employers as employee benefits. Some of these plans provide broad and comprehensive protection, while others list numerous exclusions, hence quality of products is varied. Important health insurance products are, for example, hospital-surgical insurance, long-term care insurance, disability-income insurance, and major medical insurance. Group insurance creates insurance of many persons under a single contract. A master contract defines the terms between insurer and group policyowner for the eventual benefits of individual members. Employers often offer health and life insurance of various kind to their employees through their group life and group health insurance contracts.

Property and casualty insurance for individual and households is for risk transfer related with property and physical assets owned. Other than fire, tornado, etc., causing damage to property or marine insurance, all other property related insurance falls under the broad category of casualty insurance. Therefore, insurance of individual and household interest, such as automobile, burglary and theft, etc. falls under casualty insurance. Probably the most important property insurance individuals buy in their life is homeowners insurance. Homeowners insurance is a package policy, which combines several separate coverages into a single policy. This bundling helps in reducing gaps in coverages, and the premium can become more economical. Some personal liability insurance is also included in homeowners policies, such as personal liability and medical payment up to certain amounts.

The property and casualty insurance needs of firms can be quite different from those of individuals and households, and they can also be quite varied depending on the nature of a firm's operations and assets. Commercial package policy combines many coverages into a single policy for firms. These coverages can include commercial property, commercial crime, commercial auto, farm coverage, boiler, and machinery coverage etc. If both property and liability line are combined in a single policy, it is called a multi-line policy. Ocean marine and inland marine insurance are transportation insurance for protection of goods transported over water and land, respectively, which can be very valuable for firms.

Earlier we made a distinction between particular and fundamental pure risks, where fundamental risk is one that affects the entire economy or a large number of persons or groups within an economy. Due to this property, private insurers are mostly incapable for insuring against this risk. Governments of nations and states must bear the responsibility of lowering the burden of these risks. Government provided insurance, such as social security benefits, Medicare, unemployment insurance, workers compensation, and disability insurance, are examples of such efforts.

11.3.1 Benefits and Cost of Insurance to Society

There are many direct benefits to individuals and firms from utilizing all the variety of insurance types available for the management of different pure risks. Other than the direct benefits, there are also indirect social and economic benefits of insurance.

- Indemnification of losses is the obvious benefit of insurance, which is the sole primary purpose for the contract's creation.

- Having a risk transfer and assurance of indemnification in place leads to less worry, anxiety, and fear, thus allowing firms and individuals to focus on their core activities.

- One side-effect of premium accumulation with the insurers, which they don't immediately need, is a collection of funds that must be invested until indemnification needs arise. Therefore, insurers are a large source of investment funds.

- A significant component of risk management using insurance is loss prevention. It is in the insurer's interest to create incentives for the insured to do active risk reduction.

- Lastly, due to availability of indemnification, insureds have lowered tail risk due to pure risk events. This helps in enhancing their creditworthiness and debt capacity.

The above benefits come at a cost to the society. The first obvious cost is the cost of doing the business by an insurance provider. This includes setting up the operations and workforce to offer and provide insurance and coverage. An expense loading is added to the premium to cover the expense incurred in the company's daily operations. The other costs which can prove to be an undesirable burden on the society are due to fraudulent claims. Moral hazard and morale hazard we defined earlier can bring severe inefficiencies in the insurance market, leading some providers to close shop in extreme cases. A lesser, but nevertheless damaging, burden comes from inflated claims.

Finally, if insurance providers expand their business into territories of risk where they don't have complete understanding and expertise, this can lead to serious cost to the society due to unfulfilled contracts and cascading effect on other firm liabilities. In Chapter 3, we had reviewed the historical evolution of regulation of the insurance sector, and noted the principles underlying the government control of the insurance industry. A poorly designed or implemented regulatory environment that fails to assure fair contracts and robust business practices is an additional cost burden on the society.

11.4   Risk Management Framework for Pure Risk

Enterprise risk management (ERM), with it modern scope beyond the traditional view restricted to pure risk, is the process of developing a risk management strategy by following a similar sequence of steps as given in Figure 2.1 of Chapter 2. In ERM, an enterprise follows the flowchart to develop a strategy for risk management of all the following components of its risk exposures.

- Pure risk

- Speculative risk

- Strategic risk

- Operational risk

The risk categories may need to be further refined for getting a handle on the firm's multitude of risk exposures, and how they may affect the firm. It will also be necessary to consider interactions between category of risk types and specific risk exposures in order to not get too restricted by silos created to aid risk management process, since in reality risks don't evolve in isolation.

Our focus in this section is to delve deeper into consideration of techniques for the management of pure risk. In this section, our reference entity for development of risk management strategy is any enterprise or household, while in the next section, we shift our attention to insurance providers. We consider the major risk management considerations an insurer must adopt for effectively managing the large pools of pure risk it collects by nature of its business.

Pure risk, by definition, has no upside. Therefore, the most obvious and dominant response to managing pure risk should be to avoid it. However, despite our best efforts, we can't rid ourselves of a large set of pure risks. The trick is through a formal risk management process, we are made aware of the pure risks we have chosen to be exposed to, and have actively developed a strategy for response should adverse events occur. Methods adopted for handling risk, with now insurance available as an added tool in our toolkit for risk management, are as follows.

- Avoidance

- Loss control

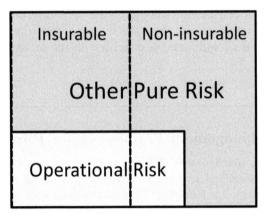

FIGURE 11.2: Relation between pure risk, operational risk, insurable versus non-insurable pure risk.

- Retention

- Non-insurance transfers

- Insurance

These are not too different from what was given in the flowchart of Figure 2.1.

In developing risk management strategies, insurance serves as a tool for the transfer of risk, much like derivatives we have studied in previous chapters. Insurance, however, pertains specifically to pure risk. It is interesting to compare insurance as a new mechanism for risk transfer we are adding beyond the other hedging instruments we have already investigated for risk transfer. While both insurance and other hedging instruments result in transfer of risk to a counterparty by means of a formal contract, they aren't the same. There are some key differences. An insurance transaction needs pooling of risks, and hence involves the transfer of only an insurable risk. Other hedging instruments did not have such requirement for the underlying risk being transferred. Insurance reduces the objective risk of the insurer, again due to the pooling of risks and the application of the law of large numbers. The counterparty of other hedging instruments may not necessarily experience risk reduction due to pooling.

We have seen several components of speculative risk in great detail, and considered the avoid-mitigate-transfer-keep response to each of them. In the last chapter, we had developed the assessment of operational risk, which was our first foray into the pure risk domain. The development of risk management for operational risk had shed some light on how one would progress with management of pure risk in general using insurance. However, operational risk is only a subset of a larger set of risks that fall under pure risks. We expand our attention to a larger set of pure risks, bearing in mind the ideas and guidelines brought forth in the context of operational risk, and how they

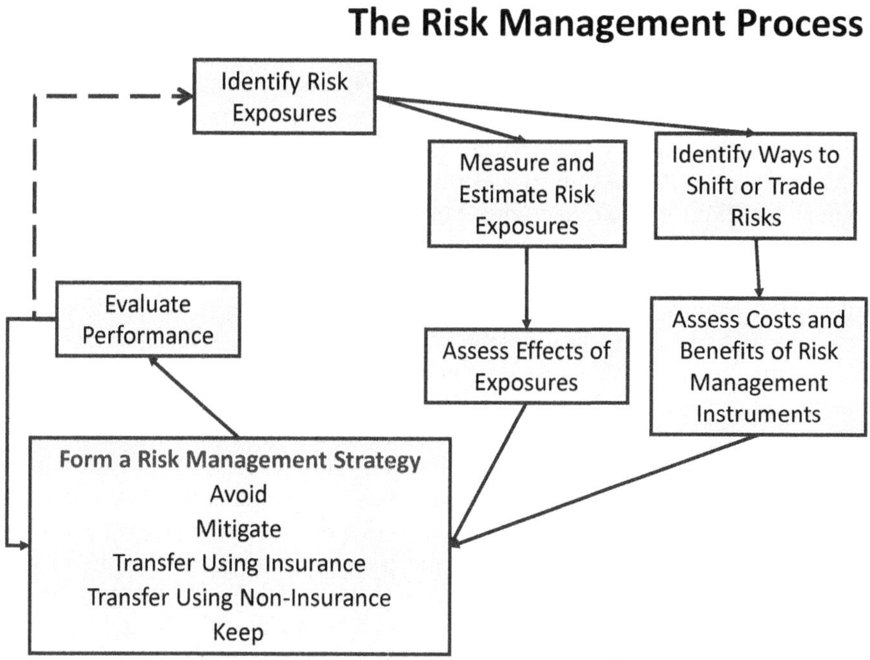

The Risk Management Process

FIGURE 11.3: The overall flowchart for the risk management process for pure risk.

may be useful here. Figure 11.3 is a reproduction of Figure 2.1 with minor modifications to highlight the use of insurance for pure risk transfer, and the fact that not all pure risks are insurable. Figure 11.2 highlights the relationship between operational risk and pure risk, and the differentiation of pure risk into insurable versus non-insurable.

Following the flowchart in Figure 11.3, we need to first identify potential pure risk exposures, and identify the loss-levels possible due to each pure risk, in the personal, property and liability categories. This is the crucial step of risk identification. The next step is to evaluate and measure the impact of each risk exposure for the potential losses it can cause. This involves estimating both **loss frequency**, i.e., the probable number of loss events that may occur during a given period of time, and **loss severity**, i.e., the probable size of loss that may occur with each event. This is the risk evaluation step. Finally, at the bottom of the flowchart, the best responses must be generated for each risk exposure. No risk management strategy is complete without feedback loops for assessment and maintenance of the strategy for its relevance and effectiveness.

Risk identification is a non-trivial exercise, especially so for the large and complex structure of a large corporation. However large or small an enterprise, or even for an individual or a household, a formal process for identifying the potential pure risk exposures is strongly recommended. It involves using a range of tools and techniques to elicit the possible losses a firm or a household may suffer from different divisions of its organization or different facets of function. Risk identification tools include risk analysis questionnaires, exposure checklists, insurance policy checklists, and expert systems. Risk analysis questionnaires must be designed as 'fact finders' for discovery of both insurable and non-insurable pure risks through a series of detailed and penetrating questions. Exposure and insurance policy checklists are often produced by insurance providers, therefore would have bias to insurable risks and specific products the provider offers, but nevertheless may be a good guide to risk discovery. Finally, expert systems can combine the insight of questionnaires and checklists to help firms identify their potential pure risk exposures.

Techniques for risk identification include orientation, analysis of documents, interviews, and inspections. The risk management decision makers, risk managers, should be well oriented to the firm's organization, its operations and goals. A detailed evaluation of the firm's annual reports, financial statements, loss reports, various sales, service lease agreements, building and other property appraisals, etc., is revealing in terms of exposing risk exposures. Finally, interviews with key personnel and physical inspection of property may be ideal to ground data in reality. The classifications developed on several themes, such as personal, property and liability; particular versus fundamental; perils and hazards can also serve as a road map to elicit all relevant pure risks of a firm. Lastly, risk identification points to measures and data to use to evaluate the impact of pure risks.

11.4.1   Pure Risk Evaluation

Assessing the risk exposures for their loss impact is crucial to developing a risk management strategy. In this assessment, while frequency is important, severity is more important, since a single catastrophic loss can wipe out the entire firm. As a preliminary step of risk evaluation, it helps to have guidance by considering upper-limit of frequency and severity. *Maximum Possible Loss* is the worst loss that could happen, whereas *Maximum Probable Loss* is the loss that is most likely to happen, or the mode of a unimodal loss distribution. For a risk exposure, it may be possible that the maximum possible loss is a very large quantity, however its likelihood is minuscule, while the maximum probable loss level is not very devastating. For instance, in Figure 11.4, both the loss distributions have the same maximum possible loss level, but the maximum probable loss level of one is much lower than that of the other. Therefore, these two quantities provide good preliminary guidance on the general shape of the loss distribution.

There are both direct and indirect losses when pure risk events occur, both of which should be accounted for while assessing loss distributions. We had made a similar distinction while assessing impact of operational risk in Chapter 10, where operational losses were computed in terms of three key terms, cost to fix, write-downs, and resolution. Direct loss is the loss due to

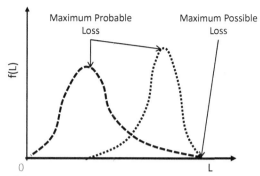

FIGURE 11.4: Maximum possible loss and maximum probable loss of two distributions. The positively skewed distribution is not as risky as the negatively skewed one.

physical damage to real property, machinery, equipment, furnishings, as well as raw materials and inventories. Indirect losses or consequential losses consist of other financial losses resulting from the damage in the form of opportunity cost.

Being able to be as quantitative as possible in risk evaluation is desirable so that the risk management decisions can be made based on objective assessments of all pure risk exposures. Models for risk evaluation must rely on availability of sufficient data, and where relevant must have the ability to make a long-term assessment of the risks. As we discussed earlier, many pure risks are long-horizon and strategic in nature, therefore their impact would only be known when prediction of the risk can be done for the long-term.

We break the task of risk evaluation into three steps; first our goal is to construct distributions for frequency of occurrence of each risk exposure anticipated for the firm. The second step would develop a conditional view of, when an event occurs, what is the loss impact of the event, both in direct and indirect terms. The third step combines the two pieces to construct the annual loss distribution due to risk exposures, or present value of future loss exposure if the impact of the risk will only be revealed in the long-term future.

The first step involves probabilistic analysis of frequency of occurrence of loss events under given conditions. For each risk exposure identified in the risk identification stage, estimates must be created of how likely it is for loss events to occur for that risk exposure under the status quo, as well as when some mitigative actions were taken. Loss frequency distributions, using standard distribution models, such as Poisson, binomial, negative binomial, normal, etc., can be fitted for each case depending on the summary statistics and empirical distribution of available data. In case of limited data, best judgment on the shape of the distribution may be needed, with key observations such as maximum and minimum frequency of events guiding the actual construction of the distribution.

The second step develops loss severity distribution, i.e., when an event occurs, what is the level of loss incurred. Some of the standard distribution models considered suitable for severity distribution include lognormal, exponential, gamma, and Pareto. We had studied all these distributions in Chapter 1, except Pareto distribution. Pareto distribution is related with the exponential distribution; if X is exponentially distributed with parameter λ, then $Y = \alpha e^X$ is Pareto-distributed with minimum α and index λ. Therefore, Pareto distribution is related to exponential distribution in the same way as lognormal distribution is related to the normal distribution.

A note of caution regarding picking a particular distribution for no other reason than the familiarity with it. There should be some empirical justification for the choice of distributions for the exercise to be sincere to reality. In the case of both frequency distribution and severity distribution, for lack of more suitable alternatives, empirical distributions constructed based on observed data may also be used. Finally, regression modeling can prove to be very useful for predicting losses when loss can be determined as a causal or correlated relation with much more readily observed quantities. For instance, there may be a meaningful relation between number of employees on payroll and the number of workers compensation claims made in a year. Using such loss predictors by capturing correlation and/or causal relations for predicting future losses can be an effective tool.

Finally, the models for loss frequency and loss severity must be combined to construct the total annual loss distribution or discounted life-time loss distribution. Combining of frequency distribution with severity distributions can be achieved through convolutions of their individual distribution functions. The probability distribution of the sum of two or more independent random variables is the convolution of their individual distributions. If L_i is the loss severity at each occurrence of a loss event, and in a period of time $(0, t]$, number of loss events experienced is, N_t, then the total loss experienced in the given time period is,

$$TL = \sum_{i=1}^{N_t} L_i. \tag{11.4}$$

We encountered a convolution of this kind in Chapter 7 in the context of the impact of sequence of jumps on the stock price in a jump-diffusion model (Section 7.2.6.4). For a general choice of frequency distributions for N_t among Poisson, binomial, negative binomial, normal distributions, and choice of loss severity distribution for L_i among lognormal, exponential, gamma, and Pareto distributions, analytically obtaining the distribution of total loss is non-trivial. However, estimation using simulation is straightforward. Moreover, the risk evaluation framework set up in this section will also help assess efficacy of risk management responses for development of a risk management strategy.

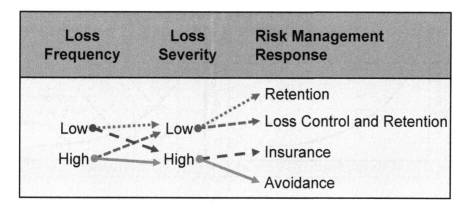

FIGURE 11.5: Risk management guideline for loss event frequency and severity for pure risk exposure.

11.4.2 Risk Management Strategies for Pure Risk

The bottom box of the risk management flowchart of Figure 11.3 has decisions regarding avoid, mitigate, transfer using insurance versus non-insurance, and keep, which must be made for each pure risk exposure of the firm. In the risk identification and risk evaluation stages, all the pure risk exposures would have been identified and assessed for their loss impact. This facilitates the next step of developing a risk management strategy for each of these risk exposures.

Risk evaluation focuses on loss frequency and loss severity, which make for the two crucial dimensions for developing a risk management response. On these two dimensions, by identifying two coarse levels of 'high' and 'low,' a preliminary understanding of risk management response can be developed. These cases and responses are summarized in Figure 11.5. The preliminary response can be broken down into the following four guidelines depending on the combination of loss frequency and severity anticipated from the pure risk. If the loss frequency and severity from the pure risk events are both anticipated to be low, it may be quite possible to retain the risk. Retention of the risk implies that if and when events happen by that pure risk, the firm must indemnify itself. Therefore, the firm needs to determine the level of loss the firm should be prepared to withstand, and the amount of funds to allocate for the purpose of indemnification of retained risk. This is also called 'self-insurance.'

If the loss frequency is low, but severity of loss can be high when loss events occur, this would be a good candidate for considering a transfer of the risk. If insurance is available for this pure risk, the firm should consider this mechanism for risk transfer. If insurance is not available, then the firm must explore opportunities for non-insurance transfer, or in fact consider retaining the risk. Funding this retained risk could be much more costly, hence some loss control or prevention would be necessary. Non-insurance transfers are

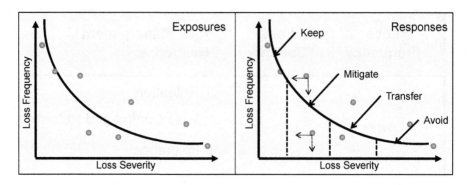

FIGURE 11.6: Firms' risk tolerance can be represented as a loss severity versus frequency curve. Avoid-mitigate-transfer-keep decisions are made relative to this risk tolerance curve in a systematic and quantitatively supported way.

methods other than insurance by which pure risk is transferred to another party, for example, outsourcing, contracts, leases, etc. For pure risks that are frequent, but less severe, risk control is possibly the best response. This will consist of developing methods for loss prevention or reduction. Finally, the high frequency and high severity pure risks should be avoided whenever possible.

A preliminary sketch of a risk management response can be developed into a complete rigorous analysis. A firm should construct its pure risk tolerance profile in terms of a risk frequency and severity frontier. Each pure risk exposure is marked on this frequency-severity grid, where the frequency is a summary statistic of the frequency distribution, N_t, constructed in the risk evaluation stage, and the severity is a key summary statistic of the severity distribution, L_i. In the left panel of Figure 11.6, all the firm's risk exposures are plotted on the frequency-severity grid. The exact location of each risk on the grid prompts the development of an appropriate response.

Further to the guideline developed based on the categories of Figure 11.5, now each risk can be quantitatively assessed for what response may be most appropriate against the firm's risk tolerance. The left-most risks are lowest in severity and fall below the risk tolerance curve, therefore they are perfect candidates to be retained, and the loss from these events is considered regular cost of conducting business.

Moving right on the severity axis are risk exposures with relatively higher loss severity, with one much lower frequency of occurrence, while the other has a higher frequency than the risk tolerance frontier would permit. The latter is a perfect candidate to conduct a complete assessment of loss control and prevention response. Based on the framework for total loss developed in the risk evaluation stage, an economic assessment of the loss control investment decision must be conducted. If the difference between expected baseline total loss, $E[TL]$, and the expected total loss after investments on loss prevention, $E[TL(p)]$ is greater than the investment made towards loss prevention, after accounting for time value of money wherever relevant, then loss prevention is a suitable response for these risks.

The next set of risks of higher loss severity are candidates for risk transfer. Even when their frequency-severity is below the firm's risk tolerance, it may not be appropriate to self-insure these risks since that may not be the best use of the firm's capital and would be an unnecessary digression from the firm's core activities. Finally, the far right risks at the highest severity level are catastrophic, and must be avoided whenever possible. It should be noted that the scale of the severity axis is relative to a firm's size, since what is severe for one firm may not be severe for another larger firm in absolute terms.

Insurance is the most relevant mode for pure risk management when the pure risk causes a high severity of loss, but with low probability of occurrence. This happens to be the case when the pure risk may be insurable. When implementing a risk management decision of transferring a pure risk by means of an insurance contract, the following additional five key issues must be addressed.

- Choice of insurance as a mechanism for risk transfer entails additional decisions regarding the exact coverage needed under the insurance contract. Pricing of insurance, and therefore cost of risk transfer, depends on the coverage and deductibles of an insurance contract. Selection of appropriate coverage should depend on the assessment of the extent of losses expected, while choice of deductible essentially points to tolerance of degree of risk retention.

- Along with the features of an insurance contract, a selection of an insurer must also be done. Given there are literally thousands of insurance providers for different kinds of personal, property, and liability insurance, this choice may be non-trivial. Several important factors should be considered while making this decision, such as, financial strength of the insurer, risk management services offered by the insurer that cater to specific customer needs, cost and terms of protection, and after-sale quality of service.

- As in the case of any bilateral contract the terms of negotiation should not be neglected. The language and meaning of the contractual provision should be clear to both parties, and everyone affected by the contractual provisions. Depending on the negotiation power of the insured, premium, coverage, etc., may be negotiable.

- Once the contract is in place, it is imperative that all the relevant information concerning the insurance coverage is properly disseminated. Efficient and effective implementation of risk management using insurance requires employees to know what services the insurer provides, and what precautions and preventions are necessary.

- As stated earlier, no risk management strategy is complete without feedback loops for assessment and maintenance of the strategy for its relevance and effectiveness. Therefore, periodic review of insurance program is necessary.

In the discussion above for developing risk management strategy for pure risk, we have focused on issues from a firm's perspective. Even though the development was done with firms as reference entities, the process is just as applicable for individuals as for institutional risk management. We consider a specific example for personal insurance decisions.

11.4.3　Modeling Individual Mortality Risk

The highest level classification of insurance is done as life versus non-life insurance. This is justified, since according to The Geneva Association, life insurance accounts for US$2.54 trillion annual worldwide premium volume, which is 58.44% of total worldwide premium income. We use this context to develop a dynamic evolution model of health risk, which creates a basis for a framework where important personal insurance decisions may be made. An explicit health evolution model is calibrated to assess an individual's lifespan distribution [37]. The most commonly used health model in health economics literature is the Grossman's [34] health evolution model.

In this framework, health is viewed as a durable stock of capital, whose inputs are medical care and time, and outputs are consumption and investment ability, usually defined as the total number of healthy days. An individual is assumed to inherit an initial stock of health. His health capital depreciates with age at a time-varying rate, which increases with age after a certain stage in the life cycle, and the health capital can be improved by health investment. Death occurs when an individual's health capital drops to a certain level. The shadow price of health depends on many other variables, besides the price of medical care, for instance, on age and education. The model can be summarized by the following relationship [69]:

$$H_{t+1} = (1 - \delta_t) \cdot H_t + \epsilon_t + \theta_1 \cdot M_t^{\theta_3} + \theta_2 \cdot \epsilon_t \cdot M_t^{\theta_3'}, \qquad (11.5)$$

where H_t and M_t stand for health capital and expenditures on health, respectively, in period t and δ_t is the health depreciation rate. The health depreciation rate is assumed to be time-dependent and can be described as an increasing function of age, such as ae^{bt}, where a and b are parameters. θ_1, θ_2, θ_3, θ_3' are parameters specified to relate the effect of health expenditures on health capital.

The above model can be calibrated for specific risk class of individuals, and used for addressing strategic decisions regarding annuity [36], life insurance, and long-term care [35] products from an individual's or household's perspective.

11.5 Risk Management by Insurers

Enormous volume of risk is collected by insurers through the transfer of risks via insurance contracts underwritten by the insurers. The 2009 figures of insurers' global financial assets under management of US $22.6 trillion, with worldwide insurance premium volume standing at US $4.34 trillion is a testimony to this fact. Insurers not only must manage these funds, but more importantly, must honor the terms of the underwritten contracts and deliver on their liabilities underlying the contracts. Moreover, as noted in Chapter 3, firms in banking and insurance sectors hold vast sums of money in trust for the public, therefore are subject to government regulation due to their fiduciary nature. Regulations are designed to ensure insurers are well-prepared to respect their fiduciary responsibilities, and provide their services fairly and reliably. Therefore, risk management must be at the core of every single activity of an insurer's business. In this section, we discuss the risk management considerations of an insurance provider, from pricing of insurance contracts, to asset management in support of the liabilities, and securitization and reinsurance activities of insurers.

11.5.1 Pricing, Investment, and Asset-Liability Management

In insurance, unlike many other products and services, the cost of production for the service to a specific insured is not known in advance. This is because the exact loss exposure due to that insured is not known. Therefore, setting price for insurance contracts that recover all the costs and possibly yield a profit is a difficult task. Pricing in insurance is a non-trivial endeavor for which specialized, well-trained professionals are engaged. These highly skilled mathematicians specialize in the discipline of actuarial science, and are called actuaries. The Society of Actuaries (SOA), established in 1889, which is now a 22,000 strong membership organization, defines an actuary as follows: 'An actuary is a business professional who analyzes the financial consequences of risk. Actuaries use mathematics, statistics, and financial theory to study uncertain future events, especially those of concern to insurance and pension programs. They evaluate the likelihood of those events, design creative ways to reduce the likelihood and decrease the impact of adverse events that actually do occur.' Professional certification as an actuary is attained by passing a series of examinations administered by the SOA, which makes them a Fellow of the society.

Pricing

An actuary determines the rates for different types of insurance products, in the personal, property, and liability categories using a range of appropriate mathematical and statistical tools. In fact as suggested by the SOA definition of an actuary, actuaries are involved not just in pricing of insurance contracts, they are engaged in all phases of an insurance company's operations. The objectives of pricing or premium determination for an insurance contract are many fold. Prices are determined to allow paying claims and expenses as they occur, satisfy any regulatory requirements, to make the business profitable, and to enable the provider to be competitive with other insurers. Therefore, the objectives of pricing can be classified as regulatory objectives and business objectives, both of which must be incorporated into pricing decisions.

As studied in Chapter 3, regulations play a prominent role in the insurance industry. Regulatory objectives of protecting the consumers require that the rates being charged by the insurers are high enough to pay all losses and expenses. Failing which the insured and the beneficiaries are at a loss if they don't receive the benefits of the insurance. While the rates are adequate, they should not be excessive. The premium for insurance should not be so high that the insured ends up paying more than the actual value of their protection. Finally, in determining the rates, the provider should not be unfairly discriminatory, i.e., insureds that result in loss exposures of similar risk characteristics and expense should not be charged substantially different rates.

The business objectives of pricing are to build a viable and profitable enterprise. Therefore, besides assuring loss recovery, operational expenses, from a business perspective pricing should also be simple. Simple pricing is easy to understand and use both by sales agents and customers, therefore reducing training and operational risk. Prices should be responsive and stable. The responsiveness of pricing comes from its ability to respond to changing loss exposure and economic conditions, which also makes the pricing stable. Nonresponsive and rapidly changing prices can result in costly adjustments and implementation. Given insurance is targeted to pure risk, and the fact that best response to pure risk is avoidance, the pricing scheme should encourage responsible behavior towards loss control. This should also be done with an eye to reduce moral and morale hazards.

Insurance industry utilizes specific terms for the pricing or rate making of its services. A *rate* is the price per unit of insurance, while an *exposure unit* is considered the unit of measurement used in insurance pricing, which may differ depending on the type of insurance. For auto insurance, it could be one car-year of auto insurance, for liability risk, it could be $1000 worth of liability insurance coverage. Given the different objectives to satisfy by price determination, *pure premium* is the portion of the rate needed to pay losses and loss-adjustment expenses for one exposure unit. The pure premium is adjusted by a *loading*, which is the amount that must be added to the pure premium for other expenses, profit, and margin of contingencies. Combining the pure premium and loading gives the *gross rate*, and the *gross premium* charged to an insured is the number of exposure units the insured requires times the gross rate.

The most popular method for insurance pricing is class rating pricing, in which exposure units with similar characteristics are placed in the same underwriting class, and the premium rate is determined for the class. Therefore, it is important to appropriately define a class. The rate then reflects the average loss experience for the class as a whole. This is very similar in essence to the credit scoring framework for pricing in consumer credit developed in Section 9.1.1 of Chapter 9. The techniques of regression modeling, discriminant analysis, and classification trees developed in that context would be valuable here, with the main difference being that the characteristics and attributes to consider here would differ. While the consideration in credit risk for characteristics and attributes were all only financially motivated, in insurance the nature of peril and physical hazard relevant to the unit, as well as type of personal, property, or liability insurance sought will dictate the characteristics and attributes to consider in the models.

Rate determination using these methods requires data, both from the insurance provider's past loss experience, as well as industry statistics and data sources, such as from rating organizations like Insurance Services Office (ISO). A well developed class rating approach achieves most of the pricing objectives discussed above, except one: creating incentives for the insured to participate in loss control. Merit rating, as a complementary approach, is when rates may be adjusted upward or downward due to merits of a specific exposure unit, such as, some of its special features within a class, past experience, or in other words, ways to create incentive for loss control.

Investment

Given the volume of global assets under management of insurers, it is safe to say that investment is an extremely important operation in an insurance company. Premiums are paid in advance, and are accrued in the statutory accounting over a period of time. However, indemnification happens after varying degrees of lag from the time of payment of the premium. The accumulated funds must be invested until funds are needed to pay claims and expenses. The funds available for investment are derived primarily from premium income, but they are also generated from investment earnings and maturing investments that must be reinvested. Therefore, the investment activity of an insurer is often a large asset management enterprise in its own right.

As discussed in Chapter 3, the accumulated premium must be invested by the insurer, however there are regulatory restrictions imposed on the investment choices of insurers. The restrictions are imposed to ensure that insurance companies reliably fulfill their fiduciary obligations. Investment is allowed in US and Canadian government bonds, mortgage loans, certain high-grade corporate bonds, and to some extent in preferred or common stock. There are additional restrictions by the lines of insurance products an insurer offers. For instance, property and liability insurers have lesser investment restrictions than life insurers, where the latter can invest only a small percentage of their assets in common stocks.

Many insurance contracts are long-term, such as life insurance and annuities, and hence make the premium income available for investment for long durations of time. On the other hand, the value of assets must be preserved for the long-term, without exposing to significant downside risk along the way, in order to meet the long duration liabilities. As a result, life insurers have a large fraction ($> 50\%$) of funds invested in government or corporate bonds for safety of principal, and maintain only limited exposure to ($\sim 20\%$) equity and other asset classes. For shorter duration insurance contracts, such as property and liability insurance, investment needs to be more liquid. Investments are predominantly in bonds, preferred and common stocks, and not in real estate.

Our analysis of equity risk and portfolio optimization framework developed in Chapter 7, as well as hedging strategies developed in the context of equity risk to construct specific risk-return profile of equity investment are clearly applicable here. Similarly, bond portfolio analysis and interest rate risk management techniques developed in Chapter 8 are fundamentally relevant here, given the heavy weight given to investment in this asset class. Additionally, fixed income derivatives discussed in Section 8.2.2, such as bond futures, bond options, and exotic option may be utilized for risk management of insurer's investment strategies.

For insurers, investment is not just a value-preserving proposition of accumulated premium, since investment income can time and again prove to be extremely important in offsetting unfavorable underwriting outcomes. Additional benefit to society of the funds resulting from accumulated premium and investment income of insurers is that these funds serve as an important source of capital for the economy.

Asset-Liability Management

The investments in government bonds, corporate bonds, preferred and common stocks, and cash held by an insurer constitute the assets of the insurer, while the portfolio of outstanding insurance contracts underwritten by the insurer are the insurer's liabilities. The cash-flow from assets must support the cash-flow requirements of the liabilities for the insurer to not default. Additionally the value of the insurer's assets should remain higher than the expected net value of the insurer's liabilities for the insurer to remain solvent.

Ideally, the risks underlying the assets should match the risks of the liabilities of the outstanding insurance contracts. This is more difficult to achieve for an insurance provider than it is for a commercial bank, since the risk factors that drive the value of an insurer's assets, such as government or corporate bonds, stocks, etc., can be quite different from the pure risks that generate the liability cash-flow for the insurer. Part of the risk on the liability side of an insurer's balance sheet is managed by how the insurance provider prices and underwrites its insurance products. The asset-liability management framework developed in Chapter 10 is applicable here to develop an asset management strategy that best meets the asset-liability management objectives.

Tools of ALM, such as gap analysis, duration gap analysis, and stochastic gap analysis can be applied to develop insight for mismatch in levels and risk of assets and liabilities of the insurer. Simulation analysis can be applied to develop scenarios of balance sheet risks, or to develop estimates of mismatches in terms of measures like long-term Value-at-Risk (LT-VaR). Usually these insights can be translated into guidance for short-term (business) and long-term (strategic) course of action. Being able to better manage the risks in its balance sheet can give a significant competitive edge to an insurance provider.

11.5.2   Risk Management, Securitization, and Reinsurance

Insurance, like banking, functions under a rigorous regulatory framework. As discussed in Chapter 3, the objective of the regulatory framework is that the insurers are not taking on excessive risk without safeguarding the goal of reliably supporting their liabilities. Therefore insurance laws take specific note of insurer's fiduciary obligations, and require them to maintain policy reserves and risk-based capital. Insurers maintain unearned premium reserve to respect the difference in timing of premium payment and service delivery, and maintain loss reserves to respect the gap of time between occurrence of loss events and their indemnification.

Asset-liability management discussed in the previous section serves the very important risk management objective of attempting to match the cash-flow and risks of assets with those of the liabilities. The objective of ALM is to mitigate the risks in the balance sheet by such matching of assets and liabilities. However, one challenge in insurance is that the risk factors underlying the liabilities are of a varied kind, and may be quite different from the risk factors of the assets. There are new instruments emerging that can alleviate this problem.

As an alternative to investment in traditional assets, an insurer can utilize new financial instruments developed in the capital markets to help support indemnification of catastrophe losses. For instance, a property and casualty insurer can invest in instruments like catastrophe futures or catastrophe options. These catastrophe derivatives started trading in 1992 on the CBOT, however were delisted in 2010 due to limited trading. The value of catastrophe futures contract increased when catastrophe losses were high and decreased when catastrophe losses were low. The contract was designed to track catastrophe loss indices developed by Property Claim Services (PCS), which provided nine loss indices each day to the Chicago Board of Trade (CBOT) based on the estimates of insured catastrophe losses in different parts of the US.

Although catastrophic futures and options have discontinued trading on CBOT, weather derivatives, defined on temperature, hurricane, frost, snowfall, and rainfall for various geographical locations are traded on the Chicago Mercantile Exchange (CME), which serve as instruments with higher correlation with cashflow from liabilities, as well as provide geographical diversification. These instruments can help match property and casualty risk underlying an insurer's outstanding insurance contracts. In the event of a weather-related catastrophe, if losses are high, the value of the contract goes up and the insurer makes a gain that hopefully offsets whatever losses it may incur through its liabilities. The reverse is also true. If weather-related catastrophe losses are lower than expected, the value of the contract decreases and the insurer loses money.

Insurance providers can also securitize the pool of risk underlying their outstanding insurance contracts. Catastrophe bonds, also called cat bonds, are special bonds issued by insurers to help them pay for natural disaster losses, such as losses from hurricanes and earthquakes. These are usually rated below investment grade and pay relatively high yield. The payment structure of these bonds is adapted to aggregate catastrophe events. If catastrophes are below a certain level during some specified period, the bond investor receives the principal with interest. However, if losses exceed the specified level, bond investors forfeit part or all of the interest or principal, or alternatively the payment of the principal is deferred. These bonds are attractive for institutional investors seeking high-yield, fixed-income securities.

Innovations in securitization for mortality and longevity risk is also developing, which can be utilized by life insurance providers. There are numerous types of products, ones that already exist and those that are proposed, to hedge the risks of mortality-sensitive products. Mortality bonds are essentially securitization of mortality risk. The bonds are issued and coupons are paid, but the principal is preserved only if a chosen mortality index is near the base level. If the mortality index rises, the principal erodes by a given schedule of percentage rise of the mortality index relative to the base level. So far, these are available as reasonably short-term bonds, usually with 3-5 years maturity. Mortality bonds designed for extreme mortality risk, such as due to a large scale major epidemic, major natural disaster, are catastrophic mortality bonds.

Longevity bonds are longer maturity bonds with maturities ranging from 20-30 years, where the coupon payments are linked to a defined cohort survivor index. Therefore, the cash flows of longevity bonds are designed to help annuity providers and pension plans to hedge their exposure to longevity risk since they are designed to meet the providers' commitment of providing level payments to the reference population over a long time horizon, depending on the population's longevity characteristics. The bond provides the investor an annual mortality-dependent payment up until the maturity of the bond. The mortality-dependence of the annual payments may be based on a publicly available death rate index for each age-group published for the year. These bonds have no terminal repayment of the principal.

Mortality or longevity swaps are mostly over-the-counter contracts, hence have the attraction that they can be customized to the particular requirements of a user, but would have a thin secondary market resulting in low liquidity. A mortality swap is a contract to exchange one or more cash flows in the future based on the outcome of a stochastic mortality index. A vanilla mortality swap may be designed similar to a floating-to-fixed interest rate swap, in which the fixed leg is linked to a published mortality projection and the floating leg is linked to the counterparty's actual realized mortality rate. A formal framework to assess the effectiveness of hedge using these instruments can be developed for constructing an optimal hedge strategy [32, 38].

Finally, an insurance provider can seek reinsurance to manage the residual risk not accounted for by any of the mechanisms discussed thus far. Reinsurance is the shifting of part or all of the insurance originally written by one insurer to another insurer, which is the transfer response of risk management. This transfer has a flavor similar to securitization involving repackaging and transferring of risk. The insurer who initially writes the business is called the ceding company, and the insurer that accepts part or all of the insurance from the ceding company is called the reinsurer. The amount of insurance retained by the ceding company for its own account is called the net retention or retention limit. The amount of insurance ceded to the reinsurer is known as the cession.

Just as securitization, reinsurance can help increase the underwriting capacity of an insurer. By transferring some of the tail risk through reinsurance, an insurer can reduce the volatility in profits. Reinsurance can be an additional mechanism for obtaining protection against catastrophic loss. It can provide considerable protection to the ceding company that experiences a catastrophic loss, since under the reinsurance contract the reinsurer is liable to pay part or all of the losses that exceed the ceding company's retention limit.

11.6 Summary

Pure risk, as discussed in Chapter 1, affects everyone, and its impact on a household or a firm, based on the nature of exposure, can be quite catastrophic. Insurance contracts are effective tools for transfer of pure risk, therefore, in this book we have treated discussion of insurance as an integral component of developing a risk management strategy. To incorporate insurance as a tool for risk management, we first look at the basic principles behind insurance contracts, which directly relate to the types of pure risk that can be insurable. To give a further classification of pure risks, we describe types of insurance and their features available for individuals and firms to offset their pure risk, along with issues related with moral hazard that bring inefficiencies in insurance markets. A specialized risk management framework for developing a strategy for managing pure risk of a firm or household was developed, along with a discussion of some example context of application of the framework. Finally, we studied the tools available to an insurance provider to manage risks in their own turn through securitization and reinsurance.

11.7   Questions and Exercises

Review Questions

1. How does the traditional view of risk management differ from the modern view of risk management?

2. Why are the boundaries between pure and speculative risks getting less obvious and clear?

3. What is enterprise risk management?

4. What are the distinctions between personal, property, and liability pure risk?

5. What is the difference between objective and subjective assessment of pure risk?

6. How are perils distinguished from hazards?

7. Discuss the different types of hazards that can lead to loss.

8. What is the distinction between fundamental and particular pure risk? Why is particular risk more amenable to being insurable?

9. What is meant by indemnification of losses?

10. What is the basic principle behind insurance provision? What is meant by spreading the losses from pure risk events?

11. When are losses not fortuitous? For insurance to be viable, why is it necessary that losses be fortuitous?

12. State the law of large numbers (LLN), both the weak and the strong version.

13. What are the requirements for a pure risk to be insurable? Discuss the requirements in detail.

14. Why is the first order of classification of insurance in terms of life versus non-life insurance? What are the types of life insurance products?

15. What are the different types of health insurance? What are group health insurance products?

16. What are the different types of property and casualty insurance products? What kind of insurance are homeowners insurance and automobile insurance?

17. What are commercial package policies?

18. Give examples of government programs that provide insurance against fundamental pure risks.

19. What are all the benefits of insurance to the society? What are the costs of insurance to society?

20. How is hedging using derivatives different from risk transfer using insurance?

21. How does operational risk relate to the larger set of pure risks, as well as with insurable versus non-insurable risk? Given examples of each.

22. What is risk identification? What is the crucial outcome of risk evaluation?

23. Why are feedback loops for assessment and maintenance of risk management strategy important?

24. What are the tools and techniques utilized for risk identification?

25. What are the three steps of risk evaluation? How does each step get accomplished?

26. What are common distributions used to model loss frequency and loss severity?

27. What is Pareto distribution? How can it be used for modeling losses due to pure risk?

28. What is self-insurance? When is it advisable to resort to this risk management response?

29. How is the frequency-severity tolerance curve for a firm useful in developing a risk management strategy for pure risks?

30. How would you evaluate the cost effectiveness of loss control and prevention response to a pure risk?

31. When implementing a risk management decision of transferring a pure risk by means of an insurance contract, what additional issues must be considered?

32. What is actuarial science? What does an actuary do?

33. What are the regulatory and business objectives of rating making in insurance?

34. What is class rating pricing? How does it achieve the regulatory and business objective of pricing in insurance?

35. What is merit rating pricing in insurance? What objective does it achieve?

36. By what considerations must insurers develop their investment strategies?

37. How is asset-liability management important for insurers? How is ALM for insurers different from that of banks?

38. How can securitization be used by an insurer to manage risk?

39. What is reinsurance? What risk management objectives can reinsurance serve?

Exercises

1. Demonstrate the law of large numbers in MATLAB. There are two pure risks being assessed for their insurability. Based on data available for the risks, the first risk is summarized as a Chi-square distribution, $L_1 \sim \chi^2(\nu)$ with degrees of freedom $\nu = 10$, and the second risk is, $L_2 \sim$ Weibull(a, b) with scale parameter $a = 5$ and shape parameter $b = 0.8$.

   (a) Show that the average loss converges to the mean of individual loss distribution as the pool size increases.

   (b) If the range of acceptable premium per exposure unit is in the range, [$5, $7], do the two pure risks appear to be insurable?

2. Give and discuss three examples of pure risks that you think should be insurable, and why. Also develop three examples of pure risk that you assess to be non-insurable, and explain why.

3. Perform a risk identification for the following stylized contexts. Identify and list all the typical pure risk exposures you anticipate the entity to have. Conduct necessary online research to explore the possible risk exposures and their impact on the entity.

   (a) A household of five individuals, consisting of two adults active in the labor market and three school-going kids.

   (b) A small service enterprise engaged in offering small business services of photocopying, digital printing, signs and graphics, and professional finishing services.

   (c) A large shipping company that offers various shipping services for the delivery of packages and freight.

4. For each of the pure risk exposures identified in Exercise 3, conduct a risk evaluation and assessment. Determine what the possible direct loss and indirect loss may be in each case. Create a rough assessment in each case of the maximum probable loss and maximum possible loss.

5. Consider the following distributional fits for loss frequency and loss severity of 5 pure risks identified through the risk identification process.

   (a) $N_{1t} \sim Po(\lambda)$, Poisson distribution with $\lambda = 10$ per year; $L_1 \sim$ Weibull(a, b) with scale parameter $a = 5$ and shape parameter $b = 0.8$

   (b) $N_{2t} \sim Po(\lambda)$, Poisson distribution with $\lambda = 15$ per year; $L_2 \sim \chi^2(\nu)$ Chi-square distribution with degrees of freedom $\nu = 25$

   (c) $N_{3t} \sim Bin(n, p)$, Binomial distribution with $n = 50$ and $p = 0.05$; $L_3 \sim$ Lognormal distribution with mean $\mu_L = 100$ and standard deviation $\sigma_L = 15$

   (d) $N_{4t} \sim NegBin(r, p)$, Negative Binomial distribution with $r = 20$ and $p = 0.13$; $L_4 \sim$ Gamma distribution with shape parameter $a = 5$ and scale parameter $b = 5$

   (e) $N_{5t} \sim Bin(n, p)$, Binomial distribution with $n = 5$ and $p = 0.2$; $L_5 \sim$ Weibull(a, b) with scale parameter $a = 500$ and shape parameter $b = 0.8$

   Construct a quantitative assessment of the total annual loss of each pure risk, as well as a grand total annual loss from all the pure risks combined.

6. In the frequency-severity dimensions of loss events, if we consider two coarse levels - high-low, give examples of risks in each of the following cases.

   (a) Risk Profile: (low, low); Risk Management Response: Retain.
   (b) Risk Profile: (low, high); Risk Management Response: Insurance or Non-insurance transfer.
   (c) Risk Profile: (high, low); Risk Management Response: Loss prevention and control.
   (d) Risk Profile: (high, high); Risk Management Response: Avoid.

7. For a severity-frequency curve given by the following relation:

$$\frac{\bar{N}^2}{2300 * \bar{L}^{0.2}} \tag{11.6}$$

make the avoid-mitigate-transfer-keep decisions for the risks in Exercise 5. $\bar{N}$ is expressed as number of occurrences per year and $\bar{L}$ is the total dollar loss per occurrence.

Part IV

Advanced Simulation

Chapter 12

Advanced Simulation Topics

All tasks of risk management benefit from simulation analysis. However, every problem solved using simulation must deal with the fact that in simulation tasks are accomplished by generating samples of observations. Estimates for the quantities of interest are obtained by applying the chosen estimators to the samples generated. In all the risk management chapters, from Chapter 7 through Chapter 11, simulation was applied for risk assessment, risk monitoring and control, and risk management.

For pricing of derivatives, defined for equity, interest rate, commodities, exchange rates, and credit risks, when the choice of model for the underlying risk doesn't readily yield an analytical solution, simulation offers an alternative for price estimation. Simulation analysis is also useful when a variety of derivatives for all the risks we have studied thus far are used to develop hedging strategies for those risks. These derivatives may be defined for single instruments, but can also be defined and utilized for hedging a portfolio of instruments, such as basket options and n-th-to-default credit swaps.

For a portfolio of instruments, whether it was market risk instruments, credit risk-sensitive instruments or pure risk instruments, we proposed using simulation analysis to assess the portfolio level risk for the purpose of mitigation, transfer or keep response of risk management. Risk measures, such as Value-at-Risk (VaR) and Conditional Value-at-Risk (CVaR), often need simulation analysis for their estimation. At the portfolio level, the models of individual risks and their interactions become complex enough that analytical solutions are rarely obtainable. Risk assessment and monitoring at the portfolio level, especially to address non-stationarity of risk factors, requires extensive scenario analysis and stress testing. The high dimensionality of these problems, due to large portfolio sizes and number of risk factors, poses significant challenge for these assessments.

Finally, the key goal of mitigation response of risk management is to construct portfolios that achieve the desired risk-reward trade-off. These portfolios could consist of a variety of instruments affected by market risks, where each instrument in the portfolio may be affected by multiple risk factors and their interactions. Moreover, in Chapter 7, we had posed these portfolio optimization problems in static as well as dynamic settings. Simulation based optimization is an area of simulation modeling and analysis that allows addressing these optimization problems using simulation analysis, especially when analytical methods are not available or efficient for solving the problems. This methodology is also applicable for determining hedging strategies, both static and dynamic ones, designed to transfer risk to achieve the desired risk-reward profile.

Using simulation to solve the above problems based on estimates of key quantities developed using finite generated samples implies there will always be some uncertainty regarding the theoretical value of these estimated quantities. As was discussed in Chapter 4, and then utilized throughout the development of the topics of this book, the uncertainty of estimates are summarized by developing confidence intervals of desired confidence level.

Construction of confidence intervals relies on the distribution of the estimator for the quantity of interest, more specifically, on the variance of the estimator. For example, if the estimator is the sample mean estimator, $\bar{X}$, then we utilize the fact that $\bar{X} \approx N(\mu, \frac{\sigma}{\sqrt{N}})$, under certain conditions. From this fact, the confidence interval for the theoretical value of the quantity of interest is obtained as, $(\bar{X} - z_{\alpha/2}\frac{\sigma}{\sqrt{N}}, \bar{X} + z_{\alpha/2}\frac{\sigma}{\sqrt{N}})$. A tighter confidence interval will assure higher reliability of decisions made based on simulation analysis.

One way to make the confidence interval tighter, or more accurate, is to increase the sample size, N. We do this as the first response, however there is a limitation to take this to the extreme due to time and computational resource restrictions, as discussed in Section 4.7.1.1. The other option available is to reduce σ^2, which is the variance of random sample element, X_i, used to construct the sample mean estimator. The latter response is the initial focus of this chapter, which will benefit all the tasks we have proposed to achieve using simulation in this book.

Once we have explored a variety of variance reduction techniques, we will move our attention to simulation optimization. As mentioned, optimization problems show up in a variety of risk management contexts, such as in portfolio optimization, developing hedging strategies, and in the management of strategic, business or operational risk. We will develop the principles of simulation optimization, and discuss several methods for implementing simulation optimization in the second half of this chapter.

12.1   Variance Reduction Techniques

As the name suggests, variance reduction techniques are designed to reduce the variance of the estimator by means other than simply raising the number of simulation runs conducted to increase the sample size. Variance reduction methods are designed on a variety of themes, all geared towards the same goal of improving the accuracy of quantities estimated using simulation. Although this topic is included in the 'Advanced Simulation' section of this book, in reality variance reduction is a basic need of almost all simulation analysis, since improving efficiency for better decision making is often a necessity, not an option.

Variance reduction techniques are built around two broad strategies. The first set of strategies take advantage of tractable features of a model, inter-relation between variables of the model, to adjust or correct simulation output. The other strategy adopted in developing variance reduction techniques is by reducing variability in simulation inputs. We will consider some methods of both category.

We will discuss and illustrate the following variance reduction methods in the coming sections.

Control Variate: The control variate method utilizes information regarding correlation between variables of the model to develop a new estimator, which is designed to have a lower variance.

Antithetic Variates: This method attempts to reduce variance by modifying how random variate inputs are used to generate a sample of observation for the quantity of interest.

Stratified Sampling: In this technique, the input random variates are produced in a controlled manner to reduce the variance.

Latin Hypercube Sampling: The Latin hypercube method is most advantageous for variance reduction as the dimension of the problem increases.

Importance Sampling: This method utilizes the properties of the probability distribution of the quantity of interest to design a second probability distribution which emphasizes the 'important' observations of the first probability distribution.

It is possible, where appropriate, to attempt to combine the application of more than one variance reduction method for estimation of a single quantity, or for estimation of different quantities in a decision making process.

In general, in implementing any variance reduction technique, attention is required for how the simulation study is designed. As discussed in Section 4.7.1.1, the important trade-off to construct in any simulation based estimation is between bias, variance, and compute-time for an estimator. In case of unbiased estimators, the focus narrows down to variance and compute time. We had defined τ as the compute time for each replication towards generating a sample for estimating the quantity of interest. Including a variance reduction technique can result in an increase in compute time for each replication, hence care is needed regarding the computation burden implied by the variance reduction technique.

The guideline from Section 4.7.1.1 applies, regarding comparing the product of compute time and variance of estimators. Let's say, Θ_2 and Θ_1 are two estimators for the quantity of interest, θ, when variance reduction is applied versus not, respectively. We compare the efficiency of applying variance reduction by comparing $\tau_1 V(\Theta_1) \gtrless \tau_2 V(\Theta_2)$, if the compute times, τ_1

and τ_2 are deterministic. If the compute times are stochastic, we compare $E[\tau_1]V(\Theta_1) \gtrless E[\tau_2]V(\Theta_2)$. If the application of a variance reduction technique ends up reducing the efficiency of the estimation task, it may actually become counterproductive to use one. Therefore, attention is require for the design and implementation of the variance technique, including in terms of efficiency of the code and data management for the implementation of the technique.

12.1.1   Control Variates

The control variate method is perhaps the most effective and broadly applicable technique in finance and risk management. The method exploits information about the errors in estimates of known quantities to reduce the error in an estimate of an unknown quantity. Hence use of the word 'control,' where one variable helps control the variance in estimate of another.

Let $Y_1, Y_2, \ldots, Y_n$ be output of n replications of a simulation for a random variable, Y, where the quantity of interest, $\theta = E[Y]$. For instance, $Y_i = C_i = e^{-rT}(S_i(T) - K)_+$, where $\theta = E[e^{-rT}(S(T) - K)_+]$ is the price of a European call option defined on the underlying asset, $S(t)$, with strike price K and maturity T. Y_i are independent, identically distributed (i.i.d.), and in order to estimate $E[Y]$, we utilize the standard sample mean estimator, $\Theta_1 = \sum_{i=1}^{n} \frac{Y_i}{n}$.

Suppose for each replication, Y_i, we also calculate another output X_i, where (X_i, Y_i) are i.i.d., and the expected value of X, $E[X]$, is known analytically. We will call the variable X a **control variate**, which we will use to create a new estimator of $\theta = E[Y]$. We first define a modified replication, $Y_i(b)$, as follows.

$$Y_i(b) = Y_i - b(X_i - E[X]), \tag{12.1}$$

where b is a fixed number picked appropriately. We compute the sample mean of $Y_i(b)$, $\Theta_2 = \sum_{i=1}^{n} \frac{Y_i(b)}{n}$, which would be the control variate estimator. It is clear that this estimator is unbiased, due to the way it is constructed,

$$E[\Theta_2] = \sum_{i=1}^{n} E[\frac{Y_i(b)}{n}], \tag{12.2}$$

$$= \sum_{i=1}^{n} \frac{1}{n}(E[Y_i] - bE[X_i - E[X]]), \tag{12.3}$$

$$= \sum_{i=1}^{n} \frac{1}{n}\theta. \tag{12.4}$$

It can also be shown that the estimator, Θ_2, is consistent.

We need to assess if the new estimator in fact results in variance reduction, and what makes this possible. For this purpose, we first compute the variance

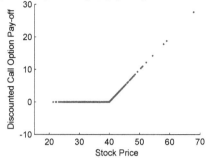

FIGURE 12.1: Scatter plot of replications of discounted European call option pay-off against replications of the stock price at option maturity. The dependence of these two quantities is as expected, which results in the strong positive correlation that the control variate method can be utilized.

of the replications, $Y_i(b)$, as follows.

$$Var[Y_i(b)] = Var[Y_i - b(X_i - E[X])], \qquad (12.5)$$
$$= \sigma_Y^2 - 2b\rho_{XY}\sigma_X\sigma_Y + b^2\sigma_X^2, \qquad (12.6)$$

where σ_Y and σ_X are the standard deviations of Y and X, respectively, while ρ_{XY} is their correlation. The variance of replications, $Y_i(b)$, is lower than the variance of the original replications, Y_i, if we can construct the control variate replication such that,

$$-2b\rho_{XY}\sigma_X\sigma_Y + b^2\sigma_X^2 < 0. \qquad (12.7)$$

Let's examine when the condition in Eqn. (12.7) can be achieved. If $\rho_{XY} \sim -1$, that is the variables X and Y are strongly negatively correlated, picking an appropriate negative value of b can satisfy Eqn. (12.7). Similarly, if $\rho_{XY} \sim +1$, picking an appropriate positive value of b can satisfy the above condition. We will be best served if we sought the best possible value of b.

The optimal choice of b is obtained by taking the first derivative of the expression on the left-hand side of Eqn. (12.7) with respect to b, and solving for its zero. We obtain the solution to be, $b = \frac{Cov[X,Y]}{Var[X]}$, which is confirmed to be the optimal by taking the second derivative of the expression on the left-hand side of Eqn. (12.7) with respect to b. In general, the theoretical value of b may not be known, therefore it must be estimated from a sample as follows,

$$\hat{b}_n = \frac{\sum_{i=1}^n (X_i - \bar{X})(Y_i - \bar{Y})}{\sum_{i=1}^n (X_i - \bar{X})^2}. \qquad (12.8)$$

Using the estimated value of b, the control variate replicates become, $Y_i(\hat{b}_n) = Y_i - \hat{b}_n(X_i - E[X])$, and the control variate estimator is, $\Theta_2 == \sum_{i=1}^n \frac{Y_i(\hat{b}_n)}{n}$. With this choice of alternate replications, if the correlation between X and Y

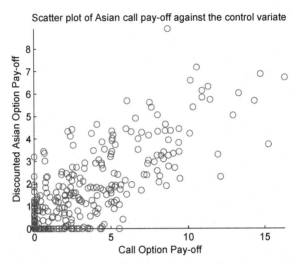

FIGURE 12.2: Scatter plot of replications of discounted Asian call option pay-off against replications of European call pay-off at option maturity. A strong positive correlation is visible, which is utilized in design of the control variate method.

is high, irrespective of its sign, we would achieve variance reduction, since the variance of the control variate estimator would be lower. We consider some examples next.

Example 1: Consider the case of pricing a European call option. We utilize the underlying asset price as the control variate for pricing the option using simulation. This choice is justified since in the risk-neutral world, we note that $E[e^{-rT}S_T] = S(0)$, therefore the theoretical mean of the control variate is known. The control variate replications are as follows,

$$Y_i(b) = Y_i - b[S_i(T) - e^{rT}S(0)]. \tag{12.9}$$

Why should this choice of control variate be successful in variance reduction? To answer this question, we need to verify the correlation between $Y_i = e^{-rT}(S_i(T) - K)_+$ and $X_i = S_i(T)$. In Figure 12.1, we display a scatter plot of Y_i and X_i. As expected, the plot shows the dependence of discounted pay-off of call option on terminal stock price, which results in the desired positive correlation.

Example 2: In this example, we go beyond the simple case of pricing a plain-vanilla European option. Pricing path-dependent options is particularly challenging, since in the path-dependent option the price of the option depends on the entire trajectory of the underlying asset price during the life of the option. Path-wise accuracy requires strong-convergence for simulation approximations, as studied in Section 6.5.2. Therefore, variance reduction can provide significant help in maintaining the accuracy and compute burden trade-off. Let's consider an arithmetic average Asian option with strike K and maturity, T.

For the purpose of pricing this Asian option using control variate, we utilize a tractable option, i.e., an option whose price is known analytically. We choose the discounted pay-off of a European call option with the same strike and maturity as our control variate, therefore $X_i = (S_i(T) - K)_+$. Under the Black-Scholes model, $C_{bls}(0, S_0; T, K, \sigma) = E[X_i]$, where C_{bls} is the Black-Scholes European call option price given in Eqn. (7.64). If the pay-off of the arithmetic average Asian option is given by,

$$Y_i = e^{-rT}(\bar{S}_A - K)_+ = e^{-rT}(\frac{1}{m}\sum_{i=1}^{m} S(t_i) - K)_+, \qquad (12.10)$$

then we create the control variate replication as, $Y_i(b) = Y_i - b(X_i - E[X])$. As stated above, the vanilla call control variate is, $X_i = (S_i(T) - K)_+$, where we know $E[X_i] = e^{rT}C_{bls}(0, S_0; T, K, \sigma)$. Similar to example 1, we can examine the degree of variance reduction obtained in this case by evaluating $\rho_{XY} = corr(e^{-rT}(\bar{S}_A - K)_+, (S_i(T) - K)_+)$. The correlation is quite visible in Figure 12.2 in which we display a scatter plot of Y_i and X_i.

In general, there is no restriction regarding the number of control variates that may be applied simultaneously. For instance, in the above example one may simultaneously apply several European call options corresponding to a range of strike prices as control variates. As seen in Figure 12.2 for a single vanilla-European call option, each of these options will have a similar correlation resulting in contribution to reduction in variance of the control variate estimator. A more detailed description and analysis of control variate based variance reduction can be found in Glasserman [30].

12.1.2 Antithetic Variables

Instead of seeking a second variable, or set of variables, which promise to have a high correlation with the quantity of interest, as utilized in control variates, we now create a negative correlation between replications themselves. This is done in a specific manner, and is successful in variance reduction under specific circumstances. Replicates are produced in pairs, where one replicate in the pair is negatively correlated with other, hence the pair is called antithetic variates.

Antithetic variates can induce a negative correlation by using a few different themes. One theme is that of generating random numbers in pairs, where the complementary random numbers are produced noting the properties of continuous uniform distribution. If $U_k \sim U(0,1)$, then automatically $1 - U_k \sim U(0,1)$, moreover if U_k is small, $1 - U_k$ is large, and vice versa. Therefore, every random number generated, U_k, is accompanied by its antithetic, $1 - U_k$, where the pair is used to generate random variates for the

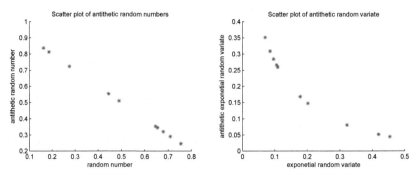

FIGURE 12.3: Scatter plot of antithetic random numbers and antithetic exponential random variates.

quantity of interest. The reader should be reminded that random numbers play a fundamental role in generating all other random variates, as presented at length in Chapter 4.

The antithetic variates method of variance reduction works when the method for generating random variates from the uniform random numbers maintains the monotonicity of the random numbers. For instance, this is achieved in the direct methods of random variates generation, such as the inverse transform method described in Section 4.4.1. An antithetic random variate pair, (X_{1k}, X_{2k}), is generated using the antithetic random number pair $(U_k, 1 - U_k)$ using the method intended to generate the random variates by the desired distribution. Figure 12.3 shows a scatter plot of antithetic random numbers, which show a perfect negative correlation. A strong negative correlation is maintained in the antithetic exponential random variates. Therefore, the use of U_k and $1 - U_k$ must be synchronized, otherwise the variance reduction may backfire.

Variance reduction is achieved from antithetic variates construction due to the following reason. Say the quantity of interest is, $\theta = E[X]$, and we generate n replicates X_{1k}, along with their antithetic pairs X_{2k}, we estimate the quantity of interest as, $\Theta_2 = \bar{X}(n) = \frac{1}{2}\left(\frac{\sum_{i=1}^{n} X_{1k}}{n} + \frac{\sum_{i=1}^{n} X_{2k}}{n}\right)$. The variance of this estimator can be computed as follows.

$$var(\bar{X}(n)) = var\left(\frac{1}{2}\left(\frac{\sum_{i=1}^{n} X_{1k}}{n} + \frac{\sum_{i=1}^{n} X_{2k}}{n}\right)\right) \tag{12.11}$$

$$= \frac{var(X_{1k})}{4n} + \frac{var(X_{2k})}{4n} + 2\frac{cov(X_{1k}, X_{2k})}{4n}, \tag{12.12}$$

$$= \frac{\sigma_X^2}{2n} + \frac{1}{2n}cov(X_{1k}, X_{2k}), \tag{12.13}$$

where σ_X is the standard deviation of each replicate, X. Therefore, the key for obtaining variance reduction is if $cov(X_{1k}, X_{2k}) < 0$. This was in fact the design for choice of antithetic pairs, therefore the antithetic estimator, $Theta_2$,

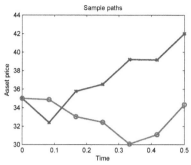
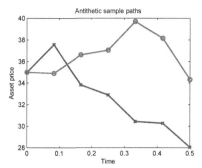

FIGURE 12.4: Sample paths for asset price evolution in antithetic pairs.

has lower variance than the usual sample mean estimator Θ_1, that is,

$$var(\bar{X}(n)) \quad = \quad \frac{\sigma_X^2}{2n} + \frac{1}{2n}cov(X_{1k}, X_{2k}), \qquad (12.14)$$

$$< \quad \frac{\sigma_X^2}{2n}, \qquad (12.15)$$

$$= \quad var(\Theta_1). \qquad (12.16)$$

The above design can also be applied if the quantity of interest is a function, $f(x)$, of the random variate, X, provided the function is monotonic. A monotonic function, whether it is non-decreasing or non-increasing, maintains the relationship between antithetic pairs, i.e., $cov(f(X_{1k}), f(X_{2k})) < 0$. Therefore, variance is reduced by applying antithetic variates.

The antithetic design of variance reduction need not be applied only to uniform random variates. In fact, it can be applied to any symmetric distribution by observing that once a random variate X_{1k} is generated by that distribution, we can generate an antithetic pair as, $X_{2k} = 2\mu_X - X_{1k}$, where μ_X is the mean (and the median) of the random variable, X. It can be shown in this case also that $cov(X_{1k}, X_{2k}) < 0$, and for a monotonic function, $f(x)$, $cov(f(X_{1k}), f(X_{2k})) < 0$. Therefore, in the numerous contexts of risk management for market risk or credit risk, where Wiener process is utilized for risk models, antithetic variates based variance reduction can be applied on this theme.

Given increments of Wiener process, $\Delta W \sim N(0, \Delta t)$, we can generate antithetic variates pairs $(\Delta W_{1k}, \Delta W_{2k}) = (\Delta W_{1k}, -\Delta W_{1k})$. Synchronizing antithetic pairs thus constructed can help generate antithetic trajectories or sample paths of various Ito processes. Figure 12.4 shows a couple of antithetic sample paths of an Ito process. Antithetic trajectories can be utilized in variance reduction, for instance, when pricing an up-and-in barrier option, if one trajectory is leading to a knock-in of the barrier option, the antithetic trajectory is likely to not lead to a knock-in, and hence have a zero pay-off. This can yield a negative correlation in replications of barrier option pay-offs, thus resulting in a variance reduction.

Feasibility and efficacy of antithetic variates is model dependent. The fundamental requirement that a model should satisfy for the antithetic variates method to work is that the random number or variate generated is transformed monotonically in the model. Therefore, implementation of antithetic variates based variance reduction can benefit from care in programming. Programming tricks that can help in the necessary synchronization for antithetic variates include random number stream dedication, using inverse-transform wherever possible, judicious wasting of random numbers, and pre-generation. In some cases, we can also seek to apply partial antithetic variates, which means only some quantities of interest in a large simulation utilize antithetically generated variates, while others are performed based on independent variates. This may be needed where full synchronization is difficult, and partial application is both feasible and effective in variance reduction.

12.1.3 Stratified Sampling

The stratified sampling method of variance reduction is built on developing a sampling mechanism that constraints the fraction of observations drawn from a specific subset or 'stratum' of the sample space. Therefore, the following constitute the two basic steps of stratified sampling.

1. Break the sample space into several strata or subsets using an appropriately defined rule.

2. Generate a sample of constrained number of observations from each stratum.

The advantage of this approach for variance reduction stems from the greater uniformity by which it represents 'sub-populations' of the entire sample space, rather than attempting to represent the entire population at once. We describe the method in detail in order to elaborate this point.

Let's say we want to estimate, $\theta = E[Y]$ for a risk factor, Y, where Y is defined on the probability space $(\Omega, \mathbf{P}, \mathcal{F})$. We identify, $A_1, A_2, \ldots, A_K$, as the disjoint subsets of the sample space, Ω, such that $\bigcup_{i=1}^{K} A_i = \Omega$. The stratified sampling based estimator is designed based on the following relationship.

$$\theta = E[Y] = \sum_{i=1}^{K} p_i E[Y|Y \in A_i], \qquad (12.17)$$

where $p_i = \mathbf{P}(Y \in A_i)$. Based on Eqn. (12.17), we define the new estimator Θ_2 for the quantity of interest, θ, as follows.

$$\Theta_2 = \sum_{i=1}^{K} p_i \bar{Y}_i, \qquad (12.18)$$

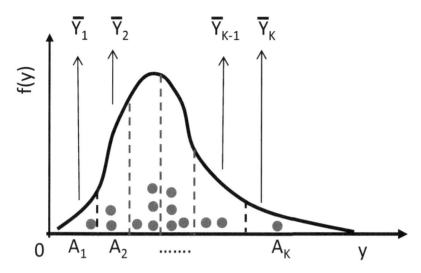

FIGURE 12.5: The sample space of the quantity of interest is broken down into strata, where samples are drawn from each stratum to create the estimator.

where $\bar{Y}_i = \frac{\sum_{j=1}^{N_i} Y_{ji}}{N_i}$ is the sample mean of observations $\{Y_{ji}\}$ generated from the subset A_i. Figure 12.5 shows the strata, sampling within each stratum, and the conditional mean, $\bar{Y}_i$, for each stratum.

In using this variance reduction estimator, a few questions need to be addressed. How does one decide the specific disjoint subsets, $A_1, A_2, \ldots, A_K$, of the sample space, Ω, to use and how does one decide the weight p_i and number of replicates to draw from each stratum, N_i, in Eqn. (12.18)? The strata are often generally defined by using a second random variable, X, as follows $A_i = \{X = x_i\}$. For instance, the random variable X may be simply defined as a K-outcome discrete random variable indicating when Y takes values in a subset A_i, i.e., $\mathcal{I}_{\{X=x_i\}} = \mathcal{I}_{\{Y \in A_i\}}$. However, using a second random variable for defining the strata allows more general definitions of the strata.

The choice of weights, p_i, and number of replicates, N_i, is not too far removed from the stratum selection. For instance, if K subsets are sought, the weight allocation to each subset may be $p_i = \frac{1}{K}$ and $N_i = Np_i$, if a total of N replicates are to be generated. This also helps define the stratum as, $A_i = (y_i, y_{i+1}]$ such that $\mathbf{P}(Y \in A_i) = p_i$. More general definitions of both weights, p_i, and the number of replicates to draw from each stratum, N_i, may also be used. For a specific choice of design of the stratified sampling based estimator of θ results in the following variance of the estimator.

$$var(\Theta_2) = \sum_{i=1}^{K} \frac{p_i^2}{N_i} \sigma_i^2, \tag{12.19}$$

where $\sigma_i^2 = var(Y|X = x_i)$, i.e., it is the conditional variance of Y for values lying within a stratum, A_i. Just as in the case of control variates, the optimum variance reduction in stratified sampling can be shown to be obtained by the following choice of N_i.

$$N_i^* = N \frac{p_i \sigma_i}{\sum_{k=1}^{K} p_k \sigma_k}, \tag{12.20}$$

where $\sigma_k^2 = var(Y|X = x_k)$, as defined earlier. Therefore, the number of observations sampled from each stratum should be proportional to the amount of variability in the quantity of interest, Y, in that stratum. In Figure 12.5, we apply stratified sampling to the normal distribution, $N(\mu, \sigma)$, where the stratum are defined by σ deviations from the mean, μ. Therefore, we have $A_1 = (-\infty, \mu - 3\sigma]$, $A_2 = (\mu - 3\sigma, \mu - 2\sigma]$, $A_3 = (\mu - 2\sigma, \mu - \sigma]$, $A_4 = (\mu - \sigma, \mu]$, $A_5 = (\mu, \mu + \sigma]$, $A_6 = (\mu + \sigma, \mu + 2\sigma]$, $A_7 = (\mu + 2\sigma, \mu + 3\sigma]$, $A_8 = (\mu + 3\sigma, -\infty)$. We define the remaining parameters for stratified sampling by first defining $p_i = \mathbf{P}(Y \in A_i)$, and then utilize the optimal sub-sample size for each stratum as per Eqn. (12.20).

If Y is instead a stochastic process, such as an Ito process defined by a Wiener process driven stochastic differential equation model, then stratified sampling can be applied if the distribution of the process is known at a specific time point, t, such as when $Y = S_t$ has the lognormal distribution. When applying stratified sampling to the simulation of sample paths of stochastic processes, where information regarding distribution of increments of the process alone may be utilized, the sample size, N, to draw in each increment of the process, Y, must be determined. This bears resemblance with simulating a multi-nomial tree of N branches at each time-step, with the difference being the N outcomes are not fixed. They are instead randomly generated from the k strata, N_i observations from the ith stratum.

For a general Ito process driven by the Wiener process, the above described stratified sampling approach for the normal distribution can be applied to the Wiener increments, $\Delta W_t \sim N(0, \Delta t)$, in order to generate sample paths for the Ito process. When N outcomes are generated for the Wiener increment using stratified sampling, stratified sampling is applied again for each of the N outcomes in order to create the next N increments per outcome for advancing the sample paths forward. Hence, the resemblance with a multi-nomial tree. Similarly, other processes for which the distribution of the process at a specific time is known or when the distribution of the increments of the process is known, stratified sampling can be adapted for generating sample paths of the stochastic process. For instance, for a homogeneous discrete- or continuous-time Markov chain, such as a homogeneous Poisson process, $Y = \Delta N_t \sim Po(\lambda t)$, stratified sampling can help reduce the variance in the increments.

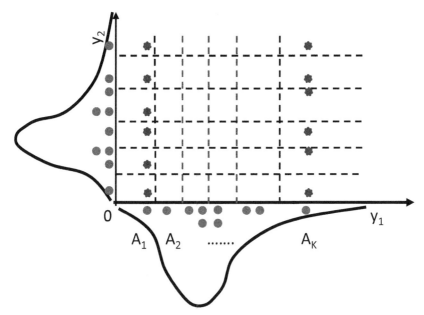

FIGURE 12.6: The two-dimensional sample space is broken down into strata constructed for marginal distribution of each coordinate. The two-dimensional realizations are constructed by randomly collating single-dimensional outcomes from single-dimensional stratified sampling.

12.1.4 Latin Hypercube Sampling

Stratified sampling is an effective method for variance reduction, however as the number of risk exposures increases that must be simultaneously simulated due to their interdependence and joint relevance to the risk management objectives, stratifying a multidimensional state space becomes prohibitively cumbersome. Latin hypercube sampling is the extension of stratification for sampling to the case of multiple dimensional random variables. Multidimensional random variables are inescapable since risks do not appear in singletons in almost any practical risk management context. Therefore, Latin hypercube sampling becomes very crucial for portfolios or multi-factor risk models, when developing stress tests for asset-liability management or VaR calculations for market or credit risk or other risk management assessments. As stated before, stratified sampling becomes infeasible in higher dimensions due to the rapid growth in the number of strata that must be sampled from in the higher dimensional sample space.

Latin hypercube sampling surmounts the curse of dimensionality by treating all coordinates of the d-dimensional sample space equally, and sampling for each coordinate of the higher dimensional space by stratified sampling. The exponential growth, K^d, of d-dimensional stratified sampling, where K is the number of strata for each dimension, is avoided by collating the outcomes generated for each coordinate. Instead of stratifying the d-dimensional space and generating outcomes from a d-dimensional hypercube, Latin hypercube method stratifies the one-dimensional marginal distributions of Y_i corresponding to the multi-dimensional joint distribution of $\mathbf{Y} = [Y_1, Y_2, \ldots, Y_d]$. It then permutes and collates each dimension to get the d-dimensional realizations. Furthermore, randomly permuting the coordinates of the d-dimensional realizations creates more realizations, thus more readily filling realizations in the K^d hypercubes.

In Figure 12.6, we display the two marginal distributions of two-dimensional risk factors, such as loss distribution from two loan portfolios. The approach developed in stratified sampling is applied to each dimension, generating the outcomes shown along the two axes. The two-dimensional outcomes are generated by randomly picking realizations from a stratum in each dimension and collating them in a two dimensional vector (indicated by stars in the two-dimensional space in the figure). As an example, say the two-dimensional joint distribution is given as,

$$f(y_1, y_2) \quad = \quad 2y_2 e^{-y_1}, \text{ for } 0 \le y_1, 0 \le y_2 \le 1 \qquad (12.21)$$
$$= \quad 0, \text{ otherwise}, \qquad (12.22)$$

then the marginal distribution of Y_1 is obtained as, $f_{Y_1}(y_1) = e^{-y_1}$, which is an exponential distribution, while the marginal distribution of Y_2 is, $f_{Y_2}(y_2) = 2y_2$. It is trivial to sample from both these marginal distributions by applying stratified sampling, followed by randomly permuting the coordinates and collating to obtain 2-dimensional samples.

As is evident from the plot in Figure 12.6, Latin hypercube sampling is also well suited for generating scenarios for scenario analysis and stress testing, especially beneficial when high-dimensional risk factors must be studied for their joint impact. Scenarios generated from single-dimensional marginal distributions that define stressed outcomes, or simply scenarios of interest for that dimension, can be picked and collated to create multi-dimensional scenarios.

12.1.5   Importance Sampling

The importance sampling method of variance reduction attempts to reduce variance of estimates of a quantity of interest regarding a random variable by changing the probability measure from which samples are generated to create the estimate. For performing this change of measure, it recruits another random variable, much like the control variate method of variance reduction.

The name 'importance' sampling is given to this method because in this method we change measure of the primary random variable of interest to try to give more weight to the *important* outcomes of the random variable, thereby increasing the sampling efficiency. Let's say we are interested in estimating θ

given below for a function, $h(.)$, of a random variable X.

$$\theta = E[h(X)] = \int_{-\infty}^{\infty} h(x)f(x)dx, \tag{12.23}$$

where X is the random variable with $f(x)$ being its probability density. We select another probability density, $g(x)$, defined on R^d that is positive wherever the original density function, $f(x)$ is positive, i.e., $g(x) > 0$ wherever $f(x) > 0$. We utilize this new probability density to make the following restatement for θ,

$$\theta = E[h(X)] = \int_{-\infty}^{\infty} h(x)[f(x)/g(x)]g(x)dx. \tag{12.24}$$

Alternatively, Eqn. (12.24) can be interpreted as follows,

$$\theta = \tilde{E}[h(X)\frac{f(X)}{g(X)}], \tag{12.25}$$

where the $\tilde{E}[]$ corresponds to taking expectation with respect to the new probability measure. The term, $\frac{f(X)}{g(X)}$, is called the *Likelihood Ratio* or the *Radon-Nikodym derivative* for the change of probability measure.

Success of importance sampling in variance reduction lies in the selection of the new probability density, $g(X)$. For variance reduction to materialize, we need to have the following relations to hold,

$$\tilde{E}[(h(X)\frac{f(X)}{g(X)})^2] = E[h(X)^2\frac{f(X)}{g(X)}], \tag{12.26}$$

$$< E[h(X)^2]. \tag{12.27}$$

This would definitely be accomplished if $f(x) < g(x)$, for all values of x. However, this would violate $g(x)$ remaining a probability density function. Therefore, one must investigate what values of x can we maintain $f(x) < g(x)$ for a more effective variance reduction while using importance sampling. One can intuitively surmise that should happen for values of random variable more likely to occur by the $f(x)$ distribution of random variable, X. Therefore $g(x) > f(x)$ for values of x more likely to occur, hence importance sampling.

12.2 Simulation-Based Optimization

Numerous problems in finance and risk management either directly involve solving an optimization problem, or indirectly precipitate into an optimization problem. Portfolio optimization, optimal asset allocation, development of optimal hedging strategy, and asset-liability management are all essentially rife with optimization problems. Indirectly, calibration of models of risk, developing forecasting models for risk, etc., involve solving some form of optimization problems. In general, optimization problems are classified by their key characteristics, which aids in developing methods for solving the problem.

A canonical optimization problem has an objective function, $f(x)$, where x are decision variables, and there may be additional constraints that the decision variables must satisfy. An optimization problem with constraints is called a constrained optimization problem, while one without constraints is an unconstrained optimization problem. Therefore, a typical optimization problem may be stated as,

$$\min_{x \in R^n} \quad f(x), \tag{12.28}$$

$$\text{such that} \quad Ax \quad \leq \quad b, \tag{12.29}$$

$$g(x) \quad \leq \quad 0, \tag{12.30}$$

$$h(x) \quad = \quad 0, \tag{12.31}$$

$$l \leq \quad x \quad \leq u. \tag{12.32}$$

where the matrix A of size $m \times n$ defines m linear constraints, $g(x)$ defines a set of non-linear inequality constraints, and $h(x)$ are a set of non-linear equality constraints. Additionally, l and u may be bounds on the decision variables. The most significant classification from the perspective of this book is that between deterministic and stochastic optimization problems. In the former, the objective function and constraints are obtainable as deterministic functions of the decision variables. In case of stochastic optimization, presence of risk factors in the definition of the objective function and/or constraints makes them a functional of random variables. Therefore, these functions must be either computed probabilistically or must be estimated statistically.

Suppose r is a set of risk factors, and the objective or constraints of the optimization problem must be stated in terms of these risk factors, along with the decision variables, x, since the risk factors in conjunction with the decision variables determine if the goals of the problem are achieved. The optimization problems must be modified as follows,

$$\min_{x \in R^n} \quad E[f(x; r)], \tag{12.33}$$

$$\text{such that} \quad E[A(r)]x \quad \leq \quad b, \tag{12.34}$$

$$E[g(x; r)] \quad \leq \quad 0, \tag{12.35}$$

$$E[h(x; r)] \quad = \quad 0, \tag{12.36}$$

$$l \leq \quad x \quad \leq u. \tag{12.37}$$

where the matrix $A(r)$ of size $m \times n$ is a function of the risk factors, r, defining m linear constraints for the decision variables, x. The set of non-linear inequality and equality constraints are defined, respectively, in terms of $E[g(x; r)]$, and $E[h(x; r)]$. A stochastic optimization problem would essentially reduce to a deterministic optimization problem if the functions, $E[f(x; r)]$, $E[g(x; r)]$, and $E[h(x; r)]$ can be computed probabilistically or analytically. For instance, this is the case for linear and quadratic functions of the risk factors, such as they appear in the mean-variance portfolio optimization problem discussed in Section 7.1.2.

Optimization problems can also be static versus dynamic, depending on whether the decision variable is independent or dependent on time, respectively. Developing investment strategies or hedging strategies, or even pricing some path-dependent options, must respond to risks evolving over time. Therefore, these involve dynamic optimization to obtain decisions that adapt to evolving risks. In Sections 7.1.1.2 and 7.1.2.2, we had developed a framework for dynamic investment strategy, which would serve as an example of a dynamic stochastic optimization problem. Exploration of dynamic hedging strategies, first developed in Section 7.3.3, can be extended utilizing dynamic optimization formulations.

An additional complexity can arise in some optimization problems when the decision variables are not continuous or are not quantitative. In some contexts for risk management decisions, such as 'education,' 'geographical region,' or 'urban-suburban-rural' characterization of clients for retail credit risk, or 'advertising or marketing channels' for business risk or 'system specifications,' 'training program choices' for operational risk, may all be non-ordinal, qualitative choices. Similarly, in other cases, while decision variables are quantitative, they may not take values in a range of the continuum. For instance, credit scores or credit rating levels for retail or commercial obligors, number of defaults to seek protection for in a loan portfolio, number of tranches designed in a CDO structure.

In any optimization problem, one must conduct a preliminary assessment of the above characteristics of the optimization problem. The set of input factors that define the problem must be identified, along with specification of whether the factors are qualitative, discrete, or continuous-valued. The input factors may be static or evolving dynamically in response to the evolving risks. Some of the input factors are controllable, while others may be uncontrollable due to a variety of reasons. Optimization can be done with respect to all the input factors, but will be meaningful to do only for the controllable ones. Moreover, there may be constraints and bounds required for the controllable input factors or decision variables. Finally, a detailed analysis is needed whether the problem is deterministic or stochastic in nature. In other words, in the presence of risk factors, can the objective and constraints functions be determined analytically or probabilistically in terms of the controllable input factors.

When the objective function or constraints can't be computed analytically is when simulation optimization is applicable and useful. We explore this in detail next.

12.2.1   Challenges of Simulation-Based Optimization

We will begin with thinking of the problem in terms of a classical mathematical optimization problem. However, the output performance measure or objective function, $R = E[f(x; r)]$, will be computed using simulation, since it can't be computed analytically. The value R takes depends on the values of input factors, $x = [x_1, x_2, \ldots, x_k]$. As stated earlier, there may be bounds on the decision variables, $l_i \leq x_i \leq u_i$, as well as other constraints. Let's first consider the unconstrained stochastic optimization problem, where further investigation of the problem will only require considering properties of the objective function.

In computational optimization, solution for the problem is sought iteratively, in each step attempting to make a better guess of the optimum. Seeking the optimal value of the objective function within the bounds for the decision variables through the iterative search is greatly facilitated by gaining some more information along the way of the objective function. If the function is continuous and differentiable, only continuous but not differentiable, or discontinuous, this can be utilized for guiding the search process. The slope or gradient of a differentiable function provides a direction of descent (or ascent, in case of a maximization problem). In case of continuous, but not differentiable, objective function, the direction of descent can be approximated using estimated slope or sub-gradients. When the objective function is discontinuous, such guidance is missing, and the search process would need to accommodate this fact.

Beyond continuity and differentiability of the problem, the objective function being linear versus non-linear qualifies the difficulty of solving the optimization problem. A linear objective function is continuous and differentiable, but it is also both convex and concave. Therefore, the optimal solution lies at the boundary of the feasible region of the problem. This knowledge can be utilized in obtaining a solution for the problem. When the objective function is not linear, i.e., it is non-linear, solving the problem may be more challenging. Determining whether a non-linear objective function is convex or concave is instructive for the existence and uniqueness of the optimal solution of the problem.

A function is convex if for any two arbitrarily picked values for the decision variables, x_1 and x_2, and for any scalar $\lambda \in [0, 1]$, we can demonstrate that,

$$E[f(\lambda x_1 + (1 - \lambda)x_2; r)] \leq \lambda E[f(x_1; r)] + (1 - \lambda)E[f(x_2; r)]. \quad (12.38)$$

If the inequality holds in the reverse for any arbitrarily picked values for decision variables, the objective function is concave. Whether an objective function is convex or concave in the entire feasible region, assuming the feasible region is a convex set, this indicates the optimization problem has a unique solution. In the case of a convex problem, the solution is an interior point, while for a concave problem, the solution lies in the boundary of the feasible region. A convex optimization problem is characterized by the gradient of the objective function, $\nabla E[f(x_0; r)]$, as follows,

$$E[f(x; r)] \geq E[f(x_0; r)] + \nabla E[f(x_0; r)]^T (x - x_0), \qquad (12.39)$$

for all x, x_0. For a convex optimization problem, the Hessian of the objective function, $H(x) = [\frac{\partial^2 E[f(x; r)]}{\partial x_i \partial x_j}]$, is positive semi-definite. Both these properties can be utilized for constructing a method for solving the optimization problem. Local convexity of an objective function, even for a non-convex problem, can be utilized for guidance in the search for the optimum.

When the objective function of a stochastic optimization problem cannot be computed analytically, there is good chance the function is not very well-behaved. Since the objective function, $R = E[f(x; r)]$, does not have a closed-form, analytical formulation, it can only be estimated with a certain degree of precision, or at a chosen confidence level. This implies that even if theoretically the objective function is continuous and differentiable, the gradient or the Hessian will also need to be estimated with some degree of accuracy. Therefore, errors in these estimates essentially create additional challenge for the solution procedures.

Since the objective is estimated with significant computational effort for the desired high accuracy, solving the optimization problem is a computationally intensive exercise. Moreover, in some cases the objective function may be highly non-linear, multi-modal, and due to the estimation error, noisy. Therefore, methods constructed to solve these optimization problems must adapt to meet these challenges. Despite the challenges, seeking even an approximately optimal decision set is a worthwhile activity in most cases.

In a constrained optimization problem, attention is needed not just to the objective function, but also to the linear or non-linear equality and inequality constraints in Eqns. (12.34), (12.35), and (12.36) respectively. These constraints define a feasible region, and remaining within the feasible region in the process of search for an optimal solution, or at the termination of the search process, is a must to satisfy the constraints. Various penalty function or barrier function approaches are developed to this end. For instance, if the problem is defined by non-linear equality constraints, $E[h(x; r)] = 0$, we can define a new objective, as follows,

$$\min_{x \in R^n} E[f(x; r)] \quad + \quad \frac{\rho}{2} E[h(x; r)]^T E[h(x; r)], \qquad (12.40)$$

$$\text{such that } l \leq \quad x \quad \leq u. \qquad (12.41)$$

where $\rho \geq 0$ is chosen as the penalty parameter. Searching for the optimal solution to this modified problem implies, depending on the size of the penalty parameter, ρ, the search may wander into the infeasible region. However, if the penalty parameter is adjusted to a large enough value, the penalty term would dominate the objective function of the modified problem, hence the solution will be feasible by the original problem. A barrier function approach modifies the objective so that the search process is not allowed to wander in the infeasible region due to the barrier introduced in the objective. For instance, for non-linear inequality constraints, $E[g(x;r)] \leq 0$, we can define a new objective, as follow,

$$\min_{x \in R^n} E[f(x;r)] \ + \ \eta \sum_{i=1}^{m_1} \ln(-E[g_i(x;r)]), \tag{12.42}$$

$$\text{such that } l \leq \ x \ \leq u. \tag{12.43}$$

where m_1 is the number of non-linear constraints and $\eta \geq 0$ is the chosen barrier parameter. In the barrier function approach the barrier parameter is initially taken to have a large enough value, but must be gradually decreased to relax the imposition of a barrier. In the above description, the penalty and barrier function approach provide an intuitive guidance for constrained optimization, however in their practical implementation additional issues may need to be resolved for convergence of search.

The necessary optimality conditions for a constrained optimization problem are defined by the Karush-Kuhn-Tucker (KKT) conditions. These are defined in terms of a Lagrangian function for the constrained optimization problem.

$$L(x,\lambda,\mu) = E[f(x;r)] + \sum_{i=1}^{m_1} \lambda_i E[g_i(x;r)] + \sum_{i=1}^{m_2} \mu_i E[h_i(x;r)]. \tag{12.44}$$

If the objective function and non-linear constraints are defined in terms of continuously differentiable functions, and if x^* is the local minimum of the constrained optimization problem, then there exist $\lambda^* \geq 0$ and μ^*, called KKT multipliers, such that,

1. Feasibility: $E[g(x^*;r)] \leq 0$ and $E[h(x^*;r)] = 0$.

2. Stationarity:
   $\nabla E[f(x^*;r)] + \sum_{i=1}^{m_1} \lambda_i^* \nabla E[g_i(x^*;r)] + \sum_{i=1}^{m_2} \mu_i^* \nabla E[h_i(x^*;r)] = 0$.

3. Complementary Slackness: $\lambda_i^* E[g_i(x^*;r)] = 0$ for all $i = 1, \ldots, m_1$.

These conditions can be utilized to develop search algorithms, where the search is done for both values of x and the KKT multipliers, λ and μ. We have provided a brief overview of challenges of constrained optimization, specifically as it is applicable for simulation-based constrained optimization. For additional discussion and development of these topics, the reader should refer to Gill et al. [28], Bertsekas [10], Fletcher [24], Luenberger and Ye [57].

12.2.2   Simulation Optimization Methodologies

Simulation optimization must rely on simulation to compute the value of the objective function and constraints, and any related constructs developed for solving the optimization problem, such as Lagrangian function, barrier or penalty function. Computational optimization or numerical optimization is an iterative process of obtaining a better solution for the problem, until the 'best' is obtained. Therefore, for simulation optimization an optimization routine has to work in tandem with a simulation routine that helps compute the required quantities that define the optimization problem. A schematic for this is provided in Figure 12.7.

With this synchronized development in view, the general steps adopted for simulation optimization as depicted in Figure 12.7 are structured as follows.

1. The process begins at the 'Start' with initializing the search process with a single or a sequence of decision choices.

2. Results from simulating earlier choices (or initial ones) are used to generate promising new directions to search in the space of possible input factor combinations.

3. Tradition or heuristic optimization search techniques may be employed. This stage will require interface with simulation routine to compute the quantities that define the problem.

4. As search progresses, the process should keep track of the best choices visited thus far. This is particularly important when objective or constraint function(s) are non-smooth, multi-modal functions.

5. Finally, the search must be terminated, which is done either when a good enough solution is obtained, significant improvements are not happening, or the process has run long enough.

Implementing a simulation optimization software from scratch is for most purposes not advisable, especially given the already available good options. In Section 12.3, we provide a list of functions available in MATLAB that may be used for computational optimization, and specifically simulation optimization, depending on the user needs and problem characteristics. In general, for a reliable, effective, and useful implementation of a simulation optimization tool, one would seek the following features.

- The foremost important feature is that the quality of solution obtained in the amount of execution time required should be efficient. For this purpose different algorithms designed for specific problem characteristics are developed.

- In many cases, as the search progresses, being able to keep track of a certain number, say 'm,' of best decision choice configurations should be available to the user. The 'm' being user-defined is an added benefit. This is particularly important in simulation optimization, where objective and constraints are computed to only a certain degree of precision.

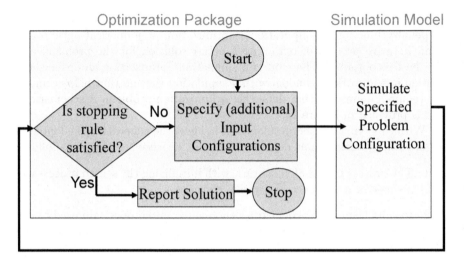

FIGURE 12.7: Typical interaction between optimization and simulation model.

- For solving a constrained optimization problem, as discussed in the previous section, accounting for constraints on decision variables besides the objective of the optimization problem is required. Specific optimization algorithm, function routines, and additional considerations must be incorporated.

- There should be several stopping rules for the termination of the search algorithm.

- Given the optimization routine is working in tandem with a simulation routine to estimate the objective and constraint functions, it is imperative that a confidence interval estimate for these performance measures corresponding to 'm' best decision configurations is available.

- Moreover, in the decision configurations for which variability in the objective function or constraints is higher, it would be desirable to do more replications in order to obtain the same degree of precision.

- Finally, in cases where the problem size is large, either due to number of variables or problem complexity, having access to parallel implementations can greatly help with simultaneous simulation runs on networked computers.

With the general framework for simulation optimization laid out and with an understanding of the desirable features for a simulation optimization implementation, we now begin a discussion of several of the important search techniques utilized for simulation optimization in the following sections.

12.2.2.1   Gradient-Based Methods

In cases where the objective function is smooth enough that its derivatives may be defined, as discussed in Section 12.2.1, this information may be gainfully utilized in the search for the optimal solution. For a continuous objective function, the gradient information, while for a continuously differentiable objective function, gradient as well as Hessian information may be employed to direct the search. In simulation optimization, however, the objective function, its gradient and Hessian must all be estimated using simulation. As opposed to analytical methods for implementing optimization algorithms, this poses an additional challenge regarding accuracy of these estimates, and how well they may guide the search. In response to this challenge, being able to construct a confidence interval around the estimated objective function value, its gradient and Hessian can prove helpful.

In these approaches, the flowchart of Figure 12.7 is primarily followed based on pursuing a single decision choice, x_k. The decision choice is iteratively improved to obtain a 'better' iterate, x_{k+1}, by taking a step, s_k, of length h. Therefore the iterations can be summarized as follows.

1. Pick a starting guess x_0.

2. Find a direction of descent for the objective function, s_k.

3. Determine an appropriate step length, $h_k \in R_+$, to move along the descent direction.

4. Update the guess, $x_{k+1} = x_k + h_k s_k$.

5. Check if termination criteria are met, or else go to Step 2.

The iterations in the search are continued until either a convergence criterion is met or an iterations limit is reached. Different gradient-based algorithms adopt specific principles to determine the directions of descent and step length. As discussed in Section 12.2.1, the above scheme can be applied to constrained optimization problems by appropriately adjusting the objective function.

The *steepest descent or gradient method* uses just gradient, i.e., first-order information in a Taylor expansion, for the direction of descent in each iteration. Therefore,

$$s_k = -\frac{\nabla E[f(x_k; r)]}{\|\nabla E[f(x_k; r)]\|}. \tag{12.45}$$

The step length h_k can be chosen by solving a one-dimensional problem $\min_{h \geq 0} l(h) = E[f(x_k + h s_k)]$. Alternatively, step length can be determined by creating a local quadratic approximation of the objective function and picking the minimum of this quadratic fit.

A first-order Taylor expansion based method can be improved by incorporating second-order information in the Taylor expansion, provided this is available. This is the basis for the Newton's method or its variant, the quasi-Newton's method. In the *Newton's method*,

$$E[f(x_k + s; r)] = E[f(x_k; r)] + \nabla E[f(x_k; r)]^T s + \frac{1}{2} s^T H(x)s, \qquad (12.46)$$

where $H(x)$ is the Hessian of the objective function. If the Hessian is positive definite, implying the function is locally strictly convex, we may find the minimizer of the second-order Taylor approximation of the objective as a solution of the system, $H(x)s = -\nabla E[f(x_k; r)]$. The solution of the linear system is taken as the descent direction, i.e., $s^* = s_k$.

Indeed in simulation optimization, as stated earlier, objective function, its gradient and Hessian must be estimated. When the gradient and Hessian information are approximated in each iteration, the method is called *quasi-Newton's method*. In fact, the gradient and/or Hessian may be adjusted iteratively as more information regarding the objective is accumulated. Moreover, in simulation optimization these algorithms will also need to be adapted for the accuracy with which these estimates are made. For instance, at an iterate, x_k, if the variance of $f(x_k; r)$ is higher, the search algorithm will need to generate a higher number of replicates for the same desired level of accuracy.

Beyond the gradient (and Hessian) based strategies for searching for the optimal decision, we also consider some heuristic strategies next. As stated earlier, gradient based optimization algorithms are feasible for objective functions that are well-behaved, i.e., those that are continuously differentiable. For objective functions that are not well-behaved, these algorithms would have limited applicability. Moreover, when the problem has discrete decision variables, or when the problem is combinatorial, by definition convexity of the feasible region is lost. In these cases, heuristic approaches may be necessary.

Additional to the cases when gradient and Hessian information of the objective is not available, if the simulation-based estimates of these quantities are too noisy, or are too computationally demanding to compute at the appropriate accuracy, heuristic methods must be considered. One need not make an 'either-or' decision in this regard, since heuristic algorithms can be combined with gradient based information to construct hybrid algorithms, where such an approach holds merit. Since these hybrid methods will venture beyond the traditional realm of theoretical convergence results of optimization algorithms, such methods would still be considered heuristics.

12.2.2.2   Simulated Annealing

The key to success of an optimization algorithm is the quality of explorative search for the optimum decisions. In gradient (and Hessian) based methods, the exploration is guided by this information. In absence of local slope and curvature information, the farthest extreme one might swing to is to pick random selection of candidate solutions in the feasible region. Repeatedly randomly picking candidate solutions from the entire feasible region may be

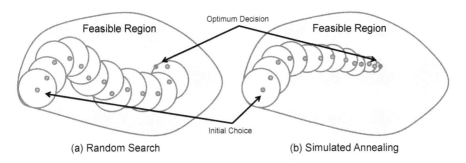

(a) Random Search   (b) Simulated Annealing

FIGURE 12.8: (a) A sample search trajectory for random search. (b) A sample search trajectory for simulated annealing.

counter-productive, therefore once an initial solution is guessed, random selection of subsequent guesses may be made from a pre-defined neighborhood of the previously guessed solutions. Moreover, a randomly selected decision in the neighborhood of a guessed solution may be allowed to graduate to be the next guessed solution only if it results in a significant improvement in the objective function value. This is the random search algorithm, which we summarize in Figure 12.8(a).

The size of neighborhood to pick the next random solution from is a critical parameter in the design of random search. If chosen to be big, it will give the opportunity to explore broad and wide from the current guessed solution. However, once the guessed solution approaches an optimal solution, a large neighborhood to select the next solution from carries the risk of wandering away from the solution. This issue can be remedied by progressively reducing the size of the neighborhood from which to select the next candidate solution. This modification results in a new algorithm, called simulated annealing.

In simulated annealing, the logic applied is that as better candidate solutions are obtained, the search space is reduced to avoid wandering away in the progress toward the optimal solution. This is shown in Figure 12.8(b). The method initializes as in random search, and progresses similarly with one important difference. Every time a new candidate solution is adopted, the admissible size of the next neighborhood to select the next candidate solution from is simultaneously adjusted. The algorithm for simulated annealing would have the following structure.

Pick an initial guess of a solution, x_0,
initial size of neighborhood, d_0,
desired reduction in function value in each iteration, ρ, and
reduction in neighborhood size in each iteration, ϵ.
    While termination criteria are not fulfilled;
        Define a neighborhood, N_i, of size d_i such that $x_i \in N_i$;
        Generate a randomly selected candidate $x_c \in N_i$;

Compute $E[f(x_c; r)]$ and $\Delta = E[f(x_c; r)] - E[f(x_i; r)]$;
If $\Delta \leq \rho$ then $x_{i+1} = x_c$, and $d_{i+1} = \max(d_i - \epsilon, \epsilon)$
Else go to Step 'Generate a randomly selected candidate...'.
End

In the above algorithm, we have made sure that the neighborhood size doesn't become zero, or worse, negative. By not changing the neighborhood size in each iteration, the algorithm would mimic a random search. Moreover, the reduction in neighborhood size can be iteratively adjusted by making it depend on the iteration, namely ϵ_i. Several variations of simulated annealing may be developed, where in one variation the size of neighborhood is not always reduced. It is in the contrary expanded in cases where reduction in objective function is large. This variation can help faster convergence, since it is adaptive to when better progress is made in the search process.

12.2.2.3   Tabu Search

In random search or simulated annealing, even though the trajectories in Figure 12.8 paint a rather favorable scenario, in reality there may be instances where the search process gets misdirected and delays the convergence to an optimum decision. Tabu search addresses this shortcoming by identifying a set of candidate solutions that have been marked either inferior or visited, and must not be revisited in future exploration. This set is called a tabu set, T, and hence the algorithm is called tabu search. The algorithmic description of tabu search, as a modification of the simulated annealing algorithm, is as follows.

Pick an initial guess of a solution, x_0,
define an initial size of neighborhood, d_0,
desired reduction in function value in each iteration, ρ,
reduction in neighborhood size in each iteration, ϵ, and
a tabu set, $T = \emptyset$.
    While termination criteria are not fulfilled;
        Define a neighborhood, N_i, of size d_i such that $x_i \in N_i$;
        Generate a randomly selected candidate $x_c \in N_i \backslash T$;
        Assign $T = T \bigcup \{x_c\}$;
        Compute $E[f(x_c; r)]$ and $\Delta = E[f(x_c; r)] - E[f(x_i; r)]$;
        If $\Delta \leq \rho$ then $x_{i+1} = x_c$, and $d_{i+1} = \max(d_i - \epsilon, \epsilon)$
        Else go to Step 'Generate a randomly selected candidate...'.
    End

An alternate version of tabu search may be developed that can help avoid converging to a local optimum. Each time search for an optimum is conducted, the local optimum obtained is included in the tabu set to avoid revisiting it in future exploration, thus forcing one to explore alternatives. This version of tabu search can be combined with random search or simulated annealing towards the later iterations, once the search has reached a promising region, to avoid convergence to a local optima. The algorithm can be structured as follows.

Pick an initial guess of a solution, x_0, and
Define a neighborhood, N_0, of size d_0 such that $x_0 \in N_0$;.
    While termination criteria are not fulfilled;
        Obtain a local optimum, $x_c = argmin\ E[f(x;r)]$ for $x \in N_0 \backslash T$;
        Assign $T = T \bigcup \{x_c\}$;
    End
Pick the best solution visited throughout as the solution, x_{sol}.

The above heuristic methods have been examples of the general optimization process described in Figure 12.7. More specifically, they have been instances of solution algorithms where a single decision choice is considered in each iteration. We next develop solution algorithms that work with a set of decision choices in each iteration, and evolve the entire set in an iteration.

12.2.2.4 Scatter Search

There are many variants of scatter search possible, the common theme of them will involve creating a set of possible decision choices, selecting the better performing ones from this set, creating a new set of solutions based on the better ones, and continuing this process until satisfactory decision choices are obtained. This general theme is depicted in Figure 12.9. Specific implementations of scatter search identify specific mechanisms for accomplishing each of the three tasks, until sufficient effort is made for the search process and good enough solutions are obtained, at which point the search is terminated. The best decision choice in the final set of decision choices is taken to be the optimum solution.

It is totally conceivable and advisable to apply tabu search after scatter search, based on treating the solution of scatter search as the initial guess for tabu search. This allows combining the benefit of scatter search, of allowing a broad and wide search of the feasible region, with the benefit of tabu search of not settling into a local optimum. We next present two biologically-inspired population-based optimization algorithms, that implement specific principles for accomplishing the three tasks of scatter search.

12.2.2.5 Evolutionary Strategies

Optimization algorithms that base search on objective function value alone, and not gradient or Hessian information, are often called the class of direct methods. Evolutionary strategies are population-based direct methods, where concepts of biological evolution of living species form the thematic

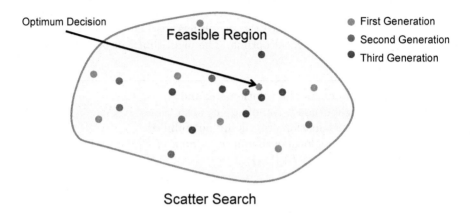

Scatter Search

FIGURE 12.9: Three generations of a population-based scatter search algorithm.

basis. In the Darwinian theory of evolution, a population of a living species adapts to its environment in order to survive. Survival favors the fittest, while the poor and weak members of the population fade away. The fittest survive, get to mate and propagate their genetic material into new generations of population. Progressively, the population gets fitter and more adapted to its environment.

Genetic Algorithm

A genetic algorithm is a specific evolutionary strategy for solving optimization problems, where the above theme of evolution is utilized to obtain better solutions. As in any population-based method, a population of decision choices are evaluated in every iteration. Considering a population of solutions increases the chances of finding the global optimal. At any stage of the iterations of the algorithm, there is a population of solutions under consideration, which is in contrast with traditional optimization approaches, such as gradient-based methods or the first two heuristic methods considered in this section.

The objective function value corresponding to each decision choice in the population indicates the fitness of that member of the population. The less fit, or higher objective function valued (in a minimization problem), decision choice is a poor solution, and is allowed to fade away. The more fit solutions are allowed to produce the next generation of a set of decision choices. This production of a new generation, in genetic reproductive terminology, involves transfer, cross-over, and mutation of genetic material of parents as it is transferred to the offspring decision choices.

The main steps of an algorithmic implementation of the genetic algorithm takes the following form.

Generate an initial population of solutions of size N randomly scattered in the feasible region;
>   While termination criteria are not fulfilled;
>>   Compute the objective value (or fitness) of each member of the population;
>>
>>   Based on fitness of the decision choices, let a fraction, ρ, of the population fade away;
>>
>>   Apply idealized-type genetic operators to the remaining ρN members of the population to produce N 'offspring' decision choices;
>   End
>
Pick the best solution from the final population as the solution, x_{sol}.

A variation in the above algorithm may be made in terms of probabilistic fading of a member of a population, where the probability is inversely proportional to fitness of the member.

The above properties of genetic algorithm make it suitable for simulation optimization. These population-based methods make no restrictive assumptions or require no prior knowledge of topology of response surface. They can work even if the objective function is high-dimensional, multi-modal, discontinuous, non-differentiable, and stochastic. Their popularity is also justified by their relative ease of use, as well as reasonable reliability, especially when combined with tabu search that hones in the final solution obtained from a genetic algorithm.

As is evident from the description of the algorithm, an evolutionary strategy, and for that matter all population-based strategies, requires a high amount of objective value evaluation. This can be somewhat tough if the simulation-based estimates of the objective function are computationally very intensive. Computation of a population's fitness is, however, very amenable to parallelization, therefore implementations of genetic algorithms are favored on parallel computing environments.

12.2.2.6 Particle Swarm Optimization

This population-based optimization algorithm is also biologically inspired, more specifically by certain zoological characteristics. If one has seen the collective movement of a flock of birds, a shoal of fish or a swarm of flying insects, this approach can be visualized as attempting to mimic this movement to achieve the optimum solution. As in any population-based approach, the algorithm begins with an initial set of decision choices. Thereafter, for moving all the decision choices of the population in the improved direction, some 'leaders' or best decision choices must be identified. The best decision choices of a single population and historically best decision choices are used to determine how each member of a population migrates to a new location. This guided movement continues until satisfactory improvement is achieved and further improvement ceases.

The general direction of improvement for each member of a population in each iteration can be summarized by a velocity vector, v_i. This is used to advance the progress to the optimum. In each iteration the velocity vector is modified by the updated information for that iteration to obtain the change in velocity, Δv_i. The particle swarm optimization algorithm can be summarized as follows.

Generate an initial population of solutions of size N randomly scattered in the feasible region;

Compute the objective value of each member of the initial population;

Define the universal best solution, x_u^*, as the best of initial population, and a vector of movement for the population, $v = 0$;

While termination criteria are not fulfilled;

Compute the objective value of each member of the population;

Based on objective function value for the decision choices, pick the best solution, x_i^*;

If $E[f(x_i^*; r)] < E[f(x_u^*; r)]$, update the historically best solution, $x_u^* = x_i^*$;

Use a linear combination of x_u^* and x_i^* to obtain the change in velocity, Δv_i;

Update every member of the population by new velocity vector, $v_i = v_i + \Delta v_i$;

End

Pick the best solution from the final population as the solution, x_{sol}.

As stated earlier, the particle swarm optimization algorithm can be appended with a single decision choice based method in the end, such as tabu search, to further fine-tune the solution. We have provided a set of example heuristic methods for simulation optimization, as well as gradient and Hessian based methods for optimization. Using the given algorithms, as well as other heuristic algorithms, hybrid methods can also be constructed to take advantage of algorithm characteristics that suit the specific problem at hand. Interested readers should access more dedicated discussion of these algorithms in Gill et al. [28], Fletcher [24], and Gilli et al. [29].

12.3 MATLAB Tools for Variance Reduction and Optimization

MATLAB mathematical software has a vast array of functions for working with random variates and methodologies in its Optimization Toolbox. We list a few of these functions here relevant for variance reduction and optimization. The reader is advised to look up the extensive help documentation available with MATLAB to see the details of these and other related functions. At the bottom of each function description in MATLAB help documentation, look for 'See Also' to explore other related functions. Resources such as MATLAB Primer [20] are also useful.

Variance reduction: transprob, transprobbytotal, lhsample

Optimization: classify, fmincon, fminsearch, linprog, quadprog

Global optimization: patternsearch, ga, simulannealbnd, gamultiobj, GlobalSearch, MultiStart, run

12.4 Summary

In the earlier chapters we developed the principles of simulation (Chapter 4) and models of risk in Chapters 5 and 6. We applied these principles and modeling techniques to solve various risk management problems in Chapters 7 to 11. In this chapter, we developed some more principles and problem solving techniques for more advanced users of simulation modeling and analysis. We began with variance reduction techniques that help in reducing the computational burden to achieve the desired accuracy in descriptive and inferential statistical analysis. Better accuracy at lower computational effort opens the door for more sophisticated usage of simulation models. In particular, it facilitates an efficient use of simulation models to perform simulation-based optimization. In Section 12.2, we developed principles of simulation-based optimization and algorithms to solve these problems. Some of these algorithms use gradient information for solving the problem, while a set of approaches presented are classified as heuristic algorithms. In today's world, as computing resources are becoming cheaper and more efficient, there is an increased appetite for the size and complexity of problems we attempt to solve using simulation.

12.5 Questions and Exercises

Review Questions

1. In what risk management tasks is simulation analysis useful?

2. What risk management problems can simulation optimization be applied to?

3. What is the fundamental challenge when solving risk management problems using simulation?

4. What are variance reduction techniques? What are the two broad strategies on which variance reduction techniques are built?

5. How should attention be paid for efficiently implementing variance reduction techniques?

6. What is the basic principle on which the control variate method for variance reduction is built?

7. How is the control variate constructed? When does it reduce variance?

8. Show that the control variate estimator is unbiased and consistent.

9. Construct examples of the control variate estimator for pricing derivatives.

10. What is the principle behind the antithetic variates method of variance reduction?

11. What role does monotonicity of functions play in the efficacy of antithetic variates variance reduction?

12. How do uniform random numbers based antithetic variates differ from those generated for other symmetric distributions?

13. How does stratified sampling achieve variance reduction?

14. What parametric choices are needed for implementing stratified sampling? What role do these parameters play in the method?

15. What is the optimal choice of sub-sample size for stratified sampling?

16. When does Latin hypercube sampling become a useful approach for variance reduction?

17. How does the Latin hypercube sampling method differ from the stratified sampling method of variance reduction?

18. How can the Latin hypercube sampling method be used for scenario analysis and stress testing?

19. What is the main principle behind the importance sampling method of variance reduction?

20. When is an optimization problem considered deterministic versus stochastic? Give examples.

21. What is a canonical constrained optimization problem?

22. How is a static optimization problem different from a dynamic optimization problem? Give an example for both.

23. Give examples of discrete variables and qualitative variables for optimization problems.

24. Why is a discrete optimization problem more challenging than a continuous optimization problem?

25. When is an optimization problem convex or concave?

26. What information do gradient and Hessian of the objective function provide in an unconstrained optimization problem? How is it useful to solve the problem?

27. What are the challenges of solving a stochastic optimization problem? When must simulation optimization be used for such problems?

28. Why are constrained optimization problems harder to solve than unconstrained optimization problems?

29. What is the penalty function approach to solving constrained optimization problems?

30. What is the barrier function approach to solving constrained optimization problems? How does this compare with the penalty function approach?

31. What is a Lagrangian function for a constrained optimization problem? What are the Karush-Kuhn-Tucker conditions for optimality?

32. What are the general steps of a simulation optimization methodology?

33. What are the features of a simulation optimization tool that provide reliability and effectiveness?

34. What are gradient-based methods for simulation optimization?

35. How can Hessian information be useful for simulation optimization? When can this information be beneficially utilized?

36. What is a random search heuristic for solving simulation optimization problems? How is simulated annealing utilized beyond random search for solving these problems?

37. What is tabu set, and how is it utilized in tabu search algorithm?

38. What are population-based methods for simulation optimization? How is scatter search performed?

39. What are evolutionary strategies? What is the basic theme behind these strategies? Give an example.

40. What is particle swarm optimization? What is this method inspired by?

Exercises

1. Consider a stock price evolving by the following model,

$$dS_t = 0.19S_t dt + 1.2S_t^{0.8}dW_t, \qquad (12.47)$$

where the current price of the stock is, $S_0 = \$20$. A discrete-average Asian option is defined on the stock evolving by the above model. The Asian call option pay-off is determined by the arithmetic average of weekly closing price of the stock, with a maturity of $T = 0.25$ years and strike price $K = 20$. Estimate the price of the option using simulation. Define a control variate for implementing a variance reduction for the price estimate, and recompute the price by applying the control variate price estimator. Compare the variance of the price estimator under no variance reduction versus under control variate variance reduction. Assume that the short-term risk-free interest rate is 2.3%.

2. Consider a stock evolving by the following Black-Scholes model

$$dS_{2t} = 0.10S_{2t} dt + 0.24S_{2t}dW_{2t}, S_{20} = \$45. \qquad (12.48)$$

Define an exchange option between this stock and the stock defined in Eqn. (12.47) with initial price of $S_0 = \$20$. Estimate the price of the exchange option, assuming that the correlation between the two driving Wiener processes is, $\rho = -0.15$ and the short-term risk-free interest rate is 2.3%. Now define a control variate to improve the performance of your estimator, apply the control variate estimator and assess its efficiency in variance reduction.

3. Consider the following distributional fits for loss frequency and loss severity of three pure risks identified through the pure risk identification process.

   (a) $N_{1t} \sim Po(\lambda)$, Poisson distribution with $\lambda = 10$ per year; $L_1 \sim$ Weibull(a, b) with scale parameter $a = 5$ and shape parameter $b = 0.8$

(b) $N_{3t} \sim Bin(n,p)$, Binomial distribution with $n = 50$ and $p = 0.05$; $L_3 \sim$ Lognormal distribution with mean $\mu_L = 100$ and standard deviation $\sigma_L = 15$

(c) $N_{5t} \sim Bin(n,p)$, Binomial distribution with $n = 5$ and $p = 0.2$; $L_5 \sim$ Weibull(a,b) with scale parameter $a = 500$ and shape parameter $b = 0.8$

Construct a quantitative assessment of the total annual loss of each pure risk, as well as the grand total annual loss from all the pure risks combined. Apply wherever possible antithetic variates based variance reduction and report the efficiency of variance reduction achieved for each pure risk, as well as the grand total annual loss.

4. The 100,000 customers in a retail credit portfolio have credit scores given by the following distribution, $270 * Beta(1.2, 1.5) + 580$. Therefore, the cutoff score used for this product is 580. The portfolio experiences a total of 5% default. The default experience by credit score in this portfolio is described by the following distribution, $\chi^2(5) + 580$. Apply stratified sampling to determine the distribution of false goods and false bads. How is your efficiency improved by the application of stratified sampling?

5. A portfolio, $\vec{w} = [10{,}000; 22{,}000; 15{,}000; 20{,}000; 30{,}000; 8{,}000; 20{,}000]$, is constructed given by number of bonds and shares of stocks. The portfolio investment is in the following bonds and stocks.

- US Treasury T-Note, Annual Coupon Rate: 3.25% (paid semi-annually); Maturity: 6 Years; Rating: AAA

- US Treasury T-Bond, Annual Coupon Rate: 7.25% (paid semi-annually); Maturity: 12 Years; Rating: AAA

- Corporate Bond, Issuer: Johnson & Johnson; Coupon Rate: 5.55% (paid semi-annually); Maturity: 5 years; Rating: AAA

- Corporate Bond, Issuer: Southwest Airlines; Coupon Rate: 7.375% (paid semi-annually); Maturity: 15 years; Rating: BBB

- Consider three stocks evolving by continuous-time stock price evolution model of the form,

$$dS_{it} = \mu_i S_{it} dt + \sigma_i S_{it} dW_{it}, \qquad (12.49)$$

for $i = 1, 2, 3$, where initial stock price is, $\vec{S_0} = [19; 53; 26]$, $\vec{\mu} = [0.09; 0.05; 0.16]$ and $\vec{\sigma} = [0.10; 0.06; 0.25]$. The three correlated Wiener processes are described by the following correlation matrix.

$$\rho = \begin{pmatrix} 1 & 0.3 & 0.1 \\ 0.3 & 1 & -0.05 \\ 0.1 & -0.05 & 1 \end{pmatrix} \qquad (12.50)$$

Apply stratified sampling where appropriate to compute the Value-at-Risk at a desired confidence level and duration of time for the above portfolio.

6. Perform a detailed stress testing of the stock-bond portfolio in Problem 5 by applying the Latin hypercube sampling for all risk factor interactions.

7. Consider three stocks with the following summary information regarding their annual returns. The mean annual return of three stocks is estimated to be, $\vec{\mu} = [0.09; 0.05; 0.16]$; the annual standard deviation of returns is $\vec{\sigma} = [0.10; 0.06; 0.25]$, and the correlation matrix is given as follows.

$$\rho = \begin{pmatrix} 1 & 0.3 & 0.1 \\ 0.3 & 1 & -0.05 \\ 0.1 & -0.05 & 1 \end{pmatrix} \qquad (12.51)$$

Assuming a planning horizon of one year and return of these stocks to be log-normally distributed, construct and analyze the following portfolios using simulation optimization.

(a) Construct the minimum variance optimal portfolio under no short-selling constraints.

(b) Choose a desired target mean return, r_{th}, which should serve as a lower bound for the optimal portfolio return. Compute the minimum variance portfolio under no short-selling constraint. Relax the no short-selling constraint and re-optimize your portfolio. How do your portfolio weights change?

(c) Construct an optimal portfolio under no short-selling constraints using a downside risk measure, such as expected shortfall. Examine how the optimal portfolio changes for different percentiles by which expected shortfall is defined.

8. Consider continuous-time stock price evolution model for three stocks of the form,

$$dS_{it} = \mu_i S_{it} dt + \sigma_i S_{it} dW_{it}, \qquad (12.52)$$

for $i = 1, 2, 3$, where initial stock price is, $\vec{S_0} = [19; 53; 26]$, $\vec{\mu} = [0.09; 0.05; 0.16]$ and $\vec{\sigma} = [0.10; 0.06; 0.25]$. The three correlated Wiener processes are described by the following correlation matrix.

$$\rho = \begin{pmatrix} 1 & 0.3 & 0.1 \\ 0.3 & 1 & -0.05 \\ 0.1 & -0.05 & 1 \end{pmatrix} \qquad (12.53)$$

Assume a monthly trading strategy for an annual planning horizon, and a chosen terminal wealth performance measure, $U(W_T) = E[u(W_T)]$, where $u(x) = \frac{x^\gamma - 1}{\gamma}$ is a constant relative risk aversion utility. Analyze different investment strategies for these three stocks using simulation for different choices of coefficient of relative risk aversion, γ. Pick a $\gamma < 0$ for degree of high risk aversion and a $\gamma > 0$ for low risk aversion. Construct an optimal trading strategy in each case using simulation optimization.

9. The Dord Motors company is considering introducing a new sports car model, named *The Racer*. The management is trying to assess the prospects for this new model. While understanding the project's profitability is a difficult task, it is an important task before the project is taken up. For this purpose, they have put together estimates for fixed and variable costs, projected future sales and prices at which they intend to sell this model. Each of these project features are described in a model, as follows.

- Fixed cost of developing *The Racer* is equally likely to be either $3 or $5 billion. At an upfront expense of $200,000, the management can acquire additional information that helps narrow the fixed cost to be equally likely at either $3.5 or $4.5 billion.

- Variable cost per car manufactured for the first three years are: For year 1 it is equally likely to be $5,000 or $8,000, for year 2 it is going to be $1.05 * (year 1 variable cost)$, for year 3 it is going to be $1.05 * (year 2 variable cost)$. Here again the management has an option to incur an upfront expense of $100,000 to narrow the first year variable cost per car to be equally likely at either $6,000 or $7,000.

- Sales projections are determined as average sales in year 1 at 200,000, with a standard deviation of 50,000 cars. A normal distribution is chosen to describe year 1 sales. Average sales of year 2 and 3 are expected to be at the sales level of the previous year, with a standard deviation of 50,000 cars. An advertising company has advised that an ad-campaign can help increase the first year sales, where it proposes a 5,000 increase in first year mean sales for every additional $100,000 spent on the campaign. After $1,000,000 spent on the ad-campaign, the market is expected to saturate and no further increase in first year sales would be possible.

- Pricing for year 1 is set at $13,000. Years 2 and 3 prices will be determined based on the previous year's price and sales. Specifically, year 2 price $= 1.05*(\text{year 1 price})+\$30*(\text{percentage by which year 1 sales exceed expected year 1 sales})$, year 3 price $= 1.05 * (\text{year 2 price})+\$30*(\text{percentage by which year 2 sales exceed expected year 2 sales})$.

Conduct a simulation optimization for the optimal design for the profitability of *The Racer*.

Bibliography

[1] G.J. Alexander and A.M. Baptista. Economic implications of using a mean-var model for portfolio selection: A comparison with mean-variance analysis. *Journal of Economic Dynamics and Control*, 26:1159–1193, 2002.

[2] G.J. Alexander and A.M. Baptista. A comparison of var and cvar constraints on portfolio selection with the mean-variance model. *Management Science*, 50:1261–1273, 2004.

[3] E.I. Altman, B. Brady, A. Resti, and A. Sironi. The link between default and recovery rates: Theory, empirical evidence, and implications. *Journal of Business*, 78(6):2203–2227, 2005.

[4] E.I. Altman and J.K. La Fleur. Managing a return to financial health. *The Journal of Business Strategy*, 2(1):31–38, 1981.

[5] M. Ammann. *Credit Risk, Valuation, Methods, Models, and Applications.* Springer, 2010.

[6] P. Artzner, F. Delbaen, J.-M. Eber, and D. Heath. Coherent measures of risk. *Mathematical Finance*, 9(3):203–228, 1999.

[7] J. Barth and J. Jahera. US enacts sweeping financial reform legislation. *Journal of Financial Economic Policy*, 2(3):192–195, 2010.

[8] D.E. Bell, H. Raiffa, and A. Tversky. *Decision Making - Descriptive, Normative, and Prescriptive Interactions.* Cambridge University Press, Cambridge, U.K., 1988.

[9] Daniel Bernoulli. Specimen theoriae novae de mensura sortis. *Comentarii Academiae Scientiarum Imperiales Petropolitanae*, 5:175–192, 1738.

[10] D.P. Bertsekas. *Nonlinear Programming, 2nd edition.* Athena Scientific, September 1999.

[11] P. Billingsley. *Probability and Measure, Third Edition.* Wiley-Interscience, 1995.

[12] F. Black and R. Litterman. Global portfolio optimization. *Financial Analysts Journal*, pages 28–43, Sept 1992.

[13] J. Y. Campbell, A.W. Lo, and A.C. MacKinlay. *The Econometrics of Financial Markets.* Princeton University Press, 1996.

[14] J. Casassus. *Stochastic behavior of spot and futures commodity prices: Theory and evidence.* PhD thesis, Doctoral dissertation, Tepper School of Business, Carnegie Mellon University, 2004. 138 pages.

[15] S. Chen and M. Insley. Regime switching in stochastic models of commodity prices: An application to an optimal tree harvesting problem. *Journal of Economic Dynamics & Control*, 36(2):201–219, Feb 2012.

[16] E.A. Coddington and J. Landin. *An Introduction to Ordinary Differential Equations.* Dover Books, 2001.

[17] B.M. Collins and F.J. Fabozzi. *Derivatives and Equity Portfolio Management.* Wiley; 1 edition, 1999.

[18] R. Cont and P. Tankov. *Financial Modelling with Jump Processes.* Chapman & Hall/CRC Financial Mathematics Series, 2003.

[19] C.L. Culp. *The Risk Management Process, Business Strategy and Tactics.* John Wiley & Sons, Inc., Wiley Finance Series, 2001.

[20] T. A. Davis. *MATLAB Primer, Eighth Edition.* CRC Press, 2010.

[21] D. Duffie and K.J. Singleton. *Credit Risk, Pricing Measurement and Management.* Princeton University Press, 2003.

[22] R. Durrett. *Probability: Theory and Examples.* Wadsworth & Brooks Cole, 1991.

[23] F.J. Fabozzi, D. Huang, and G. Zhou. Robust portfolios: Contributions from operations research and finance. *Annals of Operations Research*, 176:191–220, 2010.

[24] R. Fletcher. *Practical Methods of Optimization.* Wiley; 2 edition, May 2000.

[25] J.-P. Fouque, G. Papanicolaou, and K.R. Sircar. *Derivatives in Financial Markets with Stochastic Volatility.* Cambridge University Press, 2000.

[26] D. Foust, A. Pressman, B. Grow, and R. Berner. Credit scores: Not-so-magic numbers. *Business Week*, 4017:38–43, 2008.

[27] J. Gatheral. *The Volatility Surface: A Practitioner's Guide.* Wiley Finance, 2004.

[28] P. E. Gill, W. Murray, and M. H. Wright. *Practical Optimization.* Emerald Group Publishing Limited, January 1982.

[29] M. Gilli, D. Maringer, and E. Schumann. *Numerical Methods and Optimization in Finance.* Academic Press, January 2011.

[30] P. Glasserman. *Monte Carlo Methods in Financial Engineering.* Springer, 2004.

[31] F. Glover. Improved linear programming models for discriminant analysis. *Decision Sciences*, 21(4):771–785, 1990.

[32] H. Gründl, T. Post, and R. Schulze. To hedge or not to hedge: Managing demographic risk in life insurance companies. *Journal of Risk & Insurance*, 73(1):19–41, 2006.

[33] W.H. Greene. *Econometric Analysis, Sixth Ed.* Pearson Prentice Hall, 2011.

[34] M. Grossman. On the concept of health capital and the demand for health. *Journal of Political Economy*, 80(2):223–255, 1972.

[35] A. Gupta and L. Li. Integrating long-term care insurance purchase decision with saving and investment for retirement. *Insurance: Mathematics and Economics*, 41(3):362–381, 2007.

[36] A. Gupta and Z. Li. Integrating optimal annuity planning with consumption-investment selections. *Insurance: Mathematics and Economics*, 41(1):96–110, 2007.

[37] A. Gupta and Z. Li. Calibration of a stochastic health evolution model for optimal healthcare financing decisions using NHIS data. *Physica A*, 390(1):3524–3540, 2011.

[38] A. Gupta and H. Wang. Assessing securitization and hedging strategies for management of longevity risk. *International Journal of Banking Accounting and Finance*, 3:47–72, 2011.

[39] S.L. Heston. A closed-form solution for options with stochastic volatility with applications to bond and currency options. *The Review of Financial Studies*, 6(2):327–343, Summer 1993.

[40] C.-F. Huang and R.H. Litzenberger. *Foundations for Financial Economics.* Prentice Hall, Inc., Englewood Cliffs, New Jersey, 1988.

[41] J.C. Hull. *Options, Futures, and Other Derivatives and DerivaGem CD Package.* Prentice Hall; 8 edition, 2011.

[42] P. Jorion. On jump processes in the foreign exchange and stock markets. *Review of Financial Studies*, 1:427–445, 1988.

[43] B. Kaffel and F. Abid. A methodology for the choice of the best fitting continuous-time stochastic models of crude oil price. *Quarterly Review of Economics and Finance*, 49(3):971–1000, 2009.

[44] Y.M. Kaniovski and G.Ch. Pflug. Risk assessment for credit portfolios: A coupled Markov chain model. *Journal of Banking & Finance*, 31:2303–2323, 2007.

[45] I. Karatzas and S.E. Shreve. *Brownian Motion and Stochastic Calculus (Graduate Texts in Mathematics)*. Springer, 1991.

[46] S. Karlin and H.M. Taylor. *A First Course in Stochastic Processes, Second Edition*. Academic Press, 1975.

[47] S. Karlin and H.M. Taylor. *A Second Course in Stochastic Processes*. Academic Press, 1981.

[48] P.E. Kloeden and E. Platen. *Numerical Solution of Stochastic Differential Equations*. Springer, 2000.

[49] F.H. Knight. *Risk, Uncertainty, and Profit*. The Riverside Press, 1921.

[50] D. Knuth. *The Art of Computer Programming, Vol. 2*. Addison-Wesley, 2000.

[51] A.J. Koning and L. Peng. Goodness-of-fit tests for a heavy tailed distribution. *Journal of Statistical Planning and Inference*, 138(12):3960–3981, 2008.

[52] H. Konno. Portfolio optimization of small scale fund using mean-absolute deviation model. *International Journal of Theoretical and Applied Finance*, 6(4), June 2003.

[53] H. Konno and H. Yamazaki. Mean-absolute deviation portfolio optimization model and its applications to Tokyo stock market. *Management Science*, 37(5), May 1991.

[54] J. Lam. *Enterprise Risk Management, From Incentives to Controls*. John Wiley & Sons, Inc., Wiley Finance Series, 2003.

[55] A.M. Law and W.D. Kelton. *Simulation Modeling and Analysis, Third Edition*. McGraw Hill, 2000.

[56] E.A. Ludwig. Assessment of Dodd-Frank financial regulatory reform: Strengths, challenges, and opportunities for a stronger regulatory system. *Yale Journal on Regulation*, 29(1):181–199, 2012.

[57] D. G. Luenberger and Y. Ye. *Linear and Nonlinear Programming (International Series in Operations Research & Management Science), 3rd edition*. Springer, November 2010.

[58] B.G. Malkiel. *A Random Walk Down Wall Street: The Time-Tested Strategy for Successful Investing (Completely Revised and Updated)*. W. W. Norton & Company, 2011.

[59] L. Martellini, P. Priaulet, and S. Priaulet. *Fixed-Income Securities: Valuation, Risk Management and Portfolio Strategies.* The Wiley Finance Series, 2003.

[60] A. Mas-Colell, M.D. Winston, and J.R. Green. *Microeconomic Theory.* Oxford University Press, New York, 1995.

[61] MATLAB. *MATLAB Documentation.* MathWorks Inc., 2009.

[62] A. Melino and S.M. Turnbull. Pricing foreign currency options with stochastic volatility. *Journal of Econometrics*, 45:239–265, 1990.

[63] E.L. Melnick and B.S. Everitt. *Encyclopedia of Quantitative Risk Analysis and Assessment.* Wiley, 2008.

[64] F. Modigliani and M.H. Miller. The cost of capital, corporation finance, and the theory of investment. *American Economic Review*, 48:261–297, 1958.

[65] N. Mora. The cost of capital, corporation finance, and the theory of investment. *Economic Review - Federal Reserve Bank of Kansas City*, 2, 2012.

[66] D.R. Nance, C.W. Smith Jr., and C.W. Smithson. On the determinants of corporate hedging. *The Journal of Finance*, 48(1):267–284, 1993.

[67] B.L. Nelson. *Stochastic Modeling, Analysis and Simulation.* McGraw Hill Inc., 1995.

[68] B. Oksendal. *Stochastic Differential Equations: An Introduction with Applications (Universitext).* Springer, 2010.

[69] G. Picone, M. Uribe, and R. M. Wilson. The effect of uncertainty on the demand for medical care, health capital and wealth. *Journal of Health Economics*, 17(2):171–185, 1998.

[70] G.E. Rejda. *Principles of Risk Management and Insurance.* Prentice Hall; 11 edition, 2010.

[71] R.T. Rockafellar and S. Uryasev. Optimization of conditional value-at-risk. *Journal of Risk*, 2:21–41, 2000.

[72] S. Ross. *Stochastic Processes.* Wiley Publishers, 1995.

[73] H. Royden. *Real Analysis (3rd Edition).* Prentice Hall, 1988.

[74] W. Rudin. *Principles of Mathematical Analysis, Third Edition.* McGraw-Hill Science/Engineering/Math, 1976.

[75] W. Rudin. *Real and Complex Analysis (International Series in Pure and Applied Mathematics).* McGraw-Hill Science/Engineering/Math, 1986.

[76] D. Ruppert. *Statistics and Data Analysis for Financial Engineering.* Springer; 1st Edition, 2010.

[77] W. Schoutens. *Levy Processes in Finance: Pricing Financial Derivatives.* Wiley Series in Probability and Statistics, 2003.

[78] E. Schwartz. The stochastic behavior of commodity prices: Implication for valuation and hedging. *Journal of Finance*, 51(3):923–973, 1997.

[79] S.G. Sharma. Over-the-counter derivatives: A new era of financial regulation. *Law and Business Review of the Americas*, 17(2):279–315, 2011.

[80] W.F. Sharpe. Mean-absolute-deviation characteristic lines for securities and portfolios. *Management Science*, 18(2), 1971.

[81] S.E. Shreve. *Stochastic Calculus for Finance II: Continuous-Time Models.* Springer, 2004.

[82] T.K. Siu, H. Yang, and J.W. Lau. Pricing currency options under two-factor Markov-modulated stochastic volatility models. *Insurance: Mathematics and Economics*, 43(3):295–302, 2008.

[83] C.W. Smith and R.M. Stulz. The determinants of firms' hedging policies. *The Journal of Financial and Quantitative Analysis*, 20(4):391–405, 1985.

[84] S. Sundaresan. *Fixed Income Markets and Their Derivatives.* South-Western Pub, 1996.

[85] M. Tenenbaum and H. Pollard. *Ordinary Differential Equations.* Dover, 2001.

[86] L.C. Thomas. *Consumer Credit Models, Pricing, Profit and Portfolios.* Oxford University Press, 2009.

[87] L.C. Thomas. Consumer finance: Challenges for operational research. *Journal of the Operational Research Society*, 61:41–52, 2010.

[88] A. Tversky and D. Kahneman. Loss aversion in riskless choice: A reference-dependent model. *Quarterly Journal of Economics*, 106(4):1039–1061, 1991.

[89] H.R. Varian. *Microeconomic Theory.* W.W. Norton and Company, Inc., New York, 1992.

[90] P. Wilmott. *Paul Wilmott on Quantitative Finance.* Wiley, January 2000.

[91] G. Wright. *Behavioral Decision Making.* Plenum Press, New York, N.Y., 1985.

[92] G. Zhou. Beyond Black-Litterman: Letting the data speak. *Journal of Portfolio Management*, 36(1):36–45, 2009.

Index

Printed in the United States
by Baker & Taylor Publisher Services